Genetic Engineering

Principles and Methods

Volume 3

GENETIC ENGINEERING
Principles and Methods

Advisory Board

Carl W. Anderson
Donald D. Brown
Peter Day
Donald R. Helinski
Tom Maniatis

A Continuation Order Plan is available for this series. A continuation order will bring delivery of each new volume immediately upon publication. Volumes are billed only upon actual shipment. For further information please contact the publisher.

Genetic Engineering

Principles and Methods

Volume 3

Edited by
Jane K. Setlow
Brookhaven National Laboratory
Upton, New York

and
Alexander Hollaender
Associated Universities, Inc.
Washington, D.C.

QH442
G454
v.3
1981

Plenum Press · New York and London

The Library of Congress cataloged the first volume of this title as follows:

Genetic engineering; principles and methods. v.1-
New York, Plenum Press [1979-
v. ill. 26 cm.
Editors: 1979-   J. K. Setlow and A. Hollaender.
Key title: Genetic engineering, ISSN 0196-3716.

1. Genetic engineering—Collected works. I. Setlow, Jane K. II. Hollaender, Alexander, 1898-
QH442.G454   575.1   79-644807
                                                                 MARC-S

Library of Congress Catalog Card Number 79-644807
ISBN 0-306-40729-9

©1981 Plenum Press, New York
A Division of Plenum Publishing Corporation
233 Spring Street, New York, N.Y. 10013

All rights reserved

No part of this book may be reproduced, stored in a retrieval system, or transmitted, in any form or by any means, electronic, mechanical, photocopying, microfilming, recording, or otherwise, without written permission from the Publisher

Printed in the United States of America

PREFACE TO VOLUME 1

This volume is the first of a series concerning a new technology which is revolutionizing the study of biology, perhaps as profoundly as the discovery of the gene. As pointed out in the introductory chapter, we look forward to the future impact of the technology, but we cannot see where it might take us. The purpose of these volumes is to follow closely the explosion of new techniques and information that is occurring as a result of the newly-acquired ability to make particular kinds of precise cuts in DNA molecules. Thus we are particularly committed to rapid publication.

Jane K. Setlow
Alexander Hollaender

ACKNOWLEDGMENT

The Editors are particularly thankful to Laine McCarthy, who helped with the editing and did the final processing of the manuscripts.

CONTENTS

CONSTRUCTED MUTANTS USING SYNTHETIC OLIGODEOXYRIBONUCLEOTIDES
AS SITE-SPECIFIC MUTAGENS 1
    M. Smith and S. Gillam

EVOLUTION OF THE INSERTION ELEMENT IS1 THAT CAUSES GENETIC
ENGINEERING OF BACTERIAL GENOMES IN VIVO. 33
    E. Ohtsubo, K. Nyman, K. Nakamura and H. Ohtsubo

APPLICATIONS OF MOLECULAR CLONING TO SACCHAROMYCES. 57
    M.V. Olson

CLONING RETROVIRUSES: RETROVIRUS CLONING? 89
    W.L. McClements and G.F. Vande Woude

REPEATED DNA SEQUENCES IN DROSOPHILA. 109
    M.W. Young

MICROBIAL SURFACE ELEMENTS: THE CASE OF VARIANT SURFACE
GLYCOPROTEIN (VSG) GENES OF AFRICAN TRYPANOSOMES. 129
    K.B. Marcu and R.O. Williams

MOUSE IMMUNOGLOBULIN GENES. 157
    P. Early and L. Hood

THE USE OF CLONED DNA FRAGMENTS TO STUDY HUMAN DISEASE. 189
    S.H. Orkin

PHYSICAL MAPPING OF PLANT CHROMOSOMES BY IN SITU
HYBRIDIZATION 207
    J. Hutchinson, R.B. Flavell and J. Jones

MUTANTS AND VARIANTS OF THE ALCOHOL DEHYDROGENASE-1 GENE
IN MAIZE. 223
    M. Freeling and J.A. Birchler

DEVELOPMENTALLY REGULATED MULTIGENE FAMILIES IN
DICTYOSTELIUM DISCOIDEUM 265
    R.A. Firtel, M. McKeown, S. Poole, A.R. Kimmel,
    J. Brandis and W. Rowekamp

COMPUTER ASSISTED METHODS FOR NUCLEIC ACID SEQUENCING. . . . 319
    T.R. Gingeras and R.J. Roberts

INDEX. 339

CONSTRUCTED MUTANTS USING SYNTHETIC OLIGODEOXYRIBONUCLEOTIDES AS

SITE-SPECIFIC MUTAGENS

M. Smith and S. Gillam

Department of Biochemistry, Faculty of Medicine
University of British Columbia, 2075 Wesbrook Place
Vancouver, B.C. V6T 1W5, Canada

INTRODUCTION

Classical genetics provides a bridge between biology and the chemical structure of the gene. In the past, isolation of spontaneous or induced mutants by screening or selection for a changed phenotype was followed by genetic mapping and, more recently, by determination of the structure of the gene of interest. The advent of rapid methods for DNA sequence determination (1,2) has created a new situation whereby whole genomes, or parts of genomes, can have their sequences determined in advance of detailed genetic analysis. On occasion it is possible to assign genetic function to a DNA sequence by inspection. For example, in the case of genes J and K of the small coliphage ØX174 and G4, knowledge of the protein sequences allowed assignment of gene function to DNA in the absence of mutants (3,4). Assignment of tRNA genes to particular tracts of mitochondrial DNA sequence was achieved by using computer analysis of possible secondary structure (5). However, such parallel information is not usually available. This has led to the development of methods for constructing mutants in vitro by modifying DNA of known sequence. Examination of the in vivo properties of the modified DNA allows assignment of particular functions to specific DNA sequences. An early example of this approach is the construction of specific deletions in simian virus 40 (6). Allied methodologies include strategies for introducing deletions of variable length at specific or random sites (7,8) and specific or random insertion of short segments of duplex DNA (9,10). These all are powerful techniques for demonstrating the location of genetic functions and establishing their approximate boundaries. However, full understanding of genetic functions requires mutations that modulate gene activity, e.g., mutations that increase or decrease the efficiency of origins of replication, of transcription or of translation and mutations that establish the reading frame for the triplet code which determines the sequence of a protein. In general, this requires point mutations.

A number of strategies for making localized point mutations have been developed. These include chemical modification of a specific DNA fragment followed by reintegration into genomic DNA (11), mutations induced by chemically modified RNA (12), chemical modification of a specific single-stranded gap in duplex DNA (13) and enzymatic incorporation at a specific site of a mutation-inducing nucleotide analog (14). While the general locality of the changes introduced by any of these methods is known, full characterization of mutant DNA requires biological cloning and DNA sequence determination of many clones. In addition, the methods do not allow the production of all types of point mutation, i.e., transitions, transversions, insertions and deletions. In this article, we describe a strategy for production and isolation of specific point mutations, even if they are phenotypically silent. The method uses a synthetic oligodeoxyribonucleotide, different from wild-type in a defined position, as a specific mutagen and involves in vitro integration of the oligodeoxyribonucleotide into genomic DNA (15-18). At first sight, the method appears to be more laborious than others because of the requirement for a synthetic oligodeoxyribonucleotide. However, newer methods of synthesis are reducing the time it takes to make oligodeoxyribonucleotides (see below). The unique features offered by this approach are that it is not constrained to the neighborhood of a restriction endonuclease cleavage, a specific site in the DNA is targeted and the precise, desired change at the target is programmed by the synthetic oligodeoxyribonucleotide. The initial development of the methodology used the small single-stranded circular DNA of the coliphage ØX174. Because most cloning of recombinant DNA is carried out in E. coli, ØX174 is a good model for other minigenomes. Consequently, extension of the method to other single or double-stranded circular DNA is straightforward; not only can bacteriophage genomes be mutated specifically, but so can plasmid and recombinant DNAs. With the in vitro selection techniques that have been developed, it is possible to mutate eukaryote DNA that is not expressed in the host microorganism used for cloning. All these features of the oligodeoxyribonucleotide mutagenic method make it a uniquely powerful tool for defining DNA function.

THE MUTAGENIC OLIGODEOXYRIBONUCLEOTIDES

The concept of a polynucleotide as a specific mutagen is not a new one; it was discussed by J. Lederberg in 1960 (19). However, as mentioned above, rapid DNA sequence determination has only recently become available and accurate DNA sequencing is a prerequisite of the method. In addition, convenient methods for oligodeoxyribonucleotide synthesis are required together with precise information on the specificity and stability of duplex structures involving oligodeoxyribonucleotides.

Recent developments in the chemical synthesis of oligodeoxyribonucleotides by the phosphotriester method have greatly facilitated synthesis and subsequent studies on solid-phase supported synthesis have raised the real possibility of fully automated chemical synthesis of oligodeoxyribonucleotides (20,21). Enzymatic synthesis of oligodeoxyribonucleotides of defined sequence is an attractive alternate route for biochemists and molecular geneticists because of their familiarity with the methodologies and isolation techniques. The deoxynucleotide-5' triphosphate polymerizing enzyme, terminal transferase, can be used in this type of synthesis (22). We have concentrated our efforts on the development of a method using E. coli polynucleotide phosphorylase. In the presence of Mn^{+2}, this enzyme will catalyze the primer-dependent stepwise synthesis of oligodeoxyribonucleotides (24-32). This method of synthesis has been used for all the oligodeoxyribonucleotides employed in the development of the mutagenic methods (15-18). Enzymatic synthesis has the attractive feature that it will incorporate nucleotide analogs (31). The yields of desired products are not always high because the reaction is controlled kinetically (32); however, the mutagenic reactions are carried out on pmole amounts of DNA and require only minute amounts of the synthetic oligodeoxyribonucleotide.

Studies with defined oligodeoxyribonucleotides covalently bound to cellulose provided the basic information on duplex stability and specificity of interactions with complementary oligodeoxyribonucleotides (33-36). Pertinent data related to the use of oligodeoxyribonucleotides are summarized in Figure 1. From the point of view of specific mutagenesis, the important principle arising from these experiments is that short oligodeoxyribonucleotides can form a stable duplex with one mismatched nucleotide even though the mismatch is in the middle of the oligodeoxyribonucleotide. Clearly, it ought to be possible to use such structures, involving relatively accessible synthetic oligodeoxyribonucleotides, to induce specific point mutations.

The length, n, required for an oligodeoxyribonucleotide to recognize a unique sequence in a DNA strand is such that 4^n = number of nucleotides in the DNA strand (34,37). Thus for small bacteriophage, virus, plasmid and recombinant DNAs, which contain from 5,000 to 10,000 nucleotides, n is 6 to 7. Since this is a statistical calculation, certain sites may require a slightly longer oligodeoxyribonucleotide for unique recognition. However, the similarity of the length of oligodeoxyribonucleotide demanded by the physical chemistry and by the genetic complexity is both remarkable and gratifying.

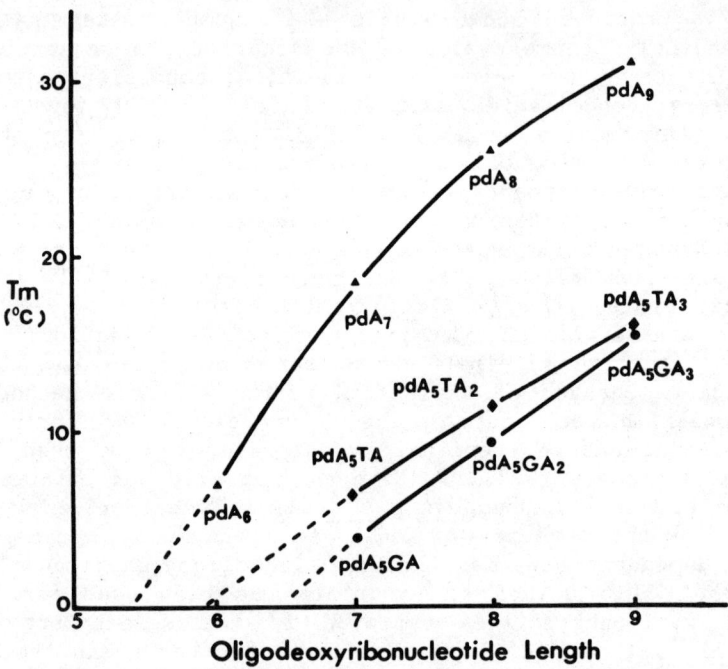

Figure 1. The relationship between melting temperature (Tm) and oligodeoxyribonucleotide length for matched and mismatched oligodeoxyribonucleotides interacting with cellulose-pT$_9$ in 1 M NaCl at pH 7.5 (36).

STRATEGY FOR OLIGODEOXYRIBONUCLEOTIDE MUTAGENESIS

The most simple method for incorporation of a single-stranded DNA fragment into genomic DNA is the marker-rescue method. In this procedure, the fragment is annealed with complementary genomic length DNA and the hybrid is used to transform cells (38, 39). However, genomic DNA derived from the fragment is obtained with a very low efficiency and the method does not work at all with very short DNA fragments (39).

As mentioned above, the initial target used for developing this mutagenic method was ØX174 viral DNA. This was chosen because its sequence is known and because the bacteriophage is well-defined genetically and biologically (40,41). Many studies

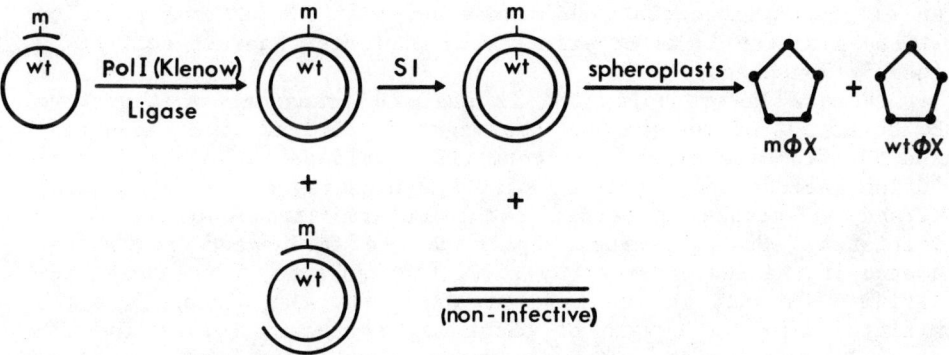

Figure 2. Synthesis of a specific mutation (m) in ØX174 DNA starting from a mutant oligodeoxyribonucleotide primer (m) and wild-type ØX174 DNA (wt). The primer is extended by E. coli DNA polymerase I, large fragment [Pol I (Klenow)], and the resultant double-stranded DNA converted to closed-circular DNA with T4 DNA ligase (Ligase). Infective molecules containing single-stranded regions are destroyed by S1 endonuclease (S1) and the DNA is then used to transfect E. coli spheroplasts. The mutant is isolated by phenotypic selection or screening or by selecting or screening for mutant DNA.

have been carried out on in vitro synthesis of double-stranded ØX174 DNA using viral DNA as template for E. coli DNA polymerase I together with DNA ligase (42-45)'. At 20°C, oligodeoxyribonucleotides containing nine or more nucleotides are optimal primers for DNA polymerase I, although there is detectable priming with tri- and tetranucleotides (44). These results suggested that an oligodeoxyribonucleotide mismatched at one nucleotide with complementary ØX174 DNA could be used as a primer for E. coli DNA polymerase I and then could be integrated into circular duplex DNA with DNA ligase to yield a molecule that is a heteroduplex at one position. Transfection with this DNA should produce progeny derived from both strands. The basic strategy is diagramed in Figure 2. It was evident from the earlier studies on oligodeoxyribonucleotide priming that the 5'-exonuclease activity of E. coli DNA polymerase I edits out most of the primer before ligation (43). It is also evident that E. coli DNA polymerase I is one of the most effective DNA synthesizing enzymes when it is necessary to convert a long single-stranded template to double-stranded DNA (46). Hence, the large polymerizing fragment of E. coli DNA polymerase I (Klenow enzyme) from which the 5'-exonuclease had been removed proteolytically, was used in these experiments (47). Other experiments have confirmed that the native E. coli DNA polymerase I requires a rather long 5'-extension on an oligodeoxyribonucleotide mismatched with DNA at one position if the mismatch is to be effectively protected against editing by the 5'-exonuclease (48).

When a heteroduplex DNA is used to transfect or transform E. coli, 50% of the progeny DNA might be derived from one strand and 50% from the other. Factors that could perturb this distribution include the asymmetry of DNA synthesis; in ØX174, a viral strand is completed before complementary strand synthesis is initiated (49). Mismatch repair can be influenced by the sequence of the DNA under study (50), the genetics of the host bacterium (50) and, in the case of ØX174, by the distance of the mismatch from the origin of phage DNA synthesis (51). The 3'-exonuclease activity, which is intrinsic to the polymerase activity of E. coli DNA polymerase I, could edit out the mismatch of the oligodeoxyribonucleotide before priming takes place. In the case of ØX174, one further factor that could influence the yield of mutant progeny is the high infectivity of single-stranded DNA (45,52). Therefore, a step involving treatment with the single-strand specific S1 endonuclease is used to inactivate residual single-stranded bacteriophage DNA (Figure 2).

The statistical calculation discussed above predicts that an oligodeoxyribonucleotide of the appropriate length should recognize only one site in a complementary strand of genomic DNA. However, it is useful to be assured that there is not an above average frequency of occurrence of the sequence of interest. This can be readily established for a known genomic sequence by using computer sequence searching techniques (53-58). It can also be experimentally established that an oligodeoxyribonucleotide interacts with only one site and that this site is the desired one (see below).

SPECIFICITY OF OLIGODEOXYRIBONUCLEOTIDE MUTAGENS

First experiments to test the possibility of using synthetic oligodeoxyribonucleotides as site-specific mutagens were directed at inducing the two types of transition AT → GC and GC → AT, since the mismatched nucleotide pairs required, AC and GT, should least perturb a DNA duplex (59). The target was nucleotide 587 of the ØX174 genome (15,40). In wild-type bacteriophage DNA, this nucleotide is G, and when it is replaced by A an amber codon is generated in the reading frame of gene E, the lysis function (Figure 3). The oligodeoxyribonucleotides used (Figure 3) each contained 12 nucleotides. There is no sequence in ØX174 bacteriophage DNA complementary to these dodecadeoxyribonucleotides except at positions 582 to 593 (excluding the mismatch at 587 in the case of the mutant complementary oligodeoxyribonucleotide). This was established by computer search of the ØX174 DNA sequence (53,58). The fact that the oligodeoxyribonucleotides would prime

wild-type complementary
oligodeoxyribonucleotide   3' A-A-A-C-A-C-C-C-T-A-T-Gp 5'

mutant complementary oligodeoxyribonucleotide   3' A-A-A-C-A-T-C-C-T-A-T-Gp 5'
                                                              A
                                                              ↑
wild-type ØX174
5' -T-G-C-T-T-A-T-G-T-A-C-G-C-T-G-A-C-T-T-T-G-T-G-G-A-T-A-C-C-C-T-C-G-C-T- 3'
              570   580   590   ↑   600

gene D - Cys - Val - Tyr - Gly - Thr - Leu - Asp - Phe - Val - Gly - Tyr - Pro - Arg -

gene E   (Met) - Val - Arg - Trp - Thr - Leu - Trp - Asp - Thr - Leu - Ala -
                                                    ↓
                                                   Ter

Figure 3. Sequence of part of ØX174 DNA coding for the overlapping genes D and E in the vicinity of position 587 and the synthetic dodecadeoxyribonucleotides complementary to wild-type (G at 587) and mutant (A at 587) ØX174 DNA. The changes G→A and A→G at position 587 interconvert Trp and am codons in the reading frame of gene E, but do not change the amino acid, Val, coded in the reading frame of gene D (15,17). Ter, termination (= amber).

specifically at the desired site on ϕX174 DNA was confirmed in a pulse-chase experiment that accurately measures the number of priming sites and their distance from adjacent downstream restriction endonuclease cleavages (15). Only one site was detected; it was situated the exact number of nucleotides from an adjacent HaeIII cleavage predicted by the ϕX174 DNA sequence (40). It is of interest that when the priming efficiencies of the homologous and heterologous oligodeoxyribonucleotides on wild-type ϕX174 DNA template were compared, the heterologous primer was detectably less efficient (15). This difference in reactivity could be the basis of a screening procedure (see below).

An even more precise test of the target specificity of an oligodeoxyribonucleotide mutagen is to use it as a primer for DNA sequence determination by the terminator method (1,60). Specific priming at the desired site results in an unambiguous sequence pattern of the adjacent DNA (60).

It is always wise to carry out the priming specificity test prior to attempting the biological part of the mutation experiment. For example, when a synthetic undecadeoxyribonucleotide was made to induce a T → G transversion at position 5276 of ϕX174 DNA (see below), it was found that the oligodeoxyribonucleotide primed at an addition site at the ribosome binding site for gene E because the 3'-terminal octadeoxyribonucleotide is complementary to the sequence upstream of gene E. It was possible to devise a strategy for specific mutation at position 5276 by carrying out the priming in two steps. At low temperature, in the presence of only three selected deoxyribonucleoside-5' triphosphates, the primer was extended preferentially; then, at a higher temperature which would dissociate a short duplex, extension was completed in the presence of four deoxyribonucleoside-5' triphosphates (16).

In another case, the deletion of nucleotide 2925 (see below), the mutagenic oligodeoxyribonucleotide primed specifically and efficiently at the desired site (60). However, the mutant was obtained in low yield. The demonstration of efficient priming indicates that the low yield of mutant was most likely the consequence of mismatch repair. Another case is when the mismatched oligodeoxyribonucleotide does not prime. This can be due to an error in the determination of the template DNA sequence, an error in the sequence of the oligodeoxyribonucleotide or because the target region of the template DNA is not accessible because it is part of a DNA secondary (hairpin) structure (see below). Clearly, it is essential to establish that these potential reasons for an unsuccessful experiment are not present before proceeding further. It is axiomatic that the known sequence of the target DNA be accurate and should be searched for potential secondary structures. Likewise, the synthetic oligodeoxyribonucleotide should be rigorously characterized (15,28,32).

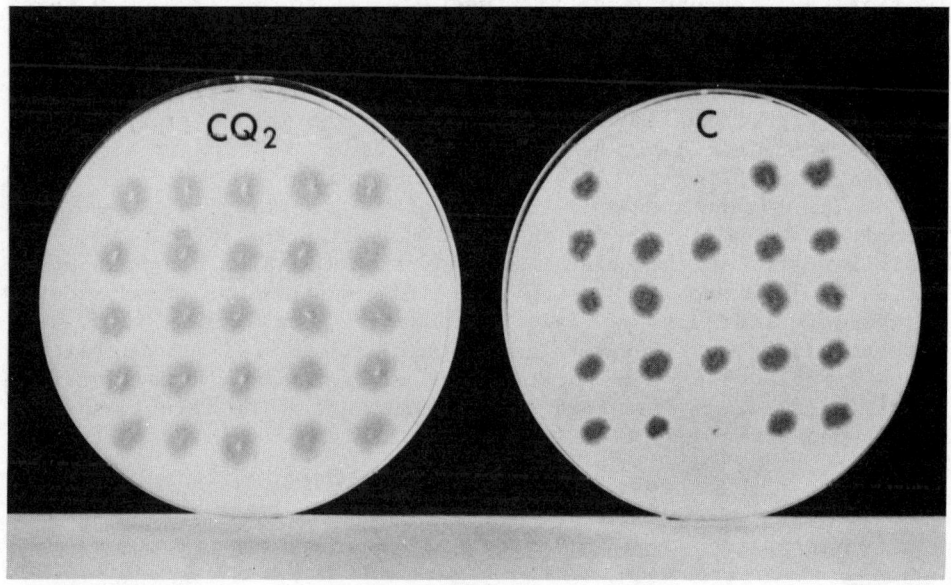

Figure 4. Progeny phage from mutation at position 587 (G · A) of ØX174 DNA using pGTATCCTACAAA as mutagen and a priming temperature of 25°C. The phage plaques obtained after plating phage (on suppressor⁺ E. coli) resulting from spheroplast transfection by mutagenized ØX174 DNA were replated on suppressor (CQ_2) and non-suppressor (C) E. coli. The missing plaques on the nonsuppressor host indicate the fraction of mutant molecules.

TRANSITION MUTATIONS

The strategy diagramed in Figure 2, when used with the oligodeoxyribonucleotide pGTATCCTACAAA as a primer at 25°C on wild-type ØX174 viral DNA template (15), induced the G→A transition at position 587 with an efficiency of about 15% (Figures 3 and 4). The oligodeoxyribonucleotide pGTATCCCACAAA induced the reversion, A→G, with a similar efficiency (15). In these first experiments, S1 endonuclease treatment had a very beneficial effect, increasing the fraction of mutated progeny from about 0.1% to about 15% (15). Presumably, this indicates incomplete elongation of the primer and/or incomplete ligation of the elongated product. In more recent experiments, with the same primers at the same temperature but a different buffer and a more active T4 DNA ligase preparation, the efficiency of mutant production prior to S1 endonuclease treatment was 2% to 10% (17). In these experiments, the fraction of progeny that were mutant after S1 endonuclease treatment was 15% to 22% (17), which is similar to the overall yield obtained in the first experiment. This result suggests that the effect is due to more efficient production of closed-circular duplexes rather than to a greater efficiency of production of the specific nucleotide change.

The production or reversion of an am mutation is easy to monitor using a suppressor host microorganism (Figure 4). A similar strategy has been used to introduce an am codon at position 2401 in gene G of ϕX174 (52); the mutagen was an octadeoxyribonucleotide (Figure 5). Presumably because the relatively short oligodeoxyribonucleotide recognizes several priming sites, the efficiency of this mutation was low even though the two-stage priming strategy described earlier was used. The relationship of oligodeoxyribonucleotide length to efficiency of mutation is discussed in the following section.

OLIGODEOXYRIBONUCLEOTIDE LENGTH AND EFFICIENCY OF MUTATION

The temperature (25°C) at which priming of DNA synthesis was carried out in the early experiments (15) was close to the melting temperature of the oligodeoxyribonucleotide duplexes that have a mismatch (Figure 1). Therefore, it was logical to investigate the effect of oligodeoxyribonucleotide length on the efficiency of mutant production (17). The results of a series of experiments directed at introducing the transition G→A at position 587 of ϕX174 DNA (Figure 3) are shown in Table 1. The series of oligodeoxyribonucleotides used in these experiments was pGTATCCTACAAA down to pGTATCCT; the underlined T residue is the mismatched nucleotide. Computer search showed that, except for pGTATCCT, the oligodeoxyribonucleotides would interact only with the desired site. The efficiency of mutant production falls off rapidly with decreasing oligodeoxyribonucleotide length (Table 1). This could be attributable solely to decreasing stability of the primer template complex. However, a second factor affecting efficiency of mutant production is editing out of the mismatched primer nucleotide by 3'-exonuclease degradation prior to priming of DNA synthesis; 3'-exonuclease is an intrinsic property of the large fragment of E. coli DNA polymerase I (61). The fact that the total yield of progeny phage, both wild-type and derived mutant, remains reasonably constant suggests that the major cause of loss of efficiency of mutant production lies in 3'-exonuclease editing rather than in reduced primer efficiency. The experiments of Table 1 and more especially those described in the next section (on the effect of temperature of priming on efficiency of mutation) suggest that a relatively short 3'-extension on the oligodeoxyribonucleotide is sufficient to provide complete protection against 3'-exonuclease editing.

mutant (am) complementary octadeoxyribonucleotide   3' C-A-A-A-A-T-C-Tp 5'

                                                          T(am)
                                                          ↑

wild-type viral DNA   5' -G-T-T-A-A-T-C-A-T-G-T-T-T-C-A-G-A-C-T- 3'

                                  2390   2400

gene G   (Met)- Phe - Gln - Thr -
                                             ↓
                                         Ter (am)

Figure 5. Sequence of part of ØX174 DNA coding for gene G in the vicinity of position 2401 and the synthetic octanucleotide complementary to mutant (T at 2401) ØX174 DNA (wild-type DNA has C at 2401). The change C → T at position 2401 changes a Gln codon to am in the reading frame of gene G (52). Ter, termination (= amber).

Table 1

Effect of length of oligodeoxyribonucleotide on the efficiency of the mutation G → A at position 587 of ϕX174 DNA with priming of the large fragment of E. coli DNA polymerase I (Klenow enzyme) at 25°C (17).

Primer*	Total ($\times 10^{-6}$) and mutant (%) phage			
	−S1 endonuclease		+S1 endonuclease	
	Total	Mutant	Total	Mutant
pGTATCCTACAAA	1.9	9.5	0.3	22.0
pGTATCCTACAA	0.4	6.5	0.2	20.0
pGTATCCTACA	1.3	4.0	0.2	18.0
pGTATCCTAC	1.5	4.5	0.2	8.0
pGTATCCTA	0.8	0.0	0.13	0.7
pGTATCCT	0.2	0.0	0.05	0.0

*Mismatch is underlined

Protection of the oligodeoxyribonucleotide against 5'-exonuclease editing is not required if the larger fragment of E. coli DNA polymerase I is used in the mutagenic strategy depicted in Figure 2. Studies with intact E. coli DNA polymerase I directed at the A → G transition at position 587 of ϕX174 DNA have defined the length of oligodeoxyribonucleotide required to protect against the 5'-exonuclease activity of this enzyme (48). A comparison of the results of these experiments with those of experiments with the Klenow enzyme (17) are summarized in Table 2. It is clear that a 5'-extension of greater than 12 nucleotides will be required to give an equivalent efficiency of mutant production with intact E. coli DNA polymerase I when compared to that obtained with the hexanucleotide 5'-extension of pGTATCCCACA under similar conditions with the large fragment (lacking 5'-exonuclease activity) of E. coli DNA polymerase I.

TEMPERATURE OF PRIMING AND EFFICIENCY OF MUTATION

As mentioned above, the temperature at which the first experiments were carried out, 25°C, is close to the melting temperature of the oligodeoxyribonucleotide duplexes under investigation. Consequently, it was very probable that a reduction in the temperature at which the priming of DNA synthesis was carried out (the first step in the strategy depicted in Figure 2) would increase the efficiency of mutation. Experiments directed at testing this possibility for the transition G → A at position 587 of ϕX174 DNA (17) are summarized in Table 3. Reduction of the

Table 2

Efficiency of the mutation A → G at position 587 of ϕX174 DNA when the large fragment of E. coli DNA polymerase I (Klenow enzyme) and the intact DNA polymerase I (Pol I) were used to elongate mutagenic oligodeoxyribonucleotides; the priming temperature was 10 to 12°C and the transfecting DNAs were treated with S1 endonuclease (17,48). (NT = not tested).

Primer *	% Mutant	
	Klenow (17)	Pol I (48)
pGTATCCC	0.7	NT
pGTATCCCA	10.0	NT
pGTATCCCAC	19.0	NT
pGTATCCCACA	37.0	NT
pGTATCCCACAA	38.0	0.02
pGTATCCCACAAAGTCCA	NT	1.2
pAGGGTATCCCACAA	NT	1.9
pAGGGTATCCCACAAAGTCCA	NT	14.0
pGCGAGGGTATCCCACAA	NT	13.9
pGCGAGGGTATCCCACAAAGTCCA	NT	32.5

*Mismatch is underlined

Table 3

Effect of temperature of priming on the efficiency of the mutation A → G at position 587 of ϕX174 DNA. The large fragment (Klenow enzyme) of E. coli DNA polymerase I was used to elongate the primers and the transfecting DNA was treated with S1 endonuclease (17).

	Total ($\times 10^{-6}$) and mutant (%) phage							
	33°		25°		10°		0°	
Primer*	Total	Mutant	Total	Mutant	Total	Mutant	Total	Mutant
pGTATCCCACAA	0.5	2.0	6.0	15.0	20.0	38.0	27.0	39.0
pGTATCCCACA	7.7	3.0	25.0	26.0	65.0	37.0	78.0	37.0
pGTATCCCAC	5.4	0.5	20.0	13.0	52.0	19.0	67.0	25.0
pGTATCCCA	3.6	5×10^{-2}	9.0	3.0	27.0	10.0	27.0	20.0
pGTATCCC	3.6	1×10^{-3}	16.0	0.2	18.0	0.7	23.0	0.5

*Mismatch is underlined

temperature of priming from 33°C to 0°C effects a very dramatic increase in the efficiency of mutation. Clearly priming at 0°C is the preferred technique; with longer oligodeoxyribonucleotides (>10), the efficiency approaches the theoretical value of 50% (assuming no biological biases are present). Another benefit of using a lower temperature for priming DNA synthesis is that shorter oligodeoxyribonucleotides become very effective mutagens; an octadeoxyribonucleotide is as effective at 0°C as is an undecadeoxyribonucleotide at 25°C (Table 3). Similar results were obtained when the series of oligodeoxyribonucleotides pGTATCCT to pGTATCCTACAAA, used to induce the transition G→A at position 587 of φX174 DNA, were used as primers at low temperature (17).

TRANSVERSION MUTATIONS

Transition mutations, discussed above, are a class of mutational changes that can be induced with reasonable predictability, but with less target specificity using a nucleotide analog or chemical reagent (13,14). As noted earlier, a potential advantage of oligodeoxynucleotide mutagenesis, in addition to precise targeting, is the possibility of programming other types of point mutation. Hence, it was of interest to investigate the possibility of inducing transversion mutations. The target for these studies was nucleotide 5276 of φX174 DNA (16,17,40). This normally is G and can be replaced by T by constructing a heteroduplex with a purine-purine (GA) mispairing (Figure 6). Reversion of this change is effected by a pyrimidine-pyrimidine (TC) mispairing. The two oligodeoxyribonucleotides constructed to induce these changes were pCCCAGCCTAAA and pCCCAGCCTCAA, respectively; they complement nucleotides 5274 to 5284 of φX174 DNA. These oligodeoxyribonucleotides contain 11 nucleotides; the mismatched nucleotides are underlined (Figure 6). The short, dinucleotide 3'-extension and the less favorable stereochemistry of the mismatches suggests that these oligodeoxyribonucleotides provide a more stringent test of the mutagenic method than the earlier experiments at position 587. In fact, when the G→T transversion at position 5276 was attempted with a temperature of 25°C for priming DNA synthesis, no mutants were detected (16,17). The experiments described above on the effect of temperature on the efficiency of mutation were undertaken and 0°C was defined as the optimum priming temperature. When pCCCAGCCTAAA was used as a mutagen at 0°C with the general strategy of Figure 2, 13% of the progeny were the desired mutant (16,17). The results of a complete series of experiments with the series of oligodeoxyribonucleotides pCCCAGCCTAAA to pCCCAGCCTA are summarized in Table 4. Comparison with the results discussed earlier (Tables 1 and 3) shows that a dinucleotide 3'-extension at 0°C provides a similar protection against 3'-exonuclease editing with both sets of oligodeoxyribonucleotides.

wild-type complementary oligodeoxyribonucleotide 3' A-A-C-T-C-C-G-A-C-C-Cp 5'

mutant complementary oligodeoxyribonucleotide 3' A-A-A-T-C-C-G-A-C-C-Cp 5'
                                                        T
                                                        ↑

X174 wild-type 5' -G-C-A-A-A-A-A-G-A-G-A-T-G-A-C-A-G-A-T-T-G-A-G-G-C-T-G-G-A-A-A- 3'
                              ↑   ↑   ↑
                             5260   5270   5280

gene B   - Lys - Lys - Arg - Asp - Glu - Ile - Glu - Ala - Gly - Lys -
                                                              ↓
                                                             Ter

gene A   - Lys - Arg - Glu - Met - Arg - Arg - Leu - Arg - Leu - Gly -
                                                                    ↓
                                                                   Phe

Figure 5. Sequence of part of ØX174 DNA coding for the overlapping genes A and B, in the vicinity of position 5265 and the synthetic undecadeoxyribonucleotides complementary to wild-type (G at 5265) and mutant (T at 5265) ØX174 DNA. The changes G → T and T → G at position 5265 interconvert Glu and am codons in the reading frame of gene B and also interconvert Leu and Phe codons in the reading frame of gene A (16,17). Ter, termination (= amber).

Table 4

Effect of length of oligodeoxyribonucleotide on the efficiency of the mutation G → T at position 5276 of ϕX174 DNA. The large fragment of E. coli DNA polymerase I (Klenow enzyme) was used to elongate the primers and the temperature of priming was 0°C. The last oligodeoxyribonucleotide is homologous with the template and is a control (16,17).

Primer*	Total ($\times 10^{-5}$) and mutant (%) phage			
	−S1 endonuclease		+S1 endonuclease	
	Total	Mutant	Total	Mutant
pCCCAGCCTAAA	3.6	7.0	1.2	13.0
pCCCAGCCTAA	3.5	4.3	0.73	7.5
pCCCAGCCTA	3.8	0.7	0.75	0.7
pCCCAGCCTCAA	0.9	0.0	0.18	0.0

*Mismatch is underlined

When the oligodeoxyribonucleotide pCCCAGCCTCAA was used under the same conditions in an attempt to revert the nucleotide change at position 5276 (T → G), no wild-type phage were detected (16,17). The explanation for this is evident from Figure 7, showing the results of a pulse-chase experiment, described earlier, which defines the number of priming sites on ϕX174 DNA for a particular oligodeoxyribonucleotide. Thus, pCCCAGCCTAAA has one strong priming site and pCCCAGCCTCAA has two. Examination of the ϕX174 DNA sequence shows that a complement of the 3'-terminal octadeoxyribonucleotide of this sequence occurs twice: once at the desired site, and again starting at position 553. The additional priming site defined by fragment B (Figure 6) corresponds to that at position 553 since it is 125 nucleotides from a HaeIII site (40). Presumably, priming at the second site preemptively interferes with the mismatched priming at the target site or results in a structure that is difficult to ligate. Knowledge of the phage DNA sequence adjacent to the 3'-end of the desired priming site (compare Figures 6 and 8) allowed the primer at the desired site to be extended preferentially by DNA polymerase in the presence of only three deoxyribonucleoside-5' triphosphates (A, T and C). Subsequent reaction at 37°C in the presence of four deoxyribonucleoside-5' triphosphates prevented elongation at the unwanted site by dissociation of the primer-template duplex. Application of the strategy diagramed in Figure 1 then resulted in the desired mutant being produced in 19% yield (16,17). The data for the complete set of experiments on this reversion are shown in Table 5. Of particular interest is the observation that S1 endonuclease did not enhance the yield of mutant. This result

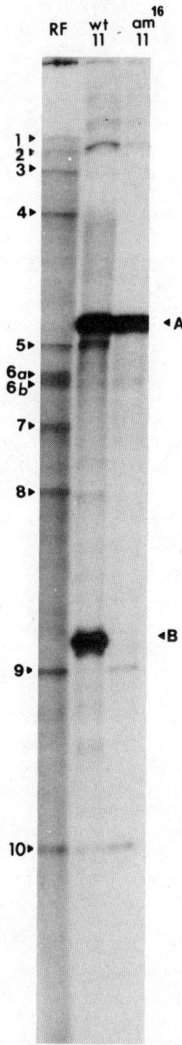

Figure 7. Gel electrophoresis of products that examine the specificity of priming by synthetic oligodeoxyribonucleotides. The pulse-chase priming (15,16) followed by HaeIII cleavage for the oligodeoxyribonucleotides pCCCAGCCTCAA (wt 11) and pCCCAGCCTAAA (am16 11) on wild-type ØX174 DNA template (see Figure 6). The fragment A corresponds to priming at the desired site in the vicinity of position 5265. The fragment B corresponds to priming at a site in the vicinity of position 557 of ØX174 (see Figure 8). The gel electrophoresis, carried out with denatured DNA in the presence of 7 M urea, has denatured HaeIII fragments of ØX174 RF DNA as size markers (nucleotide residues in brackets), 1 (1353), 2 (1078), 3 (872), 4 (604), 5 (310), 6a (271), 6b (281), 7 (234), 8 (194), 9 (118) and 10 (72).

```
mutant complementary
decadeoxyribonucleotide              3'  C-A-A-C-T-T-C-G-A-Ap  5'
                                                      A
                                                      ↑
wild-type viral DNA    5' T-G-C-G-T-G-A-G-G-C-T-T-G-C-G-T-T-T-A-T-G-G-T-A 3'
                              550            560            570

gene E                                                (Met)- Val -

gene D                    - Cys - Val - Glu - Ala - Cys - Val - Tyr - Gly -
```

Figure 8. Sequence of part of ØX174 DNA coding for gene D and the overlapping ribosome binding sequence of gene E in the vicinity of position 557 and the synthetic decadeoxyribonucleotide complementary to mutant (A at 557) ØX174 DNA (wild-type DNA has G at 557). The change G → A at position 557 changes the sequence GAGGCTT to GAAGCTT; this disrupts the ribosome binding sequence GAGG (which is also an MnlI recognition sequence) and produces a HindIII recognition sequence AAGCTT but does not change the amino acid coded in the reading frame of gene D (63).

Table 5

Effect of length of oligodeoxyribonucleotide on the efficiency of the mutation T → G at position 5276 of ØX174 DNA. The large fragment of E. coli DNA polymerase I (Klenow enzyme) was used to elongate the primers and the temperature of priming (carried out in two stages, see text) was 0°C. The last oligodeoxyribonucleotide is homologous with the template and is a control (16,17).

Primer*	Total ($\times 10^{-6}$) and mutant (%) phage			
	−S1 endonuclease		+S1 endonuclease	
	Total	Mutant	Total	Mutant
pCCCAGCCTCAA	3.6	22.0	0.9	19.0
pCCCAGCCTCA	2.4	16.0	0.7	22.0
pCCCAGCCTC	2.9	3.5	1.3	3.5
pCCCAGCCTAAA	5.4	5×10^{-3}	1.1	6×10^{-3}

*Mismatch is underlined

suggests that closed-circular duplexes were produced in vitro with essentially 100% efficiency. Also of interest, in both sets of experiments directed at producing transversions, was the mutagenic efficiency of the nonadeoxyribonucleotides where the 3'-terminal nucleotide was mismatched. Clearly, under the in vitro conditions for DNA synthesis employed in these reactions, editing by the 3'-exonuclease is not particularly efficient. This observation encourages the hope that conditions can be devised where the 3'-exonuclease is even less effective.

TRANSITION MUTATION THAT MODIFIES A RIBOSOME-BINDING SITE AND PRODUCES A RESTRICTION ENDONUCLEASE RECOGNITION SEQUENCE

The mutations discussed so far involve the production or reversion of am codons; hence, biological selection or screening is straightforward (Figure 4). Many potential mutants of interest to molecular geneticists will involve changes in DNA sequences that do not have a phenotype which is so easily selected or screened, or which are phenotypically silent. This situation is the case particularly with eukaryote genes cloned in prokaryote vectors (except in those cases where the eukaryotic gene will complement a defective prokaryote gene) (62). Consequently, methods are required for identifying mutant DNA clones directly. Of several possible methods (see below), an attractive option is to combine the construction of the desired mutant with the production or removal of a restriction endonuclease recognition site. A mutation in which the desired change in the ribosome-binding sequence of gene E of ØX174 DNA (GAGG → GAAG) also results in the production of a HindIII recognition sequence (AGGCTT →

AAGCTT) is shown in Figure 8. The desired G → A transition at position 557 of ϕX174 DNA was induced with the decadeoxyribonucleotide pAAGCTTCAAC (63). The mutation has been introduced into ϕX174 DNA in which nucleotide 587 previously had been changed from G → A (am). This change illustrates the construction of specific multiple mutants. The sequence of the mutant has been determined and the presence of the HindIII site confirmed (63). This new HindIII site is the only one present in ϕX174. The introduction (or deletion) of a unique site is particularly useful since a mixture of wild-type and mutant double-stranded DNAs can be treated with the restriction endonuclease and the circular or linear molecules separated by gel electrophoresis. This situation provides a convenient purification of the desired mutant: if it is the linear molecule, the DNA can be recircularized with DNA ligase. The introduction of a new site, particularly a unique one, provides a new entry for cloning fragments of the mutant genome and also for other strategies of constructed mutagenesis (13,14).

It might be anticipated (64) that the modification of the ribosome binding sequence (GAGG → GAAG) would affect the efficiency of production of the gene E product, the ϕX174 lysis protein. This prediction proved to be the case; lysis was suppressed under normal conditions of infection, but did occur when infected E. coli cells were grown in a lysozyme-bile salts medium.

IN VITRO SELECTION OF MUTANT DNA WITH MUTAGENIC OLIGODEOXYRIBONUCLEOTIDE

Not all desired mutants will coincidentally allow production or deletion of a restriction endonuclease site. In these cases, an alternate method for selection of the mutant DNA is required. The data shown in Figure 1 suggest a completely general method for mutant DNA selection. The perfect duplex formed by the mutating oligodeoxyribonucleotide and the complementary mutant DNA will be more stable than the mismatched duplex between the oligodeoxyribonucleotide and wild-type DNA; the enhanced stability of the duplex involving the mutant DNA should provide the basis for screening or preferably for selecting for the mutant. The difference between the stability of a perfectly matched and single nucleotide mismatched oligodeoxyribonucleotide-DNA duplex is readily detectable. This was shown in experiments on priming DNA synthesis with dodecadeoxyribonucleotide-DNA homologous and heterologous pairings (15). Single nucleotide mismatches have also been detected by direct observation of reduced duplex stability between oligonucleotides and ϕX174 or lambda DNAs (48, 65). More extended mismatches can be even more readily detected; this has been demonstrated in studies with model polydeoxyribonucleotides (66) and DNA (67). Clearly these observations provide the basis for isolation of mutants by screening.

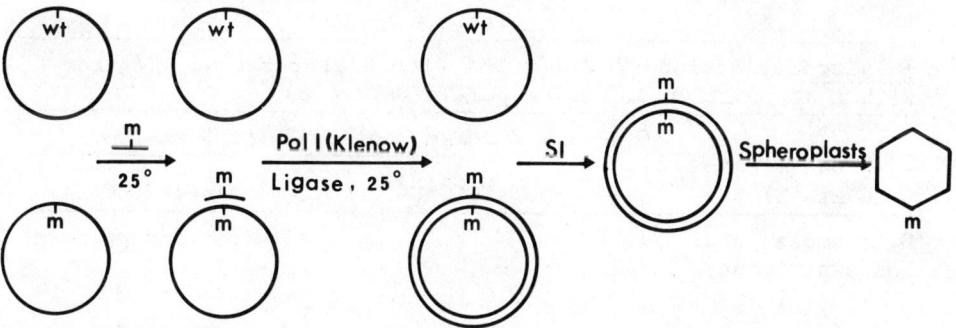

Figure 9. In vitro selection of mutant ∅X174 DNA using an oligodeoxyribonucleotide. The oligodeoxyribonucleotide can be the one used to introduce the mutation or a homolog (see Table 7). At 25°C, the mutant oligodeoxyribonucleotide (m) is a more efficient primer of E. coli DNA polymerase I, large fragment (Pol I, Klenow) on homologous (mutant) DNA template (m) than on heterologous (wild-type) DNA template (wt). Thus, the mutant DNA is selectively converted, after ligation, to closed circular duplex DNA that is resistant to S1 endonuclease. Transfection of E. coli spheroplasts then yields a phage population specifically enriched in the mutant.

An even more attractive approach, because of its simplicity and convenience, is the use of the mutating oligodeoxyribonucleotide to select for mutant DNA (18). The strategy for this is diagramed in Figure 9. The principle of the method is that the perfectly matched mutant oligodeoxyribonucleotide mutant DNA duplex should, at the appropriate temperature, be more efficiently converted to closed-circular duplex DNA and hence be resistant to the single-strand specific S1 endonuclease that will preferentially degrade wild-type DNA. Insertion of the DNA into E. coli will result in progeny enriched in mutant DNA. Results of experiments directed at determining the feasibility and efficiency of this approach are shown in Table 6; the objective was the enrichment of DNA with G at position 587 of ∅X174 DNA in a background of DNA with A at the same position. When a large excess of the mutating oligodeoxyribonucleotide was used to select for mutant DNA, an enrichment of about seven-fold was obtained with the most effective oligodeoxyribonucleotide. With a reduced excess of oligodeoxyribonucleotide, an enrichment of about 30-fold was obtained (Table 6). It is possible that the particular enrichment studied in these experiments was biologically favored because it involved the selection of wild-type DNA from an am background. Consequently the reverse selection was attempted, viz. selection

Table 6

In vitro selection of ɸX174 DNA with G at position 587 from DNA with A at position 587 using pGTATCCCACA (18).

% DNA with G at 587	% Phase Isolated with G at 587	
	-pGTATCCCACA	+pGTATCCCACA
0.25 pmole total DNA/experiment		60 pmole/experiment
0	0.0	9.0
1	1.3	18.0
10	10.0	60.0
20	17.0	80.0
50	46.0	100.0
0.76 pmole total DNA/experiment		20 pmole/experiment
0	0.0	12.0
1	1.4	33.0

of ɸX174 DNA with A at position 587 from a background with G at the same position. The results are summarized in Table 7 and shown that the selection method functions equally well in the reverse direction. It is clear that the simple procedure diagramed in Figure 9 provides a highly efficient and general method for enriching a mutant DNA by selection with the corresponding mutant oligodeoxyribonucleotide. The procedure can be carried out on a small scale (pmole amounts of DNA) and does not require any fractionation by physical or chemical means. As in all experiments directed at isolation of a specific mutant DNA, the

Table 7

In vitro selection of ɸX174 DNA with A at position 587 from DNA with G at position 587 using 20 pmole of oligodeoxyribonucleotide per 0.92 pmole of total DNA (18).

Oligodeoxyribonucleotide	% Phage Isolated with A at 587
--	5.0
pGTATCCT	16.0
pGTATCCTA	16.0
pGTATCCTAC	25.0
pGTATCCTACA	42.0

```
                                            End G    Start H
wild-type viral DNA         5' -C-C-A-C-T-T-A-A-G-[T-G-A]-G-G-T-G-A-T-T-T-[A-T-G]-T-T-T-G-G-T- 3'
                                              2911            Hph I             2930       2939
                                                              2920

mutant complementary                             3'  A-C-T-C-C-   -C-T-A-A-Ap 5'
oligodeoxyribonucleotide

                                            End G    Start H
mutant viral DNA            5' -C-C-A-C-T-T-A-A-G-[T-G-A]-G-G-A-T-T-T-[A-T-G]-T-T-T-G-G-T- 3'
```

Figure 10. Sequence of part of ØX174 DNA in the intercistronic region between genes G and H in the vicinity of position 2925. The termination codon (TGA) and the initiation codon (ATG) are boxed and the HphI recognition sequence is underlined. The synthetic decadeoxyribonucleotide is complementary to mutant (T at 2925 is deleted) ØX174 DNA. The deletion partially disrupts the ribosome binding sequence GAGGTGAT by converting it to GAGGAT, destroys a nonfunctional termination codon TGA and destroys the HphI recognition sequence, GGTGA.

ultimate test is sequence determination of the mutant DNA. The efficiency of the enrichment procedure, coupled with the fairly high efficiency of the initial oligodeoxyribonucleotide mutagenesis, ensures that sequence determination of only a very few clones of DNA will detect and characterize the desired mutant. The particular attraction of this method of isolating mutant DNA is that it makes possible the construction of mutants in phenotypically silent DNA sequences. This is particularly important in the study of eukaryote genes because they are most conveniently cloned and manipulated in prokaryote systems.

ISOLATION OF A PHENOTYPICALLY SILENT DELETION MUTANT OF ϕX174

In order to extend the scope of the method using oligodeoxyribonucleotide mutagens, a single nucleotide deletion has been constructed (60). The target was nucleotide 2925 in the intercistronic region between genes G and H of ϕX174 (Figure 10). Deletion of this nucleotide disrupts an extended sequence complementary to the 3'-terminus of 16S rRNA, eliminates a termination codon immediately upstream of the initiator ATG of gene G and disrupts an HphI recognition sequence. The mutagenic oligodeoxyribonucleotide (Figure 10) recognized only one site on ϕX174 DNA and this was shown to be the desired site by the use of the decadeoxyribonucleotide as a primer for terminator sequencing of the ϕX174 DNA (Figure 11). It was clear from these experiments that the decadeoxyribonucleotide with a deletion mismatch was as effective a primer for DNA polymerase as were oligodeoxyribonucleotides with transition or transversion mismatches. However, in the initial mutagenic experiment directed at deleting nucleotide 2925, the mutant was obtained with lower efficiency than anticipated (less than 5% rather than 10 to 40%). However, application of the in vitro selection method (three cycles) allowed isolation of the mutant with 100% efficiency (four out of four plaques were mutant phage). Figure 12, by direct sequence determination, demonstrates a 1:1 mixture of wild-type and mutant DNA after 2 cycles of selection. Extensive studies with the mutant phage demonstrated that it was completely phenotypically silent (60). This provides a very satisfactory demonstration of the potential of the in vitro selection method, particularly since the initial production of the mutant was less efficient than usual.

INTRODUCTION OF MUTATIONS INTO RECOMBINANT DNA IN PLASMID AND FILAMENTOUS PHAGE VECTORS

It was pointed out earlier that ϕX174 DNA is a good model system for most mutagenic studies that are likely to be under-

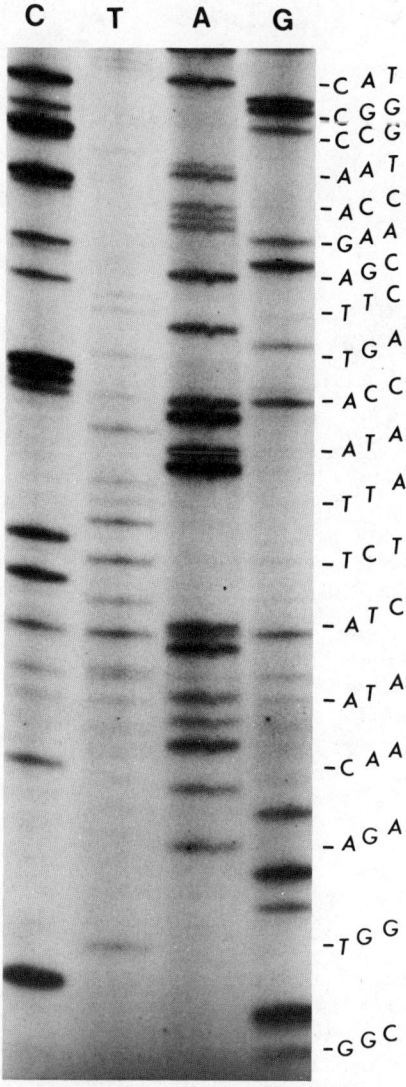

Figure 11. Sequence of part of the complementary strand of ØX174 DNA obtained by using the mutagenic oligodeoxyribonucleotide pAAATCCCTCA as a primer for Sanger dideoxyribonucleotide DNA sequencing with wild-type viral DNA as template. The sequence shown, reading from the bottom, starts at the nucleotide complementary to viral nucleotide 2912. The experiment demonstrates that the oligodeoxyribonucleotide, lacking the complement of viral nucleotide 2925, interacts specifically with the desired target.

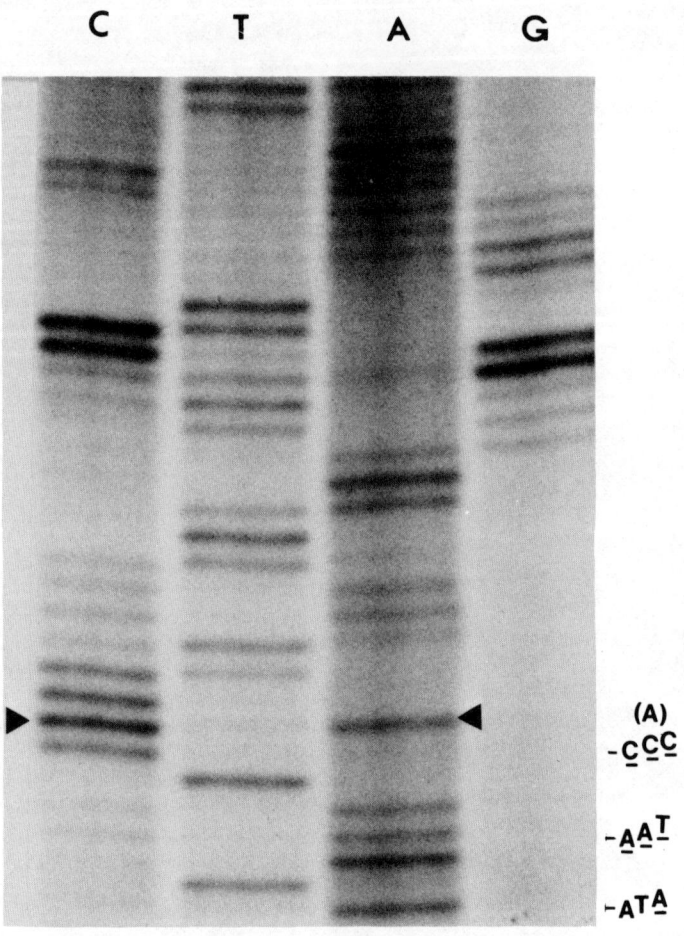

Figure 12. Sequence of mixture of wild-type and mutant DNA from which nucleotide 2925 is deleted after two cycles of enrichment of mutant DNA by the procedure depicted in Figure 9. The sequence was obtained by the enzymatic terminator method (1) using as a primer the oligodeoxyribonucleotide pAAACAATT (complementary to nucleotides 2829 to 2996 of ϕX174 DNA). The sequence (complementary to viral DNA) starting at nucleotide 2932 reads upwards from the bottom of the figure; it gives a unique pattern until the cluster of C residues and then is a series of doublets. The uppermost band in the doublet corresponds to wild-type ϕX174 DNA and the lower band to the mutant from which nucleotide 2925 is deleted. The ratio of the amounts of two bands in a doublet suggests that close to 50% of the DNA is mutant. The initial mixture of wild-type and mutant DNAs prior to enrichment contained less than 5% of mutant ϕX174 DNA (60).

taken because E. coli is the most commonly used host for recombinant DNAs. Consequently, extension of the method to recombinants carried in vectors derived from the single-stranded phage fd, fl and M13 should be quite straightforward. Studies on the use of synthetic oligodeoxyribonucleotides as primers for DNA synthesis in sequence studies have demonstrated these experiments can be carried out directly on denatured double-stranded DNA; at low temperatures (0° to 20°C) oligodeoxyribonucleotides interact rapidly with template DNA and the DNA duplex does not renature (68,69). Thus, extension of the methods for single-stranded DNA to double-stranded DNA also is straightforward.

An example of oligodeoxyribonucleotide mutagenesis applied to a clone of DNA in a single-stranded vector is a mutation in the promoter sequence of the conalbumin gene that was cloned in an fd vector (70). The mutational change is shown in Figure 3. This transversion, which changes one nucleotide in the Goldberg-Hogness sequence, also produces a XbaI site. The mutation was produced with an efficiency of about 4%.

In principle, oligodeoxyribonucleotide mutagenesis can be used to produce more extended changes than those involving a single nucleotide. Model studies on duplex stability with more extended mismatches have shown that they cause more destabilization of a duplex (66). Hence, longer mutagenic oligodeoxyribonucleotides are required than the decadeoxyribonucleotides that are effective for single nucleotide changes. A 21 base oligodeoxyribonucleotide was used to delete the 14 nucleotide intervening sequence of yeast tRNA^{Tyr} (Figure 14); the yeast gene was cloned in pBR322 (67). Thus, this experiment not only demonstrates that quite extensive changes can be made with synthetic oligodeoxyribonucleotides, but also further extends the range of recombinant vectors to which the method has been applied. The mutant was produced with an efficiency of about 4%. It is important to note that the target DNA, a tRNA gene, would have considerable secondary structure.

CONCLUSION AND PROSPECTS

This article describes the development of a highly specific method for constructing mutations in vitro with synthetic oligodeoxyribonucleotides, containing the desired change, as primers for E. coli DNA polymerase I (Klenow fragment) on circular wild-type DNA templates (15,17). Ligation, with DNA ligase, results in complete integration of the synthetic oligodeoxyribonucleotide into covalently closed-circular DNA. The resulting mutant can be screened or selected phenotypically. In the absence of a phenotype, the mutant can be detected in suitable cases where it also generates (63,70) or destroys (60) a restriction cleavage site. The mutagenic oligodeoxyribonucleotide can be used to screen for mutant complementary undecadeoxynucleotide

3' G-A-G-A-T-C-T-T-T-T-Cp 5'

5' -C-T-C-C-T-C-T-A-T-A-A-A-A-G-G-G-A- 3'
                           G
                           ↑
                 -30   -20

wild-type conalbumin DNA

Figure 13. Sequence of part of chicken conalbumin DNA, upstream of the transcription start in the vicinity of nucleotide -29 and the synthetic undecadeoxyribonucleotide complementary to mutant (G at -29) DNA (wild-type DNA has T at -29). The change T→G at position -29 changes the Goldberg-Hogness box (TATAAAA → TAGAAAA) and produces a XbaI recognition sequence (TCTAGA). The mutagenic experiment was carried out on a fragment of chicken DNA cloned in a phage fd vector (70).

---

mutant complementary 21-oligodeoxyribonucleotide

3' C-G-T-T-C-T-G-A-A-A-T-T   T-A-G-A-A-C-T-C-Tp 5'

5' -A-A-G-G-C-G-C-A-A-G-A-C-T-T-T-A-A-T-T-T-A-C-C-A-C-T-A-C-G-A-A-T-C-T-T-G-A-G-A- 3'

wild-type SUP6-o DNA

Figure 14. Sequence of center of the yeast SUP4-o gene (tyrosine-inserting ochre suppressor) spanning the 14-nucleotide intervening sequence and the 21-nucleotide oligodeoxyribonucleotide used to delete the intervening sequence. The mutagenic experiment was carried out on a fragment of yeast DNA cloned in the plasmid vector pBR322 (67).

mutant DNA by duplex hybridization (67). The most convenient and general approach to isolating a mutant is to use the mutagenic oligodeoxyribonucleotide to select for mutant DNA (18).

Mutants can be constructed with circular single-stranded or circular double-stranded DNAs that replicate in E. coli (15,17, 67,70). This, coupled with in vitro selection, allows cloned eukaryotic DNA to be mutated specifically.

Potential applications of the method are legion. Systematic changes in protein coding sequences have been carried out (15,16, 52), in regions involving transcription initiation (70) and processing (67), and in translation signal sequences (60,63). The method opens the way to studies on the role of particular amino acids in protein structure and function. Modification of DNA by generation or deletion of restriction endonuclease sites will allow more precise construction of recombinant DNAs. Studies on the sequence requirements of origins of DNA replication and even more extensive definition of transcription signals, RNA processing signals and translation signals will be possible.

Areas of uncertainty about the method lie in the mismatch repair mechanisms of the host cell. Particularly in E. coli, the mechanisms of mismatch repair are becoming better understood and mismatch repair should present no insurmountable obstacle (50). Perhaps the DNAs constructed as intermediates in the mutagenesis will prove to be useful substrates for biochemical studies on repair mechanisms.

Acknowledgments: This approach to specific mutant construction was conceived in discussions with C.A. Hutchison III during an extended study leave in 1975-1976 at the M.R.C. Laboratory of Molecular Biology in Cambridge, England, and a number of experiments have been carried in collaboration with his laboratory (15, 16,63). The study leave was supported by the Medical Research Council of Canada. Research in the authors' laboratory was supported by the Medical Research Council of Canada; M.S. is a Career Investigator of the Medical Research Council of Canada.

REFERENCES

1   Sanger, F., Nicklen, S. and Coulson, A.R. (1977) Proc. Nat. Acad. Sci. U.S.A. 74, 5463-5467.
2   Maxam, A. and Gilbert, W. (1977) Proc. Nat. Acad. Sci. U.S.A. 74, 560-564.
3   Barrell, B.C., Air, G.M. and Hutchison, C.A. III (1976) Nature 264, 34-41.
4   Shaw, D.C., Walker, J.E., Northrop, F.D., Barrell, B.G., Godson, G.N. and Fiddes, J.C. (1978) Nature 272, 510-515.
5   Barrell, B.G., Bankier, A.T. and Drouin, J. (1979) Nature 282, 189-194.

6   Lai, C.-J. and Nathans, D. (1974) J. Mol. Biol. 89, 179-193.
7   Mertz, J.E., Carbon, J., Herzberg, M., Davis, R.W. and Berg, P. (1975) Cold Spring Harbor Symp. Quant. Biol. 39, 69-84.
8   Carbon, J., Shenk, T. and Berg, P. (1975) Proc. Nat. Acad. Sci. U.S.A. 72, 1392-1396.
9   Boyer, H.B., Bettlatch, M., Bolivar, F., Rodriguez, R.L., Heyneker, H.L., Shire, J. and Goodman, H.M. (1977) in Recombinant Molecules: Impact on Science and Society (Beers, R.F. Jr. and Bassett, E.G., eds.) pp. 9-20, Raven Press, New York, NY.
10 Heffron, F., So, M. and McCarthy, B.J. (1978) Proc. Nat. Acad. Sci. U.S.A. 75, 6012-6016.
11 Borrias, W.E., Wilschut, I.J.C., Vereijken, J.M., Weisbeek, P.J. and van Arkel, G.A. (1976) Virology 70, 195-197.
12 Salganik, R.I., Dianov, G.L., Ovchinnikova, L.P., Vorinina, E.N., Kokoza, E.B. and Mazin, A.V. (1980) Proc. Nat. Acad. Sci. U.S.A. 77, 2796-2800.
13 Shortle, D., Pipas, J., Lazarowitz, S., DiMaio, D. and Nathans, D. (1979) in Genetic Engineering: Principles and Methods (Setlow, J.K. and Hollaender, A., eds.) Vol. I, pp. 73-92, Plenum Press, New York, NY.
14 Weissmann, C., Nagata, S., Taniguchi, T., Weber, H. and Meyer, F. (1979) in Genetic Engineering: Principles and Methods (Setlow, J.K. and Hollaender, A., eds.) Vol. I, pp. 133-150, Plenum Press, New York, NY.
15 Hutchison, C.A. III, Phillips, S., Edgell, M., Gillam, S., Jahnke, P. and Smith, M. (1978) J. Biol. Chem. 253, 6551-6560.
16 Gillam, S., Jahnke, P., Astell, C., Phillips, S., Hutchison, C.A. III and Smith, M. (1979) Nucl. Acids Res. 6, 2973-2985.
17 Gillam, S. and Smith, M. (1979) Gene 8, 81-97.
18 Gillam, S. and Smith, M. (1979) Gene 8, 99-106.
19 Lederberg, J. (1960) Science 131, 269-276.
20 Wu, R., Bahl, C.P. and Narang, S.A. (1978) Progr. Nucl. Acid Res. Mol. Biol. 21, 102-141.
21 Itakura, K. (1980) TIBS 5, 114-116.
22 Chang, L.M.S. and Bollum, F.J. (1971) Biochemistry 11, 536-542.
23 Hsieh, W.T. (1971) J. Biol. Chem. 246, 1780-1784.
24 Gillam, S. and Smith, M. (1972) Nature New Biology 238, 233-234.
25 Gillam, S. and Smith, M. (1974) Nucl. Acids Res. 1, 1631-1647.
26 Gillam, S., Waterman, K., Doel, M. and Smith, M. (1974) Nucl. Acids Res. 1, 1649-1664.
27 Gillam, S., Waterman, K. and Smith, M. (1975) Nucl. Acids Res. 2, 613-624.
28 Gillam, S., Rottman, F., Jahnke, P. and Smith, M. (1977) Proc. Nat. Acad. Sci. U.S.A. 74, 96-100.

29 Gillam, S., Jahnke, P. and Smith, M. (1978) J. Biol. Chem. 253, 2532-2539.
30 Trip, E.M. and Smith, M. (1978) Nucl. Acids Res. 5, 1529-1538.
31 Trip, E.M. and Smith, M. (1978) Nucl. Acids Res. 5, 1539-1549.
32 Gillam, S. and Smith, M. (1980) Methods Enzymol. 65, 687-701.
33 Astell, C. and Smith, M. (1971) J. Biol. Chem. 246, 1944-1946.
34 Astell, C.R. and Smith, M. (1972) Biochemistry 11, 4114-4120.
35 Astell, C.R., Doel, M.T., Jahnke, P.A. and Smith, M. (1973) Biochemistry 12, 5068-5074.
36 Gillam, S., Waterman, K. and Smith, M. (1975) Nucl. Acids Res. 2, 625-634.
37 Thomas, C.A. Jr. (1966) Progr. Nucl. Acid Res. Mol. Biol. 5, 315-337.
38 Weisbeek, P.J. and van de Pol, J.H. (1970) Biochim. Biophys. Acta 224, 328-338.
39 Hutchison, C.A. III and Edgell, M.H. (1971) J. Virol. 8, 181-189.
40 Sanger, F., Coulson, A.R., Friedmann, T., Air, G.M., Barrell, B.G., Brown, N.L., Fiddes, J.C., Hutchison, C.A. III, Slocombe, P.M. and Smith, M. (1978) J. Mol. Biol. 125, 225-246.
41 Tessman, E.S. and Tessman, I. (1978) in The Single-Stranded DNA Phages (Denhardt, D.T., Dressler, D. and Ray, D.S., eds.), pp. 9-29, Cold Spring Harbor Laboratory, Cold Spring Harbor, NY.
42 Goulian, M. (1968) Proc. Nat. Acad. Sci. U.S.A. 61, 284-291.
43 Goulian, M. (1968) Cold Spring Harbor Symp. Quant. Biol. 33, 11-20.
44 Goulian, M., Goulian, S.H., Codd, E.E. and Blumenfield, A.Z. (1973) Biochemistry 12, 2893-2901.
45 Goulian, M., Kornberg, A. and Sinsheimer, R.L. (1967) Proc. Nat. Acad. Sci. U.S.A. 58, 2321-2328.
46 Sherman, L.A. and Gefter, M.L. (1976) J. Mol. Biol. 103, 61-76.
47 Klenow, H., Overgaard-Hanson, K. and Patkar, S.A. (1971) Eur. J. Biochem. 22, 371-381.
48 Razin, A., Hirose, T., Itakura, K. and Riggs, A.D. (1978) Proc. Nat. Acad. Sci U.S.A. 75, 4268-4270.
49 Dressler, D., Hourcade, D., Koths, K. and Sims, J. (1978) in The Single-Stranded DNA Phages (Denhardt, D.T., Dressler, D. and Ray, D.S., eds.), pp. 187-214, Cold Spring Harbor Laboratory, Cold Spring Harbor, NY.
50 Hanawalt, P.C., Cooper, P.K., Ganesan, A.K. and Smith, C.A. (1979) Ann. Rev. Biochem. 48, 783-836.

51 Baas, P.D. and Jansz, H.S. (1978) in The Single-Stranded DNA Phages (Denhardt, D.T., Dressler, D. and Ray, D.S., eds.), pp. 215-244, Cold Spring Harbor Laboratory, Cold Spring Harbor, NY.
52 Bhanot, O.S., Khan, S.A. and Chambers, R.W. (1979) J. Biol. Chem. 254, 12684-12693.
53 McCallum, D. and Smith, M. (1977) J. Mol. Biol. 116, 29-30.
54 Staden, R. (1977) Nucl. Acids Res. 4, 4037-4051.
55 Staden, R. (1978) Nucl. Acids Res. 5, 1013-1015.
56 Staden, R. (1979) Nucl. Acids Res. 6, 2601-2610.
57 Korn, L.J., Queen, C.L. and Wegman, M.N. (1977) Proc. Nat. Acad. Sci. U.S.A. 74, 4401-4404.
58 Queen, C.L. and Korn, L.J. (1980) Methods Enzymol. 56, 595-609.
59 Lomant, A.J. and Fresco, J.R. (1975) Progr. Nucl. Acid Res. Mol. Biol. 15, 185-218.
60 Gillam, S., Astell, C.R. and Smith, M. (1980) Gene (in press).
61 Kornberg, A. (1980) DNA Replication, pp. 101-166, W.H. Freeman and Co., San Francisco, CA.
62 Petes, T. (1980) Annu. Rev. Biochem. 49, 845-876.
63 Gillam, S., Astell, C.R., Jahnke, P., Hutchison, C.A. III and Smith, M. (1980) (unpublished data).
64 Steitz, J.A. (1979) in Biological Regulation and Development (Goldberger, R.F. ed.), Vol. I, pp. 349-399, Plenum Press, New York, NY.
65 Szostak, J.W., Stiles, J.I., Tye, B.-K., Chiu, P., Sherman, F. and Wu, R. (1979) Methods Enzymol. 68, 419-428.
66 Dodgson, J.B. and Wells, R.D. (1977) Biochemistry 16, 2367-2374.
67 Wallace, R.B., Johnson, P.F., Tanaka, S., Schold, M., Itakura, K. and Abelson, J. (1980) Science 209, 1396-1400.
68 Smith, M., Leung, D.W., Gillam, S., Astell, C.R., Montgomery, D.L. and Hall, B.D. (1979) Cell 16, 753-761.
69 Kurjan, J., Hall, B.D., Gillam, S. and Smith, M. (1980) Cell 20, 701-709.
70 Wasylyk, B., Derbyshire, R., Guy, A., Molko, D., Roget, A., Teoule, R. and Chambon, P. (1980) Proc. Nat. Acad. Sci. U.S.A. 77 (in press).

EVOLUTION OF THE INSERTION ELEMENT IS1 THAT CAUSES GENETIC ENGINEERING OF BACTERIAL GENOMES IN VIVO

E. Ohtsubo, K. Nyman, K. Nakamura and H. Ohtsubo

Department of Microbiology
School of Medicine
State University of New York at Stony Brook
Stony Brook, NY 11794

INTRODUCTION

Although the joining of two different sequences or genomes can be performed in vitro with various genetic engineering techniques, the event also occurs in vivo in bacterial cells at high frequency. Recent research on a class of in vivo recombination events, called illegitimate recombination, including deletions, transpositions and inversions of DNA segments on a genome, and fusion (or cointegration) of two different genomes, has shown that most of these events are mediated by insertion elements (IS) and transposable elements (Tn or transposon). These are discrete segments of DNA that move by themselves from one site to another on the same or on different genomes.

In this chapter, we will first describe the genetic rearrangements (or genetic engineering in vivo) mediated by IS elements, focusing on the insertion element IS1. Next we will discuss our research on the distribution of the IS1 sequence in various gram-negative bacteria, the evolution of the element and our present understanding of the creation of transposable elements responsible for multiple antibiotic resistance and enterotoxin production in bacteria, the existence of which is a serious problem in medical microbiology.

THE INSERTION ELEMENT IS1 RESPONSIBLE FOR GENETIC REARRANGEMENT

IS1 is the smallest of the known active insertion elements (1-6). It was originally found in mutants of various operons as

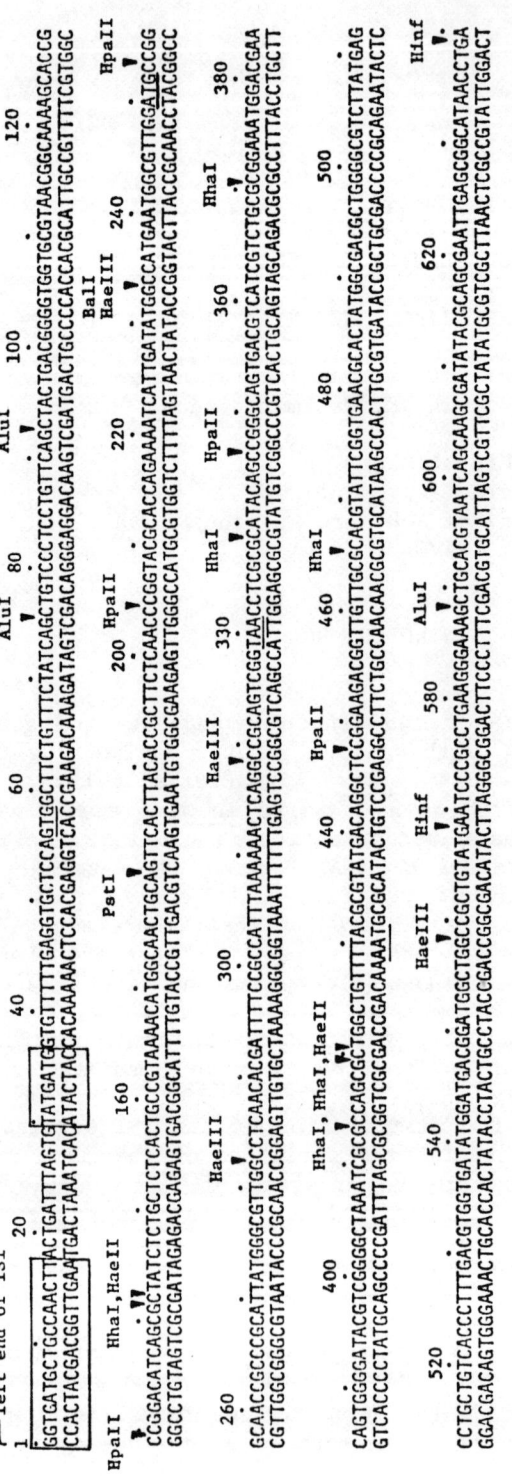

Figure 1. The entire nucleotide sequence of IS1 as determined previously (10). Possible RNA polymerase binding and recognition sequences are indicated at both end regions of IS1. Possible translatable reading frames, which may code for hypothetical polypeptides of 91, 167 and 96 amino acids, are present at 56 (G residue) to 328 (G residue), 250 (A residue) to 753 (A residue) and 719 (A residue) to 432 (G residue), respectively.

an inserted DNA sequence, causing strong polar effects on these operons (7-9). The nucleotide sequence of IS1 has been determined to be 768 base pairs (10,11) as shown in Figure 1. Insertion of the IS1 sequence into any site has been shown to generate a duplication of the nine base-pair sequence at the target site of the recipient molecule (12,13). The DNA sequence of IS1 shows that it could code for some small proteins. A sequence of about 35 base pairs at each end of the IS1 appears in an inverted orientation. Recent genetic analysis indicated that mutations occurring in the inside or end regions of IS1 decrease or abolish its translocation activity (14).

The IS1 sequence appears as repeated DNA sequences in a group of resistance plasmids, such as R100, R6 and R1, responsible for multiple antibiotic resistance (15), and in the E. coli K12 chromosome (16,17). One copy of IS1 also appears in the bacteriophage P1 (18).

Studies with the resistance plasmid R100 demonstrate the various genetic rearrangements mediated by IS1.

Figure 2 represents the genetic and physical structure of R100 (89.3 kilobases) which carries the antibiotic resistance genes for chloramphenicol (chl), sulfanilamide (sul), streptomycin (str), fusidic acid (fus) and tetracycline (tet), and the heavy metal ion mercury (mer). Note that IS1 appears on this plasmid two times in a direct orientation and flanks the resistance genes chl^r, fus^r, sul^r, str^r, and mer^r (15). It has recently been shown that this region, called the r-determinant (r-det) region, is transposable together with two direct copies of IS1 into another plasmid genome (19). It has also been demonstrated that the chloramphenicol resistance gene adjacent to one copy of IS1 can be transposed. The final structure of the transposed segment contains two copies of IS1 flanking the chl^r gene in a direct orientation (Figure 2), although the second IS1 sequence is not present in the R100 plasmid (19-21). The resulting DNA segment (Tn204 in Figure 2) is an active transposon and can be transposed into other genomes (19-21). Transposons similar to Tn204 carrying two copies of IS1 have been identified: Tn9 carries two IS1s that flank the chl^r gene in a direct orientation (22-24) and Tn1681 contains two IS1s that flank the enterotoxin-producing gene in an inverted orientation (25,26). These results indicate that DNA segments flanked by two IS1s in either a direct or an inverted orientation are transposable.

It is known that the r-determinant region can be amplified on the R100 genome under certain selective conditions in a Proteus mirabilis strain (27). It is believed that the phenomenon, called transition, is due to the ability of the IS1 element to form the transition molecules containing a tandem duplication of the r-determinant region as shown in Figure 2 (15,27,28). A phenomenon similar to transition was found to occur in any plasmid carrying a transposon with two copies of IS1 in a direct

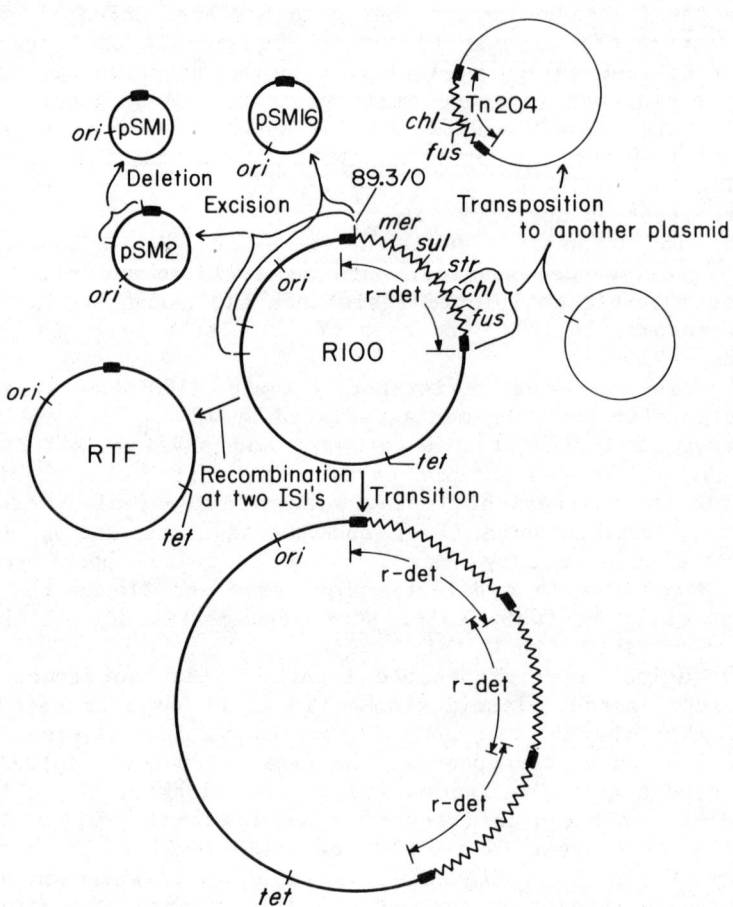

Figure 2. Physical structure of the resistance plasmid R100 and its derivatives. R100 (89.3 kilobases) has a kilobase coordinate system that starts at 0 and goes to 89.3 as indicated. Note that R100 carries direct repeats of IS1 (■), which flank the r-determinant regions (r-det) containing genes responsible for resistance to chloramphenicol (chl), sulfanilamide (sul), streptomycin (str), fusidic acid (fus) and mercury (mer). RTF deletes the r-determinant region of R100 by recombination between the two IS1 sequences and therefore carries a single copy of IS1. The pSM plasmids delete most of R100 but carry the essential region of R100 replication and a single copy of IS1.

orientation such that it can give rise to tandem duplications of the transposon in an E. coli strain (29).

In addition, the IS1 sequences are involved in the production of deletion mutants of R100. Analysis of these mutants shows that the deletion sites are usually specific in terms of involvement of the IS1 sequence. As shown in figure 2, recombination can occur at the two IS1s on R100 to generate the small-sized plasmid called RTF (28-32). Recombination between IS1 and a nonhomologous region also occurs, giving rise to various sizes of deletion mutants of R100 (see Figure 2, pSM plasmids). Note in these cases that the IS1 element is always at the site of the excisions or deletions (33,34).

The IS1 sequence also mediates the fusion of two completely different plasmid genomes by cointegration. Figure 3 shows a typical example of cointegration between pSM1, a derivative of R100, and pHS1, a derivative of the resistance plasmid pSC101. The structure of the cointegrated plasmid pMZ4 indicates that the plasmid containing one copy of IS1 is a transposon and can integrate into a second plasmid containing no IS1 sequence in a characteristic way: IS1 is duplicated at the recombinational junctions (35,36). Nucleotide sequence analysis of the resulting recombinant plasmids shows that a nine base-pair sequence at the target site appears as a direct duplication at the junctions between each IS1 and the recipient plasmid. This suggests that both the IS1 insertion event and cointegration mediated by IS1 occur by a common molecular mechanism, since IS1 insertion also generates a nine base-pair direct duplication of the target site. It has also been reported that two plasmids, both of which carry IS1, can recombine at the IS1-IS1 homology to form cointegrated plasmids (14,37).

As briefly reviewed above, IS1 can cause many different rearrangements of the R100 plasmid genome, generally in the absence of the recA function. It has been reported that IS1 causes rearrangements of not only R100 but also of other genomes carrying IS1, such as phage P1 and the E. coli K12 chromosome (19,37,38). It should be noted that other IS and Tn elements behave like IS1 and cause various genetic rearrangements of the genomes carrying them (for review see refs. 1-6).

DISTRIBUTION OF IS1 IN E. COLI AND ITS DERIVATIVES

We have examined the number of copies of IS1 in various gram-negative bacteria by the Southern blotting and hybridization techniques (41). The ^{32}P-labeled probes that specifically hybridize to the IS1 sequence were prepared from a small R100 plasmid derivative, pSM1, carrying a single copy of IS1 as follows (see Figure 2). The pSM1 DNA was cleaved with restriction endonucleases such as HaeII, HaeIII, PstI, and/or HinfI. The

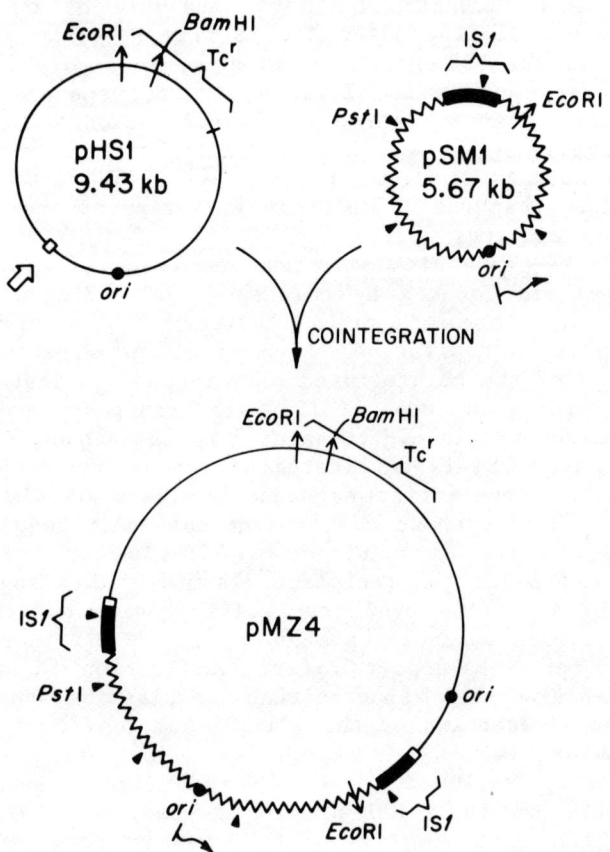

Figure 3. Physical structures of the parental plasmids (pHS1 and pSM1) and one of the cointegrate plasmids (pMZ4) (36). Cleavage sites for the restriction endonucleases EcoRI, BamHI and PstI are shown by the arrows and solid triangles. Plasmid pSM1 containing a copy of IS1 recombines with pHS1 at a site, shown by an open arrow, to give rise to the cointegrate plasmid pMZ4. The resulting pMZ4 contains parental plasmid sequences flanked by direct repeats of IS1 as shown. The nine-nucleotide sequence at the target site (□) is also duplicated in a direct orientation in pMZ4 as indicated. Approximate locations of the origin of replication of the plasmids and the direction of replication of pSM1 (39,40) are also shown.

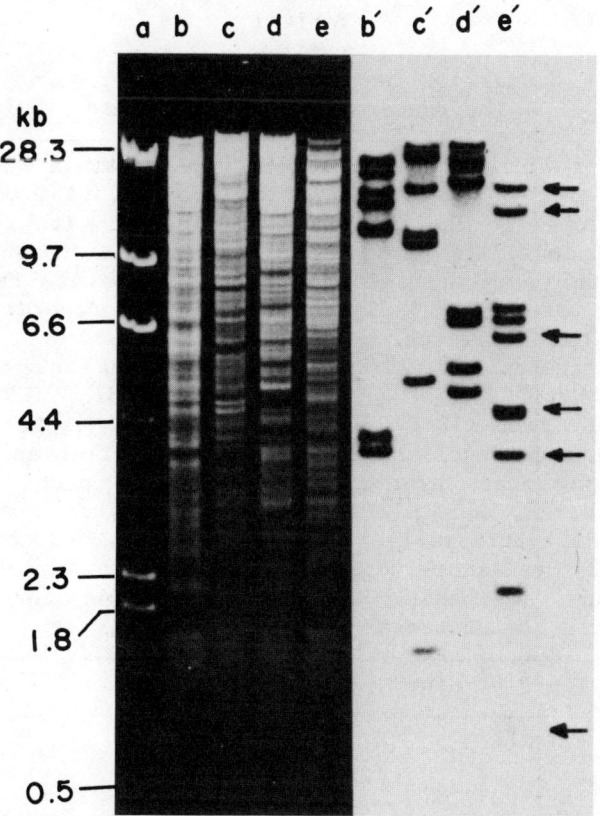

Figure 4. Ethidium bromide-stained 0.7% agarose gels and autoradiograms of the filter obtained from the gels after hybridization with ^{32}P-labeled IS1 DNA probe (β). Columns b through e are the total DNA (3 μg) of an E. coli derivative, JE5519, digested with EcoRI, HindIII, BamHI and PstI, respectively. Column a is a HindIII digest of lambda DNA used as a length standard (in kilobases) (53). Columns b' through e' are autoradiographs made after hybridization with the IS1 probe. Note that the PstI digest contained the maximum number of bands, 10, hybridizable to the IS1 probe. Six of these bands, indicated by arrows, are identical to those seen in PstI digests of the E. coli K12 derivatives EMG2, Yme1 and W1485 (see Table 1).

location of the cleavage sites for these enzymes has been determined in IS1 as well as the entire pSM1 sequence (42). With these enzymes we prepared three DNA fragments named α, β and γ from different portions of IS1: 193-233, 319-558 and 639-758, respectively (see Figure 1). The 5' ends of both strands of these fragments were labeled with ^{32}P using gamma-^{32}P-ATP and

Table 1

Summary of the Number of IS1 Copies in the Bacterial Strains Examined

The number of IS1 copies in each bacterial strain was estimated from the number of bands seen after hybridization with ^{32}P-labeled IS1 probes to the DNA that was digested with the restriction endonuclease PstI, except in Shigella dysenteriae and Shigella sonnei, which could not be digested with PstI. EcoRI and HindIII were used to digest these (43). According to Bachmann (45), the following E. coli K12 strains were obtained by one-step mutagenesis: Yme1 and W1485 from the original E. coli K12 strain; W2637 from W1485; W3110 from W2637. JE5519 was derived from E. coli K12 by at least ten steps. Note: EMG2, Yme1 and W1485 showed identical gel band patterns after examination of their PstI fragments hybridizable to the IS1 probe (see Figure 4). All other E. coli K12 derivatives contained more copies of the sequences hybridizable to IS1, in addition to the six bands seen in EMG2, Yme1 and W1485, as demonstrated by the characteristic position of the bands. See Nyman et al. (44) for sources and references for the bacterial strains listed.

Strain	IS1 copies
Escherichieae Tribe	
Escherichia coli K12 EMG2 (wild-type)	6
Yme1	6
W1485	6
W2637	8
W3110	7
W3110	8
W3110trpE$_{9914}$	9
W2252 HfrC	7
JE5519	10
Escherichia coli C	3
Escherichia coli W	0
Shigella dysenteriae SH16	>40*
Shigella sonnei	>30*
Shigella boydii	2
Shigella flexneri	>40*
Salmonella typhimurium	0
Citrobacter freundii	0
Edwardsiella tarda	0
Klebsielleae Tribe	
Enterobacter aerogenes	0
Klebsiella aerogenes	1

Table 1 (contd.)

Strain	IS1 copies
Klebsielleae Tribe (contd.)	
Serratia marcescens	2
Erwinieae Tribe	
Erwinia amylovora	0
Proteeae Tribe	
Proteus mirabilis	0
Proteus morganii	0

*The exact IS1 copy number was difficult to estimate since several of the bands appeared as doublets or more.

polynucleotide kinase, and were then separated into single strands according to the method described by Maxam and Gilbert (43). We examined whether the ^{32}P end-labeled strands could hybridize specifically with restriction fragments containing the IS1 sequence by using various plasmid DNAs such as R100 and derivatives of R100 carrying one or more copies of IS1, as listed in Figure 2. We confirmed that the ^{32}P-labeled IS1 probes hybridize specifically with the restriction fragments of these plasmids containing the appropriate portion of the IS1 sequence (44).

Chromosomal DNA was isolated from the E. coli K12 derivative JE5519 and cleaved with the restriction endonucleases EcoRI, BamHI, HindIII and PstI and examined for the presence of a sequence hybridizable to IS1 of R100. Figure 4 shows an ethidium bromide stained gel of the DNA cleaved with the four restriction endonucleases mentioned above. The DNA fragments seen in the agarose gel were denatured with alkali and then transferred onto a nitrocellulose filter according to the method described by Southern (41). The filter was hybridized with one of the ^{32}P-labeled single strands of IS1 fragment β (see above). Figure 4 (a' - c') shows an autoradiogram after hybridization with the ^{32}P-labeled probe. The results show that each enzyme generated distinctive DNA bands hybridizable to the probe. PstI, which cleaves the IS1 sequence once (see Figure 1), generated the largest number of bands, 10, the most realistic estimate of the number of IS1 copies in the chromosome, since PstI will disrupt the linkage of two directly repeated copies of IS1 and give rise to DNA fragments containing only a part of one IS1.

We examined the DNA of other E. coli K12 derivatives as well as E. coli species other than E. coli K12. These are listed in Table 1. The results of autoradiograms show that the strain EMG2, which is supposed to be a wild-type E. coli strain (45), contains six copies of a sequence hybridizable to the IS1 probe and shows the same pattern as that seen in Ymel and W1485 (see the pattern shown in Figure 4). All other derivatives listed in Table 1 contain more copies of sequences hybridizable to the IS1 probe in addition to the six bands seen in EMG2, Ymel and W1485. It is interesting to note that two W3110 strains obtained from different sources showed differences in the number of copies of the sequence hybridizable to IS1. This observation is not surprising since each bacterial clone had the same amount of time for chromosomal rearrangement and changes in the number of copies. Nevertheless, the fact that some strains of E. coli K12, such as EMG2, Ymel and W1485, showed the identical pattern of six bands hybridizable with the IS1 probe may suggest that this is the number of copies of IS1 present in the chromosome of E. coli K12 when it is originally isolated from a natural source.

E. coli C was found to contain three copies of the sequence hybridizable to IS1 of R100, whereas E. coli W contained no such sequence. This result indicates that the distribution of IS1 is unique for each species even though these species are closely related.

DISTRIBUTION OF THE SEQUENCE OF R100 HYBRIDIZABLE TO IS1 IN VARIOUS OTHER STRAINS OF THE ENTEROBACTERIACEAE

Total DNA was isolated from various strains of the Enterobacteriaceae family as listed in Table 1. Species in the family Enterobacateriaceae are grouped into tribes as described in "Bergey's Manual of Determinative Bacteriology" (46). A phylogenetic tree of the Enterobacteriaceae has recently been postulated based on an immunological comparison of enzymes (47), comparison of the 5s rRNA sequences (48) and comparison of the ribosomal proteins (49). Bacteria in the tribe Escherichieae are more closely related to those in the tribe Klebsielleae than the other tribes Erwinieae and Proteeae. Within the tribe Escherichieae, E. coli species are more closely related to Shigella species than others (47-49). These relationships are generally in agreement with various other measures of biochemical relatedness (50,51) or with the numerical taxonomic groupings (46) of the Enterobacteriaceae.

As summarized in Table 1, Klebsiella aerogenes and Serratia marcescens generated one and two DNA bands, respectively, hybridizable with the IS1 probe. Shigella boydii generated two DNA bands hybridizable with the IS1 probe, but three other Shigella strains were found to generate more than 30 bands hybridizable with the same IS1 probe. However, the pattern of ^{32}P-hybrid bands obtained was completely different in each species.

Shigella dysenteriae SH16 harbors two plasmids of approximately 2 and 30 kilobases each (52). Hybridization analysis of these plasmids shows that the 30-kilobase plasmid contains one copy of a sequence hybridizable to the IS1 probe. Shigella sonnei harbored at least four different kinds of plasmids in the 4 to 7 kilobase range (H. Ohtsubo, unpublished data). None was found to contain any sequence hybridizable to IS1. These results indicate that the IS1 sequences seen in these two Shigella species are present mostly in the bacterial chromosome. (Some of E. coli K12 strains contained the plasmid F; F did not contain any sequence hybridizable to IS1.) We have not examined whether the other bacterial species harbored plasmids or whether these plasmids, if present, carry IS1 sequences.

It is interesting to note that the distribution of IS1 sequences in various gram-negative bacteria does not correlate linearly with the taxonomic or phylogenetic relationship of these bacteria. Besides E. coli and Shigella strains, we could detect the IS1 sequence only in K. aerogeneses and S. marcescens. S. typhimurium, C. freundii and E. tarda, which did not contain IS1, are more closely related to E. coli or Shigella than K. aerogenes or S. marcescens (see Table 1). Although E. aerogenes belongs to the tribe Klebsielleae, as do K. aerogenes and S. marcescens, it did not contain IS1. Furthermore, as described previously, E. coli W did not contain IS1 even though all other E. coli species examined did contain IS1. Thus the evolution of the IS1 sequences in these bacteria did not take place simultaneously with the sequence divergence of the bulk of the bacterial DNA.

ISO-INSERTION ELEMENTS OF IS1 PRESENT IN SHIGELLA STRAINS

In the previous section, we described some Enterobacteriaceae strains, including E. coli K12, that contain sequences hybridizable to the IS1 sequence of R100. Particularly interesting are the Shigella species containing multiple copies of these sequences. Due to the large number, these repeated sequences must appear on the chromosome in both a direct and inverted orientation. In fact, many inversion loops with duplex stems of the size of IS1 have been seen after the purified DNA of Shigella dysenteriae was denatured, renatured and observed in an electron microscope (52). It has been reported that the duplex DNA stem of the inversion loops present on various resistance plasmids can be isolated to purity by digesting the single-stranded portion of the inversion loops with S1 nuclease (54). We used this method to detect the presence of inverted repeat sequences on the bacterial chromosomes, particularly in the Shigella strains. The bacterial DNA was used for denaturation and renaturation experiments without any extensive shearing, since the larger the bacterial DNA fragment, the higher the probability of inclusion of

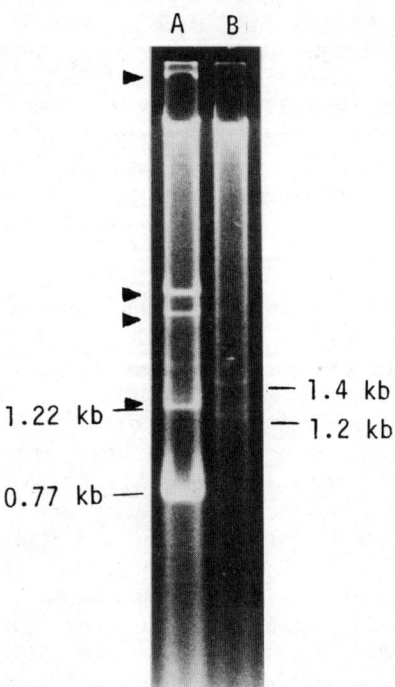

Figure 5. Electrophoresis in 1.4% agarose gels showing the generation of characteristic duplex DNA bands. (A) Shigella dysenteriae and (B) E. coli K12 (W3110) chromosomal DNA after denaturation, renaturation and S1 nuclease treatment. The four DNA bands in the gel for the Shigella dysenteriae strain (▶) are, from the top: open circular duplex configuration of a 30 kilobase plasmid, and open circular, linear and covalently closed circular duplex configurations of another 2 kilobase plasmid.

duplex regions in that strand after renaturation. After S1 treatment, to recover the duplex, we used as an additional step hydroxylapatite to concentrate duplex DNA fragments (see ref. 52 and above for further description of the method). Figure 5 shows the results of the analysis of E. coli K12 and Shigella dysenteriae DNA by gel electrophoresis. These results show that both strains produce a family of gel bands, some of which are similar in size. Both strains also produce DNA fragments similar in size to known insertion sequences such as IS1, IS2 and IS3. Note, however, that the Shigella strain produced an extremely dense band of about 770 base pairs in length which is similar in size to IS1. Shigella sonnei and Shigella flexneri also contained an unusually large amount of repeated sequences 770 base pairs in length (data not shown). We looked for inverted repeat sequences

in Proteus mirabilis and Salmonella typhimurium and found that
these strains and E. coli K12 did not contain any inverted repeat
sequences with unusually high copy numbers. Although there are
biochemical and genetic indications that the relationship between
Shigella and E. coli is very close (55-64), their chromosome
structure seems to be different when examined on the basis of the
presence of repeated sequences.

We purified the 770 base-pair inverted repeat sequences from
the Shigella strains and further examined them by restriction endonuclease cleavage to determine whether these sequences were the
insertion sequence IS1 or whether the DNA preparations contained
sequences different from IS1. Restriction endonucleases such as
HpaII, HaeII and HinfI were useful for this purpose, since these
enzymes are known to cleave the IS1 sequence at several sites,
the location of which has been determined (10).

Figure 6 shows the gel electrophoresis of purified inverted
repeat sequences of S. dysenteriae DNA after digestion with these
restriction endonucleases. The results show that there are two
major components that generated two different families of gel
bands with different intensities, as shown in schematic representations of gel bands, since the total length of each family of
DNA bands is similar in size to the original 770 base-pair fragment. There are several other minor components present in the
DNA preparation, as shown by the very faint bands close to the
top of the gel (see HaeII digests which in particular show such
bands). Of the two major components, the predominant one was
identified as a species other than IS1, since none of the bands
correspond to any of the restriction fragments of IS1 as indicated by Figure 6. On the basis of the cleavage patterns, this component has six cleavage sites for HpaII, and four cutting sites
for HaeII (see Figure 6). Another predominant component in the
DNA preparation was indeed the IS1 sequence, since it generated
fragments corresponding to those of IS1, except for the two fragments containing the ends of IS1. We named the major component
the $\nu\xi$ sequence for convenience.

The fragments isolated from S. sonnei were found to be pure
IS1 showing the expected cleavage products with several restriction endonucleases. The fragments isolated from S. flexneri
showed a different pattern from those of S. sonnei, but the
cleavage products of some endonucleases are the same, indicating
that the fragments hybridizable to the IS1 sequence have sequence
differences such that base changes occurred in restriction cutting sites to abolish the sites. We called these sequences isoinsertion sequences of IS1.

We further investigated the 770 base-pair sequence in Shigella dysenteriae by labeling its 5'-ends with ^{32}P and cleaving
with the restriction enzyme AluI. The result after gel electrophoresis and autoradiography of the digest shows two pairs of
comparatively dense bands as well as other minor components which
may or may not be cleaved with AluI, as shown in Figure 7.

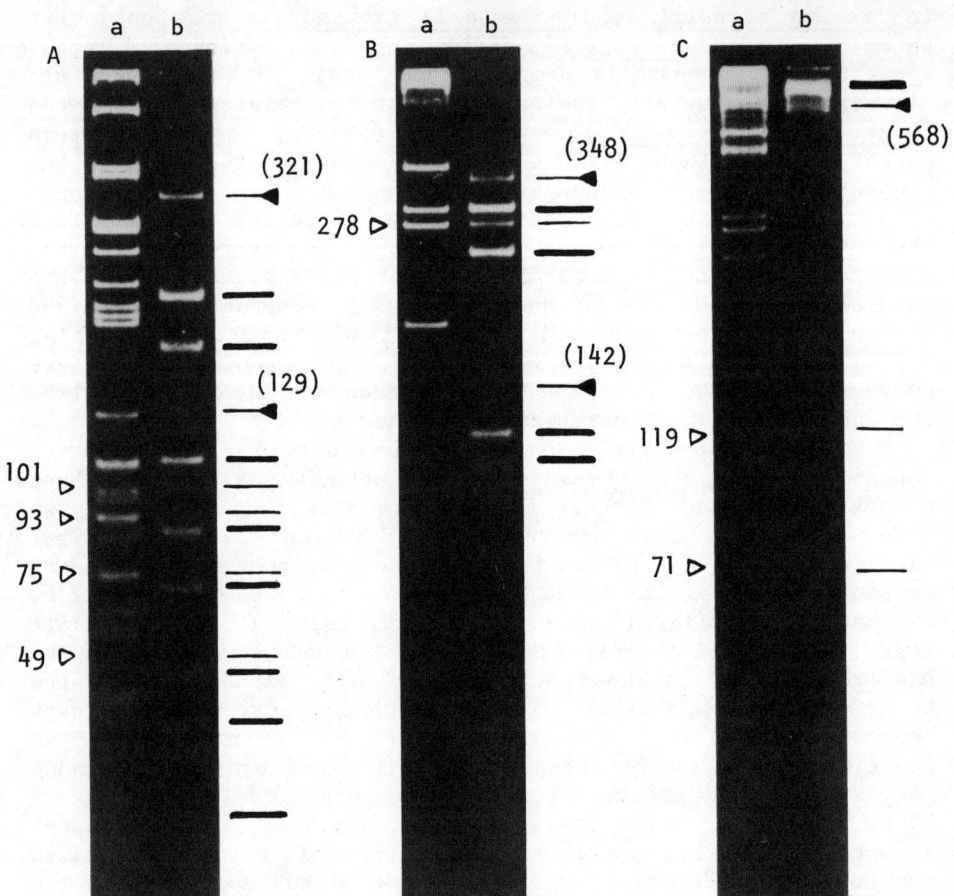

Figure 6. Polyacrylamide gel (6%) showing digests of DNA with restriction endonucleases (A) HpaII, (B) HaeII and (C) HinfI. The a columns show the plasmid DNA of pSM16 (see Figure 2) which was digested with these restriction endonucleases. The digests of pSM16 contain the DNA fragments derived from the inside of the IS1 sequence (▷, with the number of base pairs). The b columns show the restriction endonuclease digests of purified DNA fragments of the Shigella dysenteriae inverted repeat sequences 770 base pairs in length. Schematic representation of gel bands generated from the purified fragments are shown on the right of each gel. The solid triangles (▶) indicate the fragments containing the ends of IS1. The number of base pairs in these fragments is given in parenthesis.

The ratio of the two components was roughly estimated as 1:3. Our recent results show that the nucleotide sequence of one

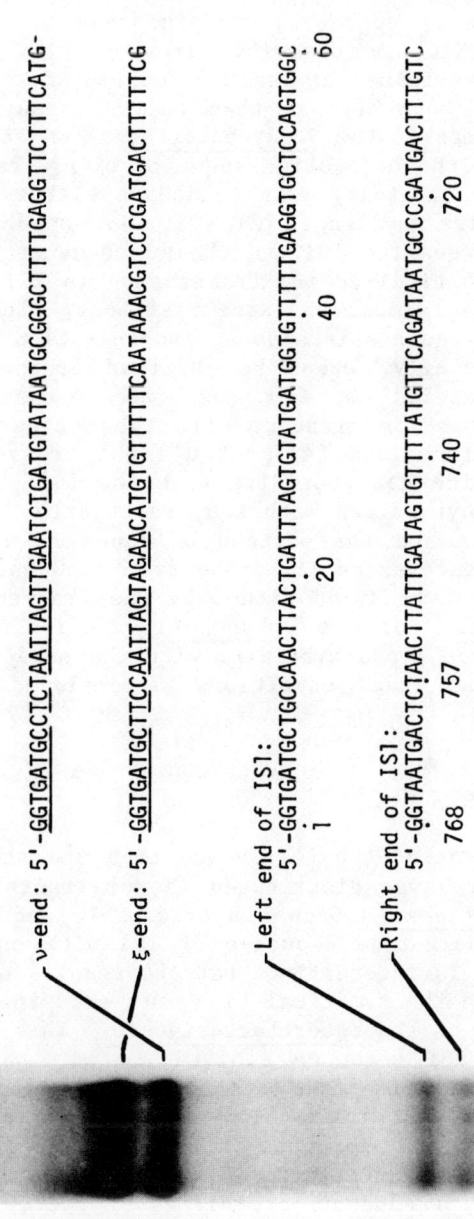

Figure 7. An autoradiogram showing two pairs of dense bands of DNA fragments seen after cleavage of the 5'-32P-labeled inverted repeat sequences of the Shigella dysenteriae chromosome with the restriction endonuclease AluI (left) and the nucleotide sequences of their ends as determined by H. Ohtsubo (unpublished data) (right).

pair of such 5'-labeled fragments was identical to the IS1 sequence except for one base pair at position 757 (see Figure 7). Another pair of fragments was found to be different but still similar to the IS1 sequence shown in Figure 7. This non-IS1 sequence (the $\nu\xi$ sequence) contains inverted repeats at its ends just as in IS1. As described previously, we estimated that S. dysenteriae contained at least 40 copies of IS1. This result and the results above suggest that the $\nu\xi$ sequence is repeated at least 120 times in the S. dysenteriae chromosome.

Most of the nucleotide sequence of $\nu\xi$ has already been determined. Preliminary results indicate that the $\nu\xi$ sequence is 766 base pairs in length (H. Ohtsubo, unpublished data). The homology between the IS1 and the $\nu\xi$ sequence is about 55%. Some of the hypothetical coding frames seen in IS1 also appear in the $\nu\xi$ sequence at almost the same positions. Therefore, we believe that the $\nu\xi$ sequence is also an iso-insertion sequence of IS1.

Interestingly, when the purified fragment of the $\nu\xi$ sequence was labeled and used as a probe to examine various Shigella strains for its presence with the Southern blotting and hybridization technique (41), it did not cross-hybridize with the sequences which can hybridize with the IS1 probe. Instead, the $\nu\xi$ sequence hybridized with many restriction fragments of S. dysenteriae DNA and the pattern of bands is completely different from that seen when an IS1 probe from R100 was used. The $\nu\xi$ sequence probe hybridized with only one fragment from S. flexneri and S. boydii (but not S. sonnei), confirming further that the probe does not cross-hybridize with the sequence hybridizable to IS1, at least under conditions we employed in this research.

EVOLUTION OF IS1

Hybridization studies showed that the sequence hybridizable to IS1 of R100 was distributed in some bacterial species of the Enterobacteriaceae. Such a hybridizable sequence seems to conserve the nucleotide sequence of IS1 with only a few base pair changes. It is interesting that the results demonstrated no correlation with the numerical taxonomic grouping or with the phylogenetic tree of the Enterobacteriaceae. This raises the question of how the IS1 sequence evolved and why only a few bacterial strains contain IS1 with such a great variation in copy number. We assume that IS1 has developed at a rather recent point in evolutionary time in one bacterial species and spread to other bacterial strains by genetic means, such as conjugation by sex factor plasmids, transduction mediated by bacteriophages and direct DNA transformation. These types of genetic exchanges are known to occur among Enterobacteriaceae (65). In fact, the resistance plasmids R100 and R6 which are originally from Shigella (65) and contain two copies of IS1 (15) have been transferred to E. coli

K12 and other bacterial strains by conjugation. Bacteriophage P1, which contains the IS1 sequence (18), has been used for genetic manipulation in the E. coli K12 system. The bacteria which have received IS1 may then increase the copy number of IS1 by further insertion into any other site on the chromosome, sometimes accompanied by rearrangement of the chromosome, as demonstrated in the analysis of E. coli K12 and its derivatives. It is therefore not too difficult to assume that some bacterial strains having a large number of IS1 sequences might have acquired them at a recent time in evolution, after divergence from other bacterial genera, increased the copy number of IS1 by translocation and subsequently transferred it onto other bacteria. It is noted, of course, that the original IS1 must have been subjected to some base changes in its sequence before or during transfer of the sequence. It is difficult to predict where IS1 originally appeared. However, the likely candidates may be those bacteria with many copies of the IS1 sequence. As described in this chapter, some Shigella strains such as S. dysenteriae, S. flexneri and S. sonnei contain multiple copies of IS1 and thus are likely candidates.

It is also possible that an IS1-like sequence without translocation activity was initially present in the chromosome of all of the Enterobacteriaceae and evolved independently in some bacteria, acquiring insertion activity to generate many copies of the sequence. Such a pro-IS1 sequence must have been subjected to drastic base changes during the evolution of bacteria, existing as a unique sequence in a chromosome. In this regard, the discovery of the $\nu\xi$ iso-insertion sequence of IS1 is very interesting. This sequence does not hybridize with the IS1 sequence of R100 under the hybridization conditions employed in this research and was actually found to have considerable differences from it at the nucleotide level. This sequence appears in the S. dysenteriae chromosome at a very high frequency, but only once in the S. flexneri and S. boydii chromosomes. If the single copy of the sequence in the latter two Shigella strains has no translocation activity, the finding of multiple copies of the $\nu\xi$ sequence in S. dysenteriae may suggest a case of independent evolution to generate an IS1-like sequence in a particular bacterial species.

Although both sequences that are very similar to IS1, with a few base changes, as well as the $\nu\xi$ sequence, are called iso-insertion sequences of IS1, it is certainly important to study all of them in great detail to examine our hypothesis for the evolution of the IS1 element.

EVOLUTION OF TRANSPOSONS AND RESISTANCE PLASMIDS

As briefly reviewed in the text, IS1, which may be dispensable for bacterial cells, can cause the rearrangement of genomes. Particularly interesting is the fact that IS1 is present in resistance plasmids and becomes a component of transposons responsible for the spread of antibiotic resistance and enterotoxin.

It is likely that the more copies of IS1 present in a bacterial chromosome, or a plasmid, the greater the chance of generating translocating segments of DNA to another site on the chromosome or into the genome of a plasmid. It is known that some transposable DNA elements containing resistance genes for antibiotics and enterotoxin genes are flanked by two IS1 sequences. Therefore, the more copies of IS1 in a genome, the greater the chance of a DNA segment flanked by two copies of IS1 to be translocated into another site on the same or a different genome.

Various kinds of resistance plasmids may have originated in this way and propagated under selective conditions, such as heavy usage of antibiotics. The fact that many resistant bacteria have different resistance spectra due to the presence of various kinds of plasmids was first noted in Shigella strains in Japan (65) particularly after antibiotic treatment. This may suggest that the elaborate recombination events responsible for the generation of resistance plasmids may have been facilitated by the large number of IS1 sequences in Shigella.

The insertion element IS1 is only one of several insertion elements, e.g. IS2, IS3, IS4, etc. It would also be interesting and important to investigate the distribution of such sequences and their iso-insertion sequences, if any, in various gram-negative bacteria to further our understanding of the evolution of IS elements. There is also a group of translocatable elements called transposons, responsible for resistance to various antibiotics, present in resistance plasmids. These elements are not always flanked by IS elements but usually by short inverted repeat sequences, for example, a characteristic 38 base-pair sequence at both ends of the transposon Tn3, which is responsible for transposition of ampicillin resistance (66-69). Recent nucleotide sequence analyses of the ends of some of the other translocatable elements have demonstrated that they contain inverted repeat sequences similar to that of Tn3 at their ends (see the most recent review by Calos and Miller, ref. 6). It is not yet known if some of these elements are iso-transposons (or "isoposons"). The investigation of the distribution of the inverted repeat sequences of transposons in gram-negative bacteria as well as genetic and physical analysis of the unique inner sequences will also be important in elucidating the origin of resistance plasmids in bacteria. Studies on the IS and Tn elements present in various bacterial species with in vitro genetic engineering techniques should improve our understanding of the evolution of these elements, which cause the spontaneous in vivo genetic engineering of bacterial genomes.

METHODS OF PROCEDURE

The following are the methods of procedure used to determine the number of copies of the IS1 sequence and to isolate inverted repeat sequences present in bacterial chromosomes. These techniques are generally very useful for the analysis of any repetitive sequence.

Southern Blot and Hybridization Technique

Three µg of chromosomal DNA was digested with restriction endonucleases to completion and subjected to 0.7% agarose gel electrophoresis on a 13 x 15 x 0.2 cm gel. The agarose gel was removed and soaked in a solution containing 0.5 M NaOH and 1.5 M NaCl for 45 min to denature the DNA in the gel. After neutralization of the gel containing the denatured DNA with 3 M NaCl and 0.5 M Tris (pH 7.0), the DNA was transferred to a sheet of nitrocellulose filter (Millipore) according to Southern (41) as follows. The agarose gel was placed on six layers of filter paper (Whatman 3 MM) on a glass plate in a baking dish half filled with 20 x saline-citrate (SSC). Four spacer bars 0.2 cm in thickness were placed 2 mm from each side of the agarose gel. A piece of nitrocellulose paper presoaked in distilled water and cut to cover exactly the gel and spacer bars was placed on top of the agarose gel. Then two layers of 3 MM paper, cut exactly the same size as the nitrocellulose paper, and paper towels (two to three inches in thickness) were placed on the nitrocellulose filter with a glass plate at the top. Two days later, the nitrocellulose filter paper was removed from the agarose gel and rinsed with 2 x SSC for 20 min, then baked at 80°C for 3 hrs in a vacuum oven. The filter paper was placed in 20 $\mu l/cm^2$ of hybridization medium (50% deionized formamide and 4 x SSC) containing the ^{32}P-labeled IS1 probe (1.5 x 10^4 cpm/ml). Wet filters were wrapped with parafilm and incubated in a sealed plastic bag at 42°C for 18 to 20 hrs. After hybridization, the filters were washed twice with 250 ml of hybridization medium at 42°C for 1.5 hrs, rinsed four times with 250 ml of 2 x SSC at room temperature for 10 min, and blotted dry. Kodak X-ray film (SF-5) was used for autoradiography.

Isolation of Inverted Repeat Sequences Present in the Bacterial Chromosome

Bacterial cells were grown and bacterial DNA was isolated according to the method described by Saito and Miura (70). Two to three mg of the DNA was dissolved in 10 ml of TE buffer (0.01 M Tris (pH 7.2) and 0.001 M EDTA). The DNA solution was titrated to pH 12.3 with 2 N NaOH at room temperature and was further incubated for 15 min. The solution containing this denatured DNA was then neutralized with 2 N HCl to pH 7.0. The [Na^+] concentration in the DNA solution was adjusted to 0.3 M with 3 M NaCl and the solution was incubated at 65°C for 1 min for renaturation of inverted DNA sequences on the DNA strands. The single-stranded DNA in the solution was digested at 37°C for 1 hr with 50 units/µg DNA of S1 nuclease in a buffer containing 40 mM sodium acetate (pH 4.6) and 4.5 mM $ZnCl_2$. After extraction of the solution with phenol saturated with TE buffer, the DNA was precipitated with ethanol and then suspended in 1 ml of TE buffer. Seven ml of 0.12 M sodium-potassium phosphate buffer (Na-P buffer, pH 6.8) containing 2 ml of hydroxylapatite (HAP) suspension (Bio-Rad) was added to the DNA solution and incubated at 65°C for 30 min. The solution was occasionally stirred with a glass rod to enhance the efficiency of adsorbing double-stranded DNA. The HAP-DNA complex was precipitated by centrifugation and resuspended in the Na-P buffer and the adsorption step above was repeated three times. Double-stranded DNA was then eluted from the HAP in 5 ml of 0.4 M Na-P buffer (pH 6.8) by incubation at 65°C for 30 min. The elution step was repeated and finally 10 ml of eluate was saved. The solution was concentrated to 2 ml in dialysis tubing which was embedded in polyethylene glycol powder (Carbowax #6000, Bio-Rad), followed by dialysis against TE buffer. After precipitation of the DNA with ethanol, the DNA was resuspended in 200 µl of TE buffer. Ten µl of the solution was loaded on a 1.4% agarose gel in electrophoresis (E) buffer containing 40 mM Tris, 20 mM sodium-acetate and 2 mM EDTA (pH 8.0).

Acknowledgments: We would like to thank J. Rosen for correcting the English in the text and Y. Tong, L. Hollmann and G. Urban for typing the manuscript. This work was supported by U.S. Public Health Service Grants GM22007 to E. Ohtsubo and GM26779 to H. Ohtsubo, and partially by support from the National Institutes of Health Training Grant CA09176.

REFERENCES

1 Cohen, S.N. (1976) Nature 263, 731-738.
2 Kleckner, N. (1977) Cell 11, 11-23.
3 Starlinger, P. and Saedler, H. (1976) Curr. Topics Microbiol. Immunol., 75, 111-152.
4 DNA Insertion Elements, Plasmids and Episomes (1977) (Bukhari, A.I., Shapiro, J.A. and Adhya, S.L., eds.), Cold Spring Harbor Laboratory, Cold Spring Harbor, NY.
5 Davidson, N., Deonier, R.C., Hu, S. and Ohtsubo, E. (1975) in Microbiology, 1974 (Schlessinger, D., ed.), pp. 56-65, American Society for Microbiology, Washington, DC.

6   Calos, M.P. and Miller, J.H. (1980) Cell 20, 579-595.
7   Hirsch, H.-J., Starlinger, P. and Brachet, P. (1972) Mol. Gen. Genet. 19, 191-206.
8   Fiandt, M., Szybalski, W. and Malamy, M.H. (1972) Mol. Gen. Genet. 119, 223-231.
9   Jaskunas, S.R. and Nomura, M. (1977) in DNA Insertion Elements, Plasmids and Episomes (Bukhari, A.I., Shapiro, J.A. and Adhya, S.L., eds.), p.487-495, Cold Spring Harbor Laboratory, Cold Spring Harbor, NY.
10 Ohtsubo, H. and Ohtsubo, E. (1978) Proc. Nat. Acad. Sci. U.S.A. 75, 615-619.
11 Johnsrud, L. (1979) Mol. Gen. Genet. 169, 213-218.
12 Calos, M.P., Johnsrud, L. and Miller, J.H. (1978) Cell 13, 411-418.
13 Grindley, N.D.F. (1978) Cell 13, 419-426.
14 Ohtsubo, E., Zenilman, M., Ohtsubo, H., McCormick, M., Machida, C. and Machida, Y. (1980) Cold Spring Harbor Symp. Quant Biol. 45 (in press).
15 Hu, S., Ohtsubo, E., Davidson, N. and Saedler, H. (1975) J. Bacteriol. 122, 764-775.
16 Saedler, H. and Heiss, B. (1973) Mol. Gen. Genet. 122, 267-277.
17 Chadwell, H., Fritz, H.-J., Haberman, P., Klaer, R., Kühn, S. and Starlinger, P. (1978) Cold Spring Harbor Symp. Quant. Biol. 43, 1187-1192.
18 Iida, S., Meyer, J. and Arber, W. (1978) Plasmid 1, 357-365.
19 Arber, W., Iida, S., Jütte, H., Caspers, P., Meyer, J. and Hänni, C. (1978) Cold Spring Harbor Symp. Quant. Biol. 43, 1197-1208.
20 Iida, S. and Arber, W. (1977) Mol. Gen. Genet. 143, 259-269.
21 Marcoli, R., Iida, S. and Bickle, T. (1980) FEBS Lett. 110, 11-14.
22 MacHattie, L.A. and Jackowski, J. (1977) in DNA Insertion Elements, Plasmids and Episomes (Bukhari, A.I., Shapiro, J.A. and Adhya, S.L., eds.), pp. 219-228, Cold Spring Harbor Laboratory, Cold Spring Harbor, NY
23 Johnsrud, L., Calos, M.P. and Miller, J.H. (1978) Cell 15, 1209-1219.
24 Alton, N.K. and Vapnek, D. (1979) Nature 282, 864-869.
25 So, M., Heffron, F. and McCarthy, B. (1979) Nature 277, 453-456.
26 So, M. and McCarthy, B. (1980) Proc. Nat. Acad. Sci. U.S.A. 77, 4011-4015.
27 Rownd, R. and Mickel, S. (1971) Nature New Biol. 234, 40-43.
28 Rownd, R., Miki, T., Applebaum, E.R., Miller, J.R., Finkelstein, M. and Barton, C.R. (1978) in Microbiology, 1978 (Schlessinger, D., ed.), pp. 33-37, American Society for Microbiology, Washington, DC.
29 Meyer, J. and Iida, S. (1979) Mol. Gen. Genet. 176, 209-219.

30 Sharp, P.A., Cohen, S.N. and Davidson, M. (1973) J. Mol. Biol. 75, 235-255.
31 Miki, T., Easton, A.M. and Rownd, R.H. (1978) Mol. Gen. Genet. 158, 217-224.
32 Chandler, M., Silver, L., Lane, D. and Caro, L. (1978) Cold Spring Harbor Symp. Quant. Biol. 43, 1223-1231.
33 Mickel, S., Ohtsubo, E. and Bauer, W. (1977) Gene 2, 193-210.
34 Ohtsubo, E., Rosenbloom, M., Schrempf, H., Goebel, W. and Rosen, J. (1978) Mol. Gen. Genet. 159, 131-141.
35 Ohtsubo, E., Zenilman, M., Rifkin, J. and Ohtsubo, H. (1979) in Abstracts of The Annual Meeting of American Society for Microbiology, p. 124.
36 Ohtsubo, E., Zenilman, M. and Ohtsubo, H. (1980) Proc. Nat. Acad. Sci. U.S.A. 77, 750-754.
37 Iida, S. and Arber, W. (1980) Mol. Gen. Genet. 177, 261-270.
38 Reif, H.-J. and Saedler, H. (1975) Mol. Gen. Genet. 137, 17-28.
39 Cabello, F., Timmis, K. and Cohen, S.N. (1976) Nature 254, 285-290.
40 Ohtsubo, E., Feingold, J., Ohtsubo, H., Mickel, S. and Bauer, W. (1977) Plasmid 1, 8-18.
41 Southern, E. (1975) J. Mol. Biol. 98, 503-517.
42 Rosen, J., Ryder, T., Inokuchi, H., Ohtsubo, H. and Ohtsubo, E. (1980) Mol. Gen. Genet. 179, 527-537.
43 Maxam, A.M. and Gilbert, W. (1977) Proc. Nat. Acad. Sci. U.S.A. 74, 560-564.
44 Nyman, K., Nakamura, K., Ohtsubo, H. and Ohtsubo, E. (1980) Nature (in press).
45 Bachmann, B.J. (1972) Bacteriol. Rev. 36, 525-557.
46 Bergey's Manual of Determinative Bacteriology, 8th Ed. (1974) (Buchanan, R.E. and Gibbons, N.E., eds.), Williams and Wilkins, Baltimore, MD.
47 Cocks, G.T. and Wilson, A.C. (1972) J. Bacteriol. 110, 793-802.
48 Hori, H. (1976) Mol. Gen. Genet. 145, 119-123.
49 Hori, H. and Osawa, S. (1978) J. Bacteriol. 133, 1089-1095.
50 Nakamura, K., Pirtle, R.M. and Inouye, M. (1979) J. Bacteriol. 137, 595-604.
51 Sanderson, K.E. (1976) Ann. Rev. Microbiol. 30, 327-349.
52 Ohtsubo, H. and Ohtsubo, E. (1977) in DNA Insertion Elements, Plasmids and Episomes (Bukhari, A.I., Shapiro, J.A. and Adhya, S.L., eds.), pp. 49-63, Cold Spring Harbor Laboratory, Cold Spring Harbor, NY.
53 Blattner, F.R., Williams, B.G., Blechl, A.E., Dennison-Thompson, K., Fabor, H.E., Furlong, L.-A., Grunwald, D.J., Kiefer, D.O., Moore, D.D., Schumm, J.W., Sheldon, E.L. and Smithies, O. (1977) Science 196, 161-169.
54 Ohtsubo, H. and Ohtsubo, E. (1975) Proc. Nat. Acad. Sci. U.S.A. 73, 2316-2320.

55 Brenner, D.J., Fanning, G.R., Johnson, K.E., Citarella, R.V. and Falkow, S. (1969) J. Bacteriol. 98, 637-650.
56 Donny, R.M. and Yanofsky, C. (1970) J. Mol. Biol. 64, 319-339.
57 Brenner, D.J., Fanning, G.R., Skerman, F.J. and Falkow, S. (1972) J. Bacteriol. 109, 953-965.
58 Creighton, T.E., Helinski, D.R., Sommerville, L. and Yanofsky, C. (1966) J. Bacteriol. 91, 1819-1826.
59 Li, S. and Yanofsky, C. (1972) J. Biol. Chem. 247, 1031-1037.
60 Luria, S.E. and Burrows, J.W. (1957) J. Bacteriol. 74, 461-476.
61 Franklin, N. and Luria, S.E. (1961) Virology 15, 299-311.
62 Sarkar, S. (1966) J. Bacteriol. 91, 1477-1488.
63 Manson, M.D. and Yanofsky, C. (1976) J. Bacteriol. 126, 668-678.
64 Manson, M.D. and Yanofsky C. (1976) J. Bacteriol. 126, 679-689.
65 Transferable Drug Resistance Factor R (1971) (Mitsuhashi, S., ed.), University of Tokyo Press, Tokyo.
66 Ohtsubo, H., Ohmori, H. and Ohtsubo, E. (1978) Cold Spring Harbor Symp. Quant. Biol. 43, 1269-1277.
67 Cohen, S.N., Casadaban, M., Chou, J. and Tu, C. (1978) Cold Spring Harbor Symp. Quant. Biol. 43,. 1247-1255.
68 Takeya, T., Nomiyama, H., Miyoshi, J., Shimada, K. and Takagi, Y. (1979) Nucl. Acids Res. 6, 1831-1841.
69 Heffron, F., McCarthy, B., Ohtsubo, H. and Ohtsubo, E. (1979) Cell 18, 1153-1163.
70 Saito, H. and Miura, K. (1963). Biochim. Biophys. Acta. 72, 619-629.

APPLICATIONS OF MOLECULAR CLONING TO SACCHAROMYCES

M.V. Olson

Department of Genetics
Washington University School of Medicine
St. Louis, Missouri

INTRODUCTION

The cloning of yeast DNA sequences in bacteria was first reported in 1976. In the short period since that time, the application of recombinant DNA techniques to Saccharomyces has burgeoned into a large and complex enterprise which already has made major contributions to our understanding of eukaryotic cells. To some extent, of course, this phenomenon simply reflects the astonishing growth of recombinant DNA methodology, a development that has affected the study of all organisms. In the case of yeast, however, it can be argued that there has been a special synergism between recombinant DNA techniques and the pre-existing strengths of Saccharomyces as an experimental organism. As a result of this synergism, the importance of yeast as a model eukaryote has increased enormously. It is no exaggeration to say that yeast is emerging from the first years of the recombinant DNA era as a kind of eukaryotic E. coli: this is not to imply that yeast is an entirely typical eukaryote or that it can or should displace more complex organisms as a focus of study. Yeast is simply becoming the most versatile single experimental system available for studies of the elementary molecular organization of eukaryotic cells.

There would appear to be four properties of yeast that largely account for the rich variety of molecular cloning experiments that have been carried out with yeast DNA in such a short period of time.

1) Yeast has a powerful genetic system. Even more than ease of growth and commercial importance, yeast's genetic system has been the organism's main experimental attraction for the past 30

years. Classical genetic studies during this period have produced a rich supply of information about many gene systems, a resource that continues to help direct yeast cloning experiments along productive lines. The genetic system is of more specific uses as well. The existence of hundreds of easily manipulated mutant loci has been essential to the development of yeast transformation (see below) and it has also made possible the identification of many cloned sequences with specific genes of known phenotypic effects. (General reviews of yeast genetics can be found in refs. 1-3.)

2) Yeast has a small genome. At 15 million base pairs, the genome complexity of yeast is only about three times that of E. coli. Single-copy yeast sequences occur in shotgun pools at a frequency between one in a few hundred and one in a few thousand clones. As a result, cloned genes can be identified by screening techniques that would be too laborious or produce too many false positives if applied to a more complex genome.

3) Many yeast genes are functionally expressed in E. coli. This property, which may be largely limited to the fungi, was of considerable historical importance. It allowed the early isolation of a number of genes coding for metabolically important enzymes by direct complementation of bacterial mutations in E. coli. It remains unclear why there are so many cases of yeast genes that are expressed in E. coli. One crude screen of promoter function -- using the ability of random eukaryotic DNA fragments to allow expression of the adjacent pBR322 tet^R gene -- suggested that promoters that can function in E. coli are much more common in yeast than in any of the other eukaryotes tested (4). It is of considerable theoretical interest that this phenomenon is reciprocal; there is now clear evidence that some bacterial genes can be functionally expressed in yeast (5).

4) Yeast is readily transformed with naked DNA. In conjunction with a well-developed genetic system, transformation is an enormously powerful tool. Its two most important consequences are that it allows the cloning of genes about which only phenotypic knowledge exists and it makes possible genetic analysis of mutants that are constructed in vitro. It is not yet possible to predict the extent to which eukaryotes in general will prove to be transformable in a way that allows normal expression of the exogenous genes. To a large extent, yeast transformation has fluorished rapidly because powerful techniques are available both to select transformants and to analyze the genotypic basis of the transformed phenotype. More intrinsic factors, such as the high level of recombination in vegetative yeast cells, and perhaps unknown characteristics of yeast's sub-cellular compartmentalization, may also be important. The initial development of the yeast transformation system was discussed in Volume 1 of this series (6).

The following sections of this review survey the results of cloning experiments carried out with yeast nuclear DNA. The cloning of mitochondrial genes, which is in itself a large topic, is not included. As a general point, this article contains almost no discussion of the specific methods that have been used to construct yeast/E. coli recombinant DNA molecules. Nearly all yeast cloning has involved plasmid and lambda cloning technology that has been expertly discussed elsewhere in this series (7,8). Readers are also directed to an article on yeast cosmid cloning, a method that shows considerable promise but has as yet seen little use (9).

An initial section examines the methods that have been used to identify specific DNA sequences within populations of recombinant clones; greatest attention is given to unusual techniques that exploit special advantages of the yeast system. The second section surveys the use of cloned sequences to study specific problems of gene organization, gene expression and genome structure. The descriptive material on methods and results is followed by a tabulation of the identified yeast DNA sequences that have now been cloned. Finally, a few comments are made about future challenges in this field.

SCREENING TECHNIQUES

In yeast cloning, in contrast to the situation in higher organisms, the use of RNA or cDNA hybridization probes for the in situ screening of colonies or plaques has been of secondary importance. Such methods, of course, have been used to clone genes that code for stable RNAs including the ribosomal RNAs and numerous transfer RNAs. However, the only genes coding for proteins that have been cloned by the direct use of cDNA probes prepared from partially purified mRNA are the genes for glycolytic enzymes such as glyceraldehyde-3-phosphate dehydrogenase (10). These mRNAs, which are highly abundant when yeast is grown on glucose, are well-suited to this technique.

Undoubtedly, the relatively sparse use of cDNA probes in yeast compared to animal cell systems is accounted for by the lack of yeast cell lines highly committed to the expression of tissue-specific genes. There is, however, an example of cDNA cloning that exploited the inducibility of specific mRNAs under specialized growth conditions. This case involved the GAL-1,7,10 cluster, which codes for the Leloir pathway enzymes galactokinase, galactose-1-phosphate uridyl transferase and uridine diphosphogalactose-4-epimerase (11,125). Lambda clones whose plaques hybridized much more strongly to cDNA probes prepared from galactose, rather than glucose-grown cells, were identified. This method is of increasing importance in the screening of cDNA shotgun pools made from heterodisperse tissue-specific mRNA populations in higher organisms. Only in an organism like yeast, with a low complexity genome, is it likely to have the specificity required to identify single-copy genomic clones.

Another hybridization screening technique that exploited the low complexity of the yeast genome was the use of a 13-base synthetic oligodeoxynucleotide to clone the gene for iso-1-cytochrome c (12). This technique, which exploited multiple advantages of yeast molecular genetics, will be difficult to generalize directly to more complex organisms. To begin with, the sequence of a short region of the iso-1-cytochrome c mRNA had previously been deduced by an elegant series of experiments involving the sequencing of the N-terminal region of several mutant proteins; protein sequences on the products of genes containing double frameshift mutations allowed the deduction of all third position nucleotides in one region of the mRNA (13). Even given this mRNA sequence, the feasibility of using a relatively short synthetic probe to locate the iso-1-cytochrome c gene in genomic DNA depended on the low complexity of the yeast genome. At a complexity of 1.5×10^7 base pairs, the largest random sequence that has a 50% chance of occurring in yeast DNA is about 12. On this basis, it was expected that the iso-1-cytochrome c-specific 13-mer would be a marginal but adequate hybridization probe. There proved to be only a narrow range between the temperature at which hybridization was relatively specific and that at which even the perfectly matched duplex melted. It was nonetheless possible to identify the EcoRI fragment coding for iso-1-cytochrome c by comparing a genomic Southern blot experiment on wild-type yeast DNA with those on a cyc1 deletion strain and a point mutant that lacks the wild-type EcoRI site in the coding region. This fragment was then cloned from a lambda shotgun pool by plaque screening with the 13-mer probe. More specific hybridization was obtained to the cyc1 gene in independent experiments with a synthetic 15-mer probe (14). As in the case of differential plaque screening with induced and uninduced cDNA probes, the main application of short synthetic oligonucleotide probes in higher organisms is likely to involve the screening of populations of cDNA clones rather than genomic shotgun pools (15).

A final type of DNA-DNA hybridization screening that deserves mention is the bootstrapping approach -- using one clone as a probe to clone a related gene. A special case of this type of cloning involves the repetitive cloning of the same gene, as when a series of mutants or natural variants are being analyzed. This approach has been applied at a number of loci -- SUP4 (16, 17), SUP-RL1 (18), SUQ5 (18), HIS3 (19,20), CYC1 and CYC7 (21). Since much more work of this kind may be expected, the efficiency of this technique is of interest. By far the largest of these projects involved the repetitive cloning of 32 SUP4 mutants (17). A key factor in making such undertakings practical is that yeast's small genome size minimizes the difficulty of preparing an all-inclusive shotgun pool from each new mutant yeast strain.

A less trivial application of bootstrapping is the use of a gene clone to screen for clones of related but not identical genes. One example of this procedure involves the cloning of the iso-2-cytochrome c gene (CYC7) by using the iso-1 gene as a probe (21). These two proteins are 85% homologous in amino acid sequence while the nucleotide sequences of the two coding regions proved to be 76% homologous. Whether or not this degree of relatedness will prove generally adequate for bootstrap cloning is problematical. It is likely that the cytochrome c experiment was greatly facilitated by a tract of 33 perfectly conserved coding region nucleotides. The question of what selective pressures have maintained third-position usage throughout this 11-codon tract is intriguing.

A rather spectacular example of efficient bootstrap cloning involved the mating type loci. In addition to the master locus at which there are two alternative sequence organizations (MATa and MATα), there are silent-copy sequences at two other sites (HML and HMR), which can themselves each exist in either an a or an α variant. Two laboratories obtained different footholds in this system by complementation cloning (see below); both groups then rapidly cloned the entire gene system by hybridization screening, exploiting what turned out to be over one kilobase of homology between all pairwise combinations of the loci (22-24). Similar techniques had been used earlier to unravel the complex relationship between Ty1 and δ sequences, yeast's two best-characterized sets of transposable repetitive sequences (25).

Partly because of yeast's phylogenetic isolation from other heavily studied organisms, there have been few efforts to clone yeast genes using the evolutionarily homologous gene from another species as a probe. The only reported case involved the yeast actin gene, which was cloned in two different laboratories using a cloned actin gene from the slime mold Dictyostelium as the probe (26-28).

As discussed in the Introduction, the most remarkable yeast cloning technique is based on the selection of yeast sequences that complement mutations in E. coli. The first published instance of a cloned eukaryotic gene coding for a protein, involved the yeast HIS3 locus, which codes for the enzyme imidazole glycerol phosphate dehydratase (29). The difficult part of this work was not the selection itself, but rather the long process of proving that the complementing yeast sequence was actually the yeast IGP dehydratase structural gene. Enzyme assays on the activity that the cloned fragment induced in E.coli showed that the properties of this activity resembled those of the yeast rather than the E. coli enzyme (19). Nonetheless, a formal proof that the cloned DNA actually coded for the yeast HIS3 gene was ultimately developed by a more complex experiment: the original complementing fragment was used as a hybridization probe to clone the homologous fragment from a yeast strain containing a his3 amber allele. This fragment, in turn, was shown to complement E. coli hisB mutants only in E. coli strains that contained amber suppressors (20).

The problem of proving that a particular yeast fragment complements an E. coli mutation by expression of the corresponding yeast structural gene has been uniformly difficult. In the case of the LEU2 gene, supporting evidence was obtained by physically mapping the complementing fragment to chromosome III, to which LEU2 maps genetically (30). The physical mapping involved comparing the intensity of hyridization of the putative LEU2 probe to DNA from haploid, diploid, chromosome III disomic (n+1) and chromosome III monosomic (2n-1) strains.

As a technical note, the original HIS3 cloning was carried out from a lambda shotgun pool with a procedure that involved the formation of double lysogens. The use of a helper phage was required since the lambda vector lacked the attachment site as well as essential integration functions. More commonly, plasmid shotgun pools have been employed; in this case, an E. coli strain that has been transformed with pool DNA can simply be plated on a selective medium (31). In the case of lambda pools, it has now been shown that lytic phage containing a complementing fragment can be selected directly by plating the phage pool on selective media with the relevant E. coli mutant as the host. Although the lawn cannot grow under the selective conditions, sufficient phage replication occurs to allow the formation of detectable plaques, the observation of which can be facilitated by ethidium bromide staining of the plate (32). Although the recent development of complementation cloning in yeast via yeast transformation has made E. coli selections obsolete, the availability of genes cloned by this method, particularly LEU2, TRP1, HIS3, URA3 and ARG4, was critical to the rapid development of yeast transformation.

One useful feature of cloned yeast genes that function in E. coli is that these systems lend themselves to the application of powerful bacterial genetic techniques for the construction of mutants that can then be further analyzed in either yeast, E. coli or in in vitro systems. A large set of his3 deletions was constructed in lambda, for example, and these deletions allowed a precise functional delimitation of the gene (33). In the case of regulatory mutants obtained in E. coli, however, more is likely to be learned about the regulation of gene expression in E. coli than in yeast. This point was emphasized by the discovery that two "up" mutations obtained by selecting for improved expression of yeast genes in E. coli were caused by the transposition of IS2 into the cloned yeast sequences (34,35).

A relative of the E. coli complementation technique, which does not require functional expression of the eukaryotic gene, is based on the use of antibody to detect the synthesis of specific yeast antigens in E. coli. The yeast PGK locus, coding for 3-phosphoglycerate kinase, was cloned with this method (36). The antibody screening technique has considerable theoretical potential but only further experience can determine its generality. As with several other methods that have been successfully used to clone yeast genes, it may be too prone to false positives and false negatives to be suitable for the screening of more complex genomic shotgun pools. It also depends, of course, on a certain degree of biochemical expression of the cloned genes. Presumably, the use of bacterial vectors that are designed to maximize the synthesis of proteins that are encoded by the inserted DNA could extend the generality of this technique to organisms whose genes do not generally function in E. coli (7).

Another method that, like antibody screening, does not require that the target gene be genetically defined, uses hybrid-selected translation to identify cloned DNA sequences that code for readily identifiable proteins. This method is of great importance in screening cDNA clones from higher organisms, but only in yeast has it been used directly to identify single-copy genomic clones. The problem with this technique is not lack of specificity but rather the amount of effort required per clone screened. Even in yeast, it would be laborious to screen complete shotgun pools with this method. Its successful application to the cloning of five ribosomal protein genes (37) and the single-copy H2A and H2B histone genes (38) exploited a prior enrichment of the pool for these genes. Translation experiments on size-fractionated yeast mRNAs had indicated that the histones and most ribosomal proteins are translated from small polyadenylated messages. Consequently, 100 colonies from a shotgun pool were chosen for further screening on the basis of their preferential hybridization to a small poly(A) RNA probe. These clones were then screened in small groups by using their DNA to select homologous mRNAs under R-looping conditions (39). These mRNAs were then translated and the translation products were examined electrophoretically for the presence of ribosomal proteins or histones. Although this strategy was rather laborious, it led to the cloning of a set of genes whose regulation is of great interest. Without any available mutants in these genes and given the low abundance of their mRNAs, there was a lack of promising alternatives.

The final method of identifying specific genes in yeast shotgun pools is to screen or select for cloned sequences that will complement mutations in yeast. The generality of this method was rapidly established following the development of yeast transformation (40). Although the first applications of complementation cloning were only published in 1979, it has already become the method of choice for the cloning of most mutationally defined yeast genes. Yeast cloning, therefore, is now likely to develop along lines similar to those pioneered in E. coli, where catalogued clone collections are rapidly encompassing the entire set of known genes (41).

The original yeast transformation experiment employed LEU2 as the selectable marker and ColE1 as the vector (40). The maximum frequency of leu$^+$ transformants was quite low, about $1/10^7$ regenerated spheroplasts. Given this low frequency, it was essential to use an extremely stable leu⁻ acceptor; for this purpose, a doubly mutant leu2 locus was constructed by intragenic recombination. The transformation frequency in this system was too low to allow complementation cloning out of total shotgun pools. The HIS4 gene was successfully cloned in a system of this kind, but it proved necessary to subdivide the E. coli/yeast shotgun pool extensively before preparation of DNA for yeast transformation (42).

A major development that paved the way for widespread applications of shotgun cloning by complementation was the discovery that transformation vectors with special sequences, either from the yeast 2μ plasmid or from certain chromosomally derived restriction fragments, will transform yeast at a frequency on the order of 10^4 times higher than vectors lacking these sequences (43,44). The property of high frequency transformation correlates with the capacity of the transforming plasmids to replicate autonomously in yeast; an appealing but still unproven hypothesis is that high frequency transformation sequences are simply DNA replication origins. A wide variety of vectors is now available in both the high frequency, autonomously replicating and the low frequency, chromosomally integrating classes (45). In general, all these vectors have been designed so that they can replicate and are subject to selection in both E. coli and yeast.

As a practical matter, the high frequency transformation vectors allow transformation of yeast at a high enough frequency to make possible the direct selection of individual genes out of total shotgun pools. In the simplest protocols, a yeast shotgun pool is prepared in E. coli from wild-type yeast DNA with a high frequency transformation vector. A mixed population of the E. coli transformants is then grown preparatively to provide the shotgun DNA sample. A yeast mutant is transformed with this DNA mixture and transformants in which the mutant phenotype has reverted are identified either by direct selection or by a suitable screening procedure (24,46). In the case of most temperature sensitive (ts) mutants and nutritional markers, it is possible simply to plate the primary yeast transformants under conditions that are restrictive for the mutant's growth.

In other instances, it is often necessary to use a replica-plating procedure to screen the primary transformants after they have been selected with the yeast marker on the vector. Since transformation is carried out using yeast spheroplasts, which regenerate well only when embedded in a soft agar overlay, replica-plating procedures must be preceded by a step in which the colonies of primary transformants are extracted from the soft agar to allow replating of the pooled cell population. An example of this procedure is provided by the original cloning of the yeast mating type genes. The cloned mating type loci were identified by means of a screen in which the primary leu⁺ transformants, which contained sterile mutations at the MAT locus, were replica-plated to a lawn of cells of the opposite mating type (23,24). Colonies whose replicas mated to the tester cells were analyzed for the presence of plasmids containing MAT locus sequences.

After yeast transformants with the desired phenotype have been identified, either by direct selection or by a screening procedure, DNA from a colony-purified transformant is used to retransform E. coli. Not uncommonly, this step can produce a variety of E. coli transformants by using DNA from a single yeast strain (46). Presumably, this phenomenon indicates that the primary yeast transformants are often co-transformed with multiple plasmids and that there is relatively slow mitotic segregation of these mixed plasmid populations. If several different plasmids are identified in separate E. coli transformants, only one of them will normally retransform yeast at a high frequency for the originally selected phenotype. At this stage, one criterion for establishing that a specific DNA sequence is responsible for the phenotypic change originally observed in a primary transformant is a 100% correlation between the phenotypic change caused by the vector's selectable marker (e.g., leu$^-$ to leu$^+$) and the phenotypic change caused by the inserted DNA. The reverse experiment can also be performed since all the autonomously replicating plasmids are mitotically unstable (43,44,47-51). If the transformants are grown without selection even for a few generations, a substantial portion of the cells will lose the plasmid. Once again, the selected phenotype caused by the inserted DNA should co-segregate with that caused by the vector.

Although the basic experimental design is quite simple, it should be pointed out that complementation cloning in yeast from shotgun pools is a new technique and further improvements in efficiency would be helpful. It is still a common practice to subdivide the original E. coli shotgun pool prior to DNA preparation, in order to compensate for the marginal efficiency of the overall experiment. A number of separate yeast transformations are then carried out, each of which involves DNA populations of less complexity than the total yeast genome. Efficiency is not a trivial problem, as indicated by the fact that three of the most important early applications of complementation cloning -- the cases of the MATα (24), HMLα (22) and CDC10 (101) loci -- hinged on the isolation of a single isolate of the desired clone. (One of the great virtues of cloning is that a single isolate is often enough.)

A final point about complementation cloning concerns methods of confirming that the cloned sequence actually corresponds to the desired locus. In one sense, this problem is simply a subset of the general challenge of confirming the identity of cloned sequences and correlating them with mutationally defined genes. This difficulty is apt to be particularly severe in many applications of complementation cloning, since the technique allows the cloning of DNA sequences whose biochemical functions are unknown. The recently cloned cell division control genes CDC10 and

CDC28, as well as the mating type genes MATa, MATα, HMR and HML, are all examples of genes whose gene products are not known. Even a gene such as the canavanine resistance locus CAN1, which is thought to code for an arginine permease, remains an essentially genetic rather than a biochemical entity. In all of these cases, even DNA sequencing of the clones would be a futile approach to clone identification. In principle, mutant alleles of these loci could be cloned by hybridization screening and the mutations located by DNA sequencing. This method is practical for small genes such as tRNA genes that correspond to nonsense suppressor loci, but it would be exceedingly laborious in the case of genes of more typical size.

In the case of the mating type genes, it was possible to correlate clones with genetic loci by taking advantage of the fact that all three loci -- HML, HMR and MAT -- can exist in different strains in a or α forms that have different restriction maps. These map variations proved to be detectable in Southern blot experiments, and their linkages to appropriate genetic markers were determined by conventional genetic mapping methods (22-24). A similar approach was used to show that the DNA sequence that complemented the can1-11 mutation actually arose from a genomic site that is tightly linked (and almost certainly corresponds) to the CAN1 locus (47). Two yeast strains were discovered whose restriction maps differed, fortuitously, in the region to which the cloned probe hybridizes. This physical variation, which again could be scored by Southern blot experiments, was shown to be tightly linked to the CAN1 locus. Restriction variation mapping, which was pioneered in yeast for rRNA (52) and tRNA (53) genes is probably a general method since interfertile laboratory yeast strains contain extensive DNA sequence polymorphisms (53). The main drawback of this method is that Southern hybridization, despite the availability of rapid yeast DNA isolation procedures, is a tedious method of scoring segregants in genetic experiments.

An elegant alternative to the use of restriction fragment size variants was employed in the CDC28 experiments, in which the TRP1-pBR322 vector YRP7 was used (46). With this vector, transformed strains in which the vector and its associated cloned sequences have become chromosomally integrated are readily selected. The selection exploits the superior growth of integrants on a tryptophan-deficient medium compared to transformants with only autonomously replicating copies; this superior growth is simply due to the stability of the trp$^+$ phenotype in integrants compared to its instability in transformants with only episomal copies. Integration can occur at the TRP1 locus by homologous recombination between the 1.4 kilobase TRP1 EcoRI fragment of YRP7 or at the chromosomal locus (or loci) sharing sequence homology with the cloned insert. In the case of the CDC28 experiments, YRP7 clones containing 6 to 8 kilobases of inserted yeast DNA, which could complement CDC28 mutants, integrated much more frequently at a single non-TRP1 site in the genome than at TRP1 itself. In these integrants the trp$^+$ phenotype associated with the integrated vector sequences was genetically mapped and shown to be very tightly linked to the CDC28 locus. In cases of clones that contain predominantly single-copy DNA, this rather simple genetic test should provide a general method of establishing sequence homology between a fragment of cloned DNA and the genetic locus to which, on the basis of mutant complementation, it is believed to correspond. The importance of such corroborating evidence must be emphasized, since bona fide complementation -- as opposed to indirect suppression -- is only one of many ways in which phenotypic reversion can occur.

An example has already been given of cloning of a sequence by complementation that did not directly correspond to the locus defined by the complemented mutant. A sterile mutant at the MAT locus acquired the ability to mate when transformed by an HMLα clone (22). Because the genetics of this system are well defined and the clones were rigorously characterized, the identity of the HMLα clone was correctly inferred. In other systems whose genetic description is less complete, the risk of outright errors in clone identification will be significant.

WHAT HAS BEEN LEARNED FROM YEAST CLONING EXPERIMENTS?

Following a description of the wide range of techniques that have been used to clone yeast genes, it is appropriate to survey the biological insights that have been gained from this work. Only a brief survey is possible since some of the most interesting cases (e.g., the mating type genes) would require major reviews in their own right. The emphasis here, therefore, will be more on the kinds of illuminating experiments that have employed yeast clones than on the details of particular gene systems. The discussion will cover four broad categories of work: studies of gene structure, gene expression, genome organization and the analysis of mutational events.

With respect to the narrow question of gene structure, the pattern of intervening sequence occurrence in yeast genes has been of major interest. The yeast tRNA^{Tyr}(16) and tRNA^{Phe}(54) genes were among the first chromosomal eukaryotic genes in which the presence of intervening sequences was demonstrated. Since then, it has been found that intervening sequences are present in some additional tRNA genes (those for tRNA^{Trp} (55), tRNA$^{Leu}_3$ (56,57) and tRNA$^{Ser}_{UCG}$ (18,58)) and absent from others (tRNA$^{Ser}_2$ (59), tRNA$^{Ser}_{UCA}$ (18), tRNA^{Arg} (60), tRNA^{Asp} (60), tRNA^{Met} (61)). A particularly striking case is the comparison between two closely related tRNA^{Ser} genes: the single gene for tRNA$^{Ser}_{UCG}$ (encoded by the SUP-RL1 locus) contains a 19 base-pair intervening sequence while that for tRNA$^{Ser}_{UCA}$ (encoded by the SUQ5 locus) has none, although the two genes differ by only three nucleotide pairs in their coding regions (18). In higher eukaryotes, intervening sequences are also an occasional but not obligatory feature of tRNA gene structure (62-65). Interestingly, there has not been evolutionary conservation of intervening sequences among genes coding particular tRNAs: for example, the yeast tRNA^{Phe} genes have intervening sequences while those from Xenopus do not (65).

Only recently has an intervening sequence been described in a yeast structural gene that codes for a protein (27,28). Unfortunately the gene in question, which codes for actin, is not associated with a known genetic locus. Nonetheless, the evidence is quite good that this split gene is not simply a vestigial actin-like gene that has been inactivated by an insertion event. Southern blot experiments indicate that the coding sequence exists in a single copy (26) and hybrid-selected message is spliced at the site predicted from protein sequence-DNA sequence comparisons (28). The splicing site fits well with the canonical sequence inferred from studies on the genes of higher organisms. The actin result is of considerable theoretical interest since it is the only evidence that yeast has a splicing activity analogous to the one that processes message precursors in higher organisms. Much has been made of the apparent fact that most yeast genes that code for proteins do not contain intervening sequences, in contrast to the situation in vertebrates. This conclusion may prove to be true, but it is still based on a limited set of observations. Most of the relevant vertebrate genes do not have yeast homologs, and in no case have genes coding for homologous products been analyzed both in yeast and in a higher organism.

A second observation about gene structure that has emerged from the analysis of cloned genes is that codon utilization patterns vary greatly from gene to gene. The only available comparisons involve two closely related glyceraldehyde-3-phosphate dehydrogenase (GAPDH) genes (67), on the one hand, and the iso-1- and iso-2-cytochrome c genes (21,68) on the other. In the two GAPDH genes, codon usage is very non-random; for example, 22/22 Arg residues are encoded by AGA (one of six Arg codons) and 49/49 Gly residues are encoded by GGT (one of four Gly codons). The cytochrome c genes tend to be biased towards the same codons favored in the GADPH genes, but the usage is not nearly as skewed; for example, from data for the two cytochrome c genes, all four Gly codons are used two or more times out of only 24 total occurrences of glycine. A reasonable starting hypothesis is that there has been overwhelming selection for translational efficiency in the case of GAPDH, yeast's most abundant single protein, but more subtle effects are also implicated.

A final case in which purely structural studies of cloned yeast genes have been highly informative concerns the mating type system. The yeast mating type genes were an immediate focus of cloning efforts after the development of high frequency yeast transformation vectors because the genetic dissection of this system had led to a provocative, testable model for yeast mating type control (69). The biological background to experiments on this gene system is as follows. Haploid yeast strains can exist in two mating types, a and α. Matings between a and α cells produce a/α diploids, which can either be vegetatively cultured or induced to undergo meiosis and sporulation. The latter process produces two a and two α haploid spores; this Mendelian segregation of mating type specificity defines a genetic locus (MAT) that maps to the right arm of chromosome III. In heterothallic strains, the mating types of haploid cell lines are quite stable, but cells that have changed mating types are nonetheless found in typical haploid cell populations at a frequency of 10^{-4} or 10^{-5}. In the presence of the single dominant gene HO (for homothallism), mating type interconversion occurs much more frequently, almost once per generation, and it becomes impossible to culture haploid cell lines since mating events lead to the rapid diploidization of the cell population after the germination of a haploid spore. A long series of elegant genetic experiments led to the hypothesis that mating type specificity is determined by the presence of one of two blocks of nonhomologous DNA at the MAT locus, MATa or MATα. In addition to the expressed block of DNA at the MAT locus, haploid strains were hypothesized to contain silent copies of the a and α gene sets at two other loci, HML and HMR. Most haploid strains contain one silent copy of each set of genes: the configuration HMLα, HMRa is particularly common in laboratory strains. Mating type interconversion was thought to involve the insertion, by an unspecified recombinational mechanism, of a set of a or α genes derived from HML or HMR at the MAT locus. The cloning of the whole set of mating type genes, originally by the identification of sequences that complement sterile mutations at MATa and MATα, led to a rapid verification of the basic predictions of the above hypothesis (23,24). Specifically, MATa proves to contain a 700 base-pair a-specific sequence that is replaced by a nonhomologous 850 base-pair sequence in MATα. Silent copies of the a and α gene sets, located at HML and HMR, differ by the identical substitution. Outside of the short nonhomologous region, all the active and silent gene sets share similar or identical sequences for at least 230 base pairs on one side and 700 base pairs on the other. Further studies aimed at relating these structural alterations to gene function are under way, but even at the present stage of analysis, the combination of classical genetic studies and molecular cloning has provided a more detailed description of this sytem than of any other reversible developmental event in eukaryotic biology.

Yeast cloning experiments have provided a number of new insights into gene expression by facilitating the analysis of gene transcripts. Two quantitative studies of mRNA levels have demonstrated that both a biosynthetic gene (URA3, encoding orotidine-5-phosphate decarboxylase) and a catabolic gene (CYC1, encoding iso-1-cytochrome c) are regulated by alterations in the rate of transcription (70,71,135).

Two other experiments, in which only steady-state message levels were measured, showed that two different regulatory loci exert their influence on genes whose expression they control by affecting mRNA levels. These experiments involved the effect of the RNA2 locus on three ribosomal protein mRNAs (37) and the GAL4 locus on four galactose-inducible transcripts (11). Both RNA2 and GAL4 behave genetically as though their gene products are positive regulators of gene expression.

Cloned hybridization probes have been equally useful in qualitative studies of RNA processing in yeast. For example, cloned yeast tRNA genes played a key role in the first direct demonstration that RNA splicing is the mechanism by which split genes are expressed. A yeast mutant (the rna1 allele ts136) was found to accumulate precursors to mature tRNAs (72). RNA-DNA hybridization with cloned tRNA genes allowed the rapid identification of specific ts136 precursors with particular tRNAs (72-74). RNA fingerprinting experiments allowed a direct proof that the precursors contained intervening sequences and that in vitro processing extracts could be used to convert them into mature tRNAs (73,74).

Three of the tRNA genes that contain intervening sequences -- $tRNA^{Tyr}$ (75-77), $tRNA^{Trp}$ (55) and $tRNA^{Leu}_3$ (57) -- represent the only cases to date in which cloned yeast genes have been used as templates in DNA-dependent transcription experiments. Intact Xenopus oocytes, oocyte extracts or somatic cell extracts were the sources of transcriptional activity. These studies allowed the analysis of steps in tRNA gene expression that precede splicing, including the nucleolytic maturation of both ends of the primary transcript, CCA addition, and a number of base modifications. Remarkably, these experiments also showed that the Xenopus splicing system could correctly splice the yeast tRNA precursors, thereby demonstrating the extreme phylogenic conservation of this RNA processing activity.

A combination of R-looping and S1 mapping techniques has allowed the precise localization of transcripts in a number of gene systems. An early result from this type of experiment concerned the transcriptional organization of the rRNA genes. In yeast, unlike higher eukaryotes, the 5S genes are interspersed among the genes for 5.8S, 18S and 25S RNA, with one gene for each RNA in each copy of the 10 kilobase rDNA repeat (78). Intriguingly, the 5S transcripts and those for the large rRNAs originate from opposite strands. Thus, there is divergent transcription by RNA polymerases I and III within the rDNA repeat. The availability of cloned sequences from the rDNA repeat has also allowed precise mapping of the site at which transcription of the precursor to the large rRNAs is initiated in vivo (79).

A more global question about rDNA concerns its genomic organization. This subject, which has a long history, has advanced rapidly only since cloned copies of the ribosomal repeat became available. Although the key experiment was an outgrowth of the analysis of cloned ribosomal RNA genes, it made little direct use of the cloned sequences themselves. In the course of mapping rDNA clones, it was observed that different haploid yeast strains have slightly different restriction maps within the rDNA repeat (80). Because of the high redundance of the rDNA, these differences can be observed directly in ethidium bromide-stained agarose gels of EcoRI-cleaved total yeast DNA. Genetic experiments were used to demonstrate that this restriction map variation segregates as a single Mendelian marker, which maps to chromosome XII (52,81,82). These experiments imply that the rRNA genes exist as a single tandem repeat of the 100 to 150 copies of the basic 10 kilobase unit, and that this entire cluster is located on chromosome XII.

The availability of suitable clones and an efficient transformation system made possible a series of experiments that have demonstrated directly unequal sister strand crossing-over within the rDNA repeat during both mitotic (83) and meiotic (84) growth. These experiments involved integrating the LEU2 gene at different sites within the ribosomal RNA gene cluster using a transformation vector that also contained rDNA sequences. The presence of this phenotypically observable gene within the rDNA cluster provided a sensitive genetic probe for recombinational events within the repeat. Although it is possible that other mechanisms of sequence homogeneity maintenance may also operate, these experiments indicate that unequal sister strand crossovers are sufficiently frequent that, by themselves, they would account for a high degree of homogeneity.

An aspect of yeast genome organization that, despite intensive study, has yet to be clarified by yeast cloning experiments, concerns the role of the 2μ plasmid. This 6 kilobase circular DNA molecule is yeast's only well defined plasmid. Recombinant DNA techniques have revealed an enormous amount of information about its sequence organization (featuring an inverted repeat across which recombinational inversion of the intervening single-copy DNA can occur (85-89)), its transcriptional map (90), the organization of its replication functions (91), the peptides whose synthesis it programs in E. coli (92) and now its entire DNA sequence (93). However, even with this knowledge, there are still no gene products or cellular functions that can be ascribed to 2μ plasmids. In a sense, the most provocative contribution of cloning techniques to the analysis of 2μ DNA was a straightforward Southern blot experiment that demonstrated that there are no sequences homologous to 2μ plasmids in chromosomal DNA, even in those strains of S. carlsbergensis that lack the plasmid altogether (87). Although, in principle, this experiment could have employed 2μ DNA itself rather than cloned 2μ sequences, its reliability benefited from the absolute sequence purification that cloning provides.

Despite the difficulty of developing any teleological rationale for 2μ DNA, much of the knowledge that has accumulated about it has been useful in designing yeast transformation vectors that incorporate the 2μ replication origin. These widely used vectors remain unique in that they appear not to support stable integration of the vector and any accompanying cloned sequences into chromosomal DNA. Further investigation of this odd phenomenon may uncover exceptions, but there is no question that there is a dramatic difference in the capacity of a cloned yeast sequence to direct relatively stable integration at its homologous chromosomal site depending on whether it is present in a plasmid that employs a chromosomally derived autonomous replication sequence (e.g., a TRP1 vector) or a 2μ-based vector.

Studies of chromosomally derived autonomous replication sequences (ARSs) are themselves a major outgrowth of yeast cloning experiments. The first ARS discovered is located on the same restriction fragment as the TRP1 gene (44,50,94) and another fortuitously discovered example was found on an ARG1 fragment (49).

The TRP1-linked ARS has been shown to be a cis-acting element that lacks any extensive homology (<20 base pairs) with other sites in the yeast genome (94). Shotgun experiments on total yeast DNA suggest that an ARS occurs on the average every 40 kilobases, a figure that is in rough agreement with estimates from electron microscopy and fiber autoradiography of the spacing between normal chromosomal replication origins (95). An important new observation is that functional ARSs can be readily cloned in yeast from the DNA of a wide variety of eukaryotic organisms, while no such sequences are found in E. coli DNA (66). Further molecular and genetic analysis of these sequences holds considerable promise for efforts to elucidate the pathways that control eukaryotic DNA replication.

As has been true in a number of organisms, yeast cloning experiments have also led to a greater insight into the nature of middle repetitive DNA. Early DNA-DNA renaturation experiments had suggested that yeast contains very little repetitive DNA beyond that accounted for by ribosomal, 2μ and mitochondrial DNA (96). It is now apparent, however, that the few percent of nuclear chromosomal DNA that does exist as dispersed repetitive sequences is a critical, if poorly understood, part of the yeast genetic system. The first such sequences that were explicitly studied were named Ty1 and δ (25). They were discovered on clones of the SUP4 tRNA^{Tyr} gene. Interest in this chromosomal region began when it was observed that the restriction map in the vicinity of SUP4 exhibited remarkable strain-to-strain variability (53). An analysis of two SUP4 clones derived from different strains showed that the differences in their restriction maps were caused, in part, by different arrangements of their middle-repetitive DNA. Both clones contained a number of copies of a

300 base-pair repetitive sequence named δ, and one strain also contained a larger element called Ty1. There are several hundred members of the δ family in the genome. These sequences are found either independently or (in approximately 35 cases) as a pair of directly repeated δs flanking a second 5 kilobase repeated element; the latter structure, which resembles a bacterial transposon, is a Ty1. More recent work, some of which is discussed below in conjunction with the analysis of mutational events, has unequivocally demonstrated that at least some Ty1 sequences are transposable and that, like bacterial transposons, their insertion leads to a short target sequence duplication (98,99) and affects the regulation of neighboring genes (97). Restriction mapping and DNA sequencing have also demonstrated that there is extensive sequence heterogeneity in both the Ty1 and δ families (97).

The discovery of the Ty1 and δ families, like many of the insights into overall genome structure that have resulted from cloning experiments, originated in fortuitous observations about individual clones. A more direct approach to studying the large scale organization of the genome is the chromosome walking technique in which a particular cloned sequence is used as a DNA-DNA hybridization probe to screen for new clones that overlap the original sequence. In this way, large contiguous stretches of particular chromosomes can be cloned and analyzed. Two projects of this kind have been reported in yeast. About 40 kilobases of DNA have been cloned near the chromosome III centromere, including the entire LEU2-CDC10 intergenic region (100,101) and a region of similar size including the CYC1 and SUP4 loci was independently analyzed (102). Estimates from these and other studies (46,103) suggest that a typical ratio of physical to genetic distance in yeast is within a factor of 2 of 3 kilobases/centimorgan. Not enough cases have been analyzed to allow any serious assessment of the uniformity of the recombination frequency from one genomic region to another.

Following discussion of applications of molecular cloning to the study of gene structure, gene expression and genome organization in yeast, the final topic is the analysis of mutational events. These experiments are grouped together rather artificially since, depending on the nature of the mutation involved, particular projects are obviously related to one or more of the above areas. They are discussed together here in order to focus attention on the range of methods that lend themselves to studying mutations that affect genes whose wild-type alleles have already been cloned and on the type of information that such studies can yield. Given the rich variety of phenotypically interesting yeast mutants now open to molecular analysis, vigorous activity in this area may be anticipated.

The most straightforward way of analyzing a mutation in a gene whose wild-type allele is already cloned is to reclone the gene from a mutant strain using the wild-type clone as a DNA-DNA hybridization probe. In the case of point mutations, DNA sequencing is then required to define the mutation. This method has seen extensive use in the analysis of tRNA genes that code for nonsense suppressors. Suppressor alleles of four tRNA^{Tyr} genes, SUP3-amber (104), SUP4-ochre (16), SUP6-ochre (105,106) and SUP11-ochre (105), have been cloned and sequenced, as have two tRNA^{Ser} genes, SUQ5-ochre (18) and SUP-RL1-amber (18). In all cases, the suppressor mutations proved to affect the expected base pairs at sites coding for the second or third nucleotide in the tRNA's anticodon. The yeast suppressor tRNA genes have attracted considerable interest, in part because they are the only known genes transcribed by RNA polymerase III that can be manipulated genetically. So far, the principal value of defining the molecular basis of suppressor mutations is that the ability to derive a suppressor allele from a particular tRNA gene provides unequivocal evidence that the gene is functional; this is not a trivial point, since nearly all eukaryotic tRNA genes belong to large multigene families.

As a technical point, the standard method of recloning the mutant tRNA genes has involved using a nick-translated restriction fragment that contains the wild-type tRNA gene as a DNA-DNA hybridization probe. In a typical case, the fragment might contain a few thousand base pairs of single-copy DNA along with roughly 100 base pairs of coding region. Although such probes will cross hybridize, by way of the coding region homology, to all the members of a particular tRNA gene family, they hybridize much more strongly to the particular locus that shares both coding region and flanking homology with the probe. Usually, there is no difficulty employing such locus-specific probes directly in plaque- or colony-screening. In some cases, however, particularly when the restriction fragment contains additional repetitive sequences besides the tRNA coding region, it may be necessary to use a single-copy subfragment as the probe (18).

In addition to the analysis of suppressor mutations, one tRNA gene, the tRNA^{Tyr}-coding SUP4 locus, has been intensively dissected by the isolation of a large collection of second-site mutations that were all recloned by the method described above and then sequenced (17). The mutations were isolated by selecting canavanine-resistant derivatives of a strain that contained both a SUP4 <u>ochre</u> suppressor and an <u>ochre</u> allele of the canavanine-resistance gene CAN1. Genetic analysis of these can^R strains identified a subset that involved complete or partial loss of SUP4 function because of second site mutations at the SUP4 locus itself. Thirty-two of these mutant genes were recloned and sequenced. Several conclusions were derived from the analysis of this mutant collection, which proved to contain exclusively point mutations (transitions, transversions and a single base-pair deletion). First, no mutations were obtained beyond the 5' or 3' termini of the region coding for mature tRNA^{Tyr}. As an indication of the good coverage of the coding region that was obtained, mutations were isolated which affected the last base pair at the 3' end and the third from the last base pair at the 5' end of the coding region. These results fit well with the emerging view that even the transcriptional specificity of RNA polymerase III genes lies within the coding region. As further support for this hypothesis, transcription experiments with the cloned mutant genes as templates indicated that some of the coding region mutations interfere with the efficiency with which SUP4 is transcribed in vitro while others enhance it (76). Only a single mutation was found within the intervening sequence and this allele proved to have a disappointingly trivial explanation: the creation of a tract of five consecutive Ts in the + strand gives rise to a premature site for transcription termination (76).

In an independent investigation of the SUP4 locus, a set of large deletions in the SUP4 region was analyzed (104). Unlike the other tRNA^{Tyr} genes that have been studied in which deletions are extremely rare, about one-third of the spontaneous loss-of-function mutations at SUP4 involve deletions of the entire coding region. It is likely, but still unproven, that the formation of these deletions is somehow mediated by the δ sequences in the vicinity of SUP4. Southern blot experiments with cloned fragments from the SUP4 region were used to define at least two classes of deletions, one of 2.1 kilobases and the other of 2.8 kilobases. Despite the relative compactness of the yeast genome, these sizable deletions are not lethal in haploid strains.

Another locus at which a rich variety of mutational events has been analyzed with the aid of cloning techniques is HIS4 (97). A principal focus of attention has been an unstable, spontaneous mutation, his4-912, which was caused by a Ty1 insertion into the promoter region. With Southern blotting experiments using the wild-type HIS4 gene as a probe, it was established that the mutation was an insertion (107). Although the mutant gene could have been cloned by a procedure analogous to that used in the SUP4 experiments, a different method was employed that exploits yeast transformation (107). A pBR322 clone of a 1.5 kilobase SalI restriction fragment from the HIS4 region was used to transform a strain bearing the his4-912 mutation. The SalI fragment contained a small part of the HIS4 coding region as well as the 5'-proximal sequences into which the insertion had been placed. Southern blot experiments confirmed that in the transformants the circular transforming DNA had integrated by homologous recombination, producing a structure in which the DNA sequences present in the 1.5 kilobase SalI fragment were duplicated on opposite sides of the pBR322 DNA. In one segment of the duplication the insertion mutation was present while the other contained the wild-type sequence. By cleavage of the transformant's DNA with a judiciously chosen restriction enzyme and ligation of the fragments into circles, molecules were produced that contained the insertion mutation, along with other adjacent DNA, in a suitable cloning site of pBR322. This molecule was readily cloned by transforming E. coli with the transformant's total mixture of cleaved and ligated DNA, and then selecting for the amp^R trait of pBR322. For well-mapped chromosomal regions, it will probably prove generally feasible to design a strategy of this kind for the recloning of a particular mutant fragment. A priori, it is not obvious whether the technique involves more or less effort than conventional recloning from a shotgun pool, and investigator preference is likely to prove the deciding factor in choosing one method or the other.

Analysis of the insertion mutation indicated that it was caused by transposition of a Ty1 sequence to a site that lacked any significant homology with the transposon (98). The insertion occurs at a site 161 base pairs upstream from the AUG translation start. Since the site of transcription initiation in the wild-type gene is not known, it is premature to speculate on the reason that the transposition prevents expression of the gene. Reversion to his$^+$ is frequent, with the most common class of such revertants involving loss of all of the Ty1 sequence except for one of its δs; a likely mechanism for this event is simply homologous recombination between the two identical, directly repeated δs that bound the Ty1 (107,108). Among the less frequent his$^+$ revertants is a bizarre variety of deletions, inversions, transpositions and translocations, not unlike events that are mediated by bacterial transposons. Although Southern blot experiments with various cloned HIS4 probes were quite helpful in the characterization of these rearrangements, the main weight of the analysis was carried by conventional genetic experiments (108). Given the complexity of the results, the prospects for the detailed analysis of transposition in eukaryotes that lack facile genetic systems are not encouraging.

A second transposition into the HIS4 promoter region, his4-917, was studied by analogous methods. It has many properties similar to those of his4-912, but the transposon itself has a substantially different sequence than the Ty1 that caused the his4-912 mutation (97).

A final example of a mutational event that was characterized with the aid of cloned DNA-DNA hybridization probes was a circular derivative of chromosome III (103). This mutant chromosome can be reproducibly selected since it results from one of the more common events that allows illegitimate matings between two haploids of mating type a. It was hypothesized that the circular chromosome was formed by homologous recombination between the silent α gene cluster at HMLα and the a cluster at MATa. These two gene clusters, which are now known to share 1.7 kilobases of homology, are present on opposite sides of the chromosome III centromere in the same orientation. Recombination between them produces a 180 kilobase circular chromosome containing a single set of mating type genes; these genes result from a HMLα-MATa fusion that produces an active set of α genes. If the haploids

Table 1

Tabulation of Cloned Yeast Sequences[a]

A. Cloned DNA Sequences That Correspond to Known Genetic Loci

Genetic Locus	Chromosome[b]	Gene Product[c]	Screening Method	References[d]
ADC1	?	alcohol dehydrogenase	complementation in yeast	134
ARG4	VIII	arginosuccinate lyase	complementation in E. coli	31, 49
CAN1	V	arginine permease	complementation in yeast	47
CDC10	III	unknown	complementation in yeast	101
CDC28	II	unknown	complementation in yeast	46
CYC1	X	iso-1-cytochrome c	DNA-DNA hybridization (synthetic oligodeoxynucleotide probe).	12, 68, 102, 136
CYC7	V	iso-2-cytochrome c	DNA-DNA hybridization (iso-1-cytochrome c probe)	21
GAL1	II	galactokinase	cDNA-DNA hybridization complementation in E. coli	11, 112, 125
GAL7	II	galactose-1-phosphate uridyl-transferase	cDNA-DNA hybridization	11, 125
GAL10	II	uridine diphosphogalactose-4-epimerase	cDNA-DNA hybridization	11, 125
HIS3	XV	imidazole glycerol phosphate dehydratase	complementation in E. coli	19, 20, 29, 33, 121
HIS4	III	phosphoribosyl-AMP cyclohydrolase, phosphoribosyl-AMP pyrophosphorylase, histidinol dehydrogenase	complementation in yeast, transformation	42, 97, 98, 107
HMLa	III	unknown	DNA-DNA hybridization (HMLα probe)	23
HMLα	III	unknown	DNA-DNA hybridization (MATα, HMLα probes) complementation in yeast	22–24
HMRa	III	unknown	DNA-DNA hybridization (MATα, HMLα probes)	23, 24
HMRα	III	unknown	DNA-DNA hybridization (HMLα probe)	23
LEU2	III	isopropylmalate dehydrogenase	complementation in E. coli	30, 43, 100, 121
MATa	III	unknown	DNA-DNA hybridization (MATα, HMLα probes)	23, 24
MATα	III	unknown	complementation in yeast DNA-DNA hybridization (HMLα probe)	23, 24

Table 1 (cont.)

Tabulation of Cloned Yeast Sequences[a]

A. Cloned DNA Sequences That Correspond to Known Genetic Loci (cont.)

Genetic Locus	Chromosome[b]	Gene Product[c]	Screening Method	References[d]
PGK1	III	3-phosphoglycerate kinase	antigen production in E. coli	36
SUP2	IV	tRNA^{Tyr}	tRNA-DNA hybridization	53, 118
SUP3	XV	tRNA^{Tyr}	tRNA-DNA hybridization	53, 104, 118
SUP4	X	tRNA^{Tyr}	tRNA-DNA hybridization	16, 17, 25, 53, 75, 76, 99, 118
SUP5	II	tRNA^{Tyr}	tRNA-DNA hybridization	53, 118
SUP6	VI	tRNA^{Tyr}	tRNA-DNA hybridization	53, 105, 118
SUP8	XIII	tRNA^{Tyr}	tRNA-DNA hybridization	16, 53, 75, 77, 118
SUP11	VI	tRNA^{Tyr}	tRNA-DNA hybridization	16, 53, 75, 105, 118
SUP-RL1	III	tRNA$^{Ser}_{UCG}$	tRNA-DNA hybridization	18, 119
SUQ5	XVI	tRNA$^{Ser}_{UCA}$	tRNA-DNA hybridization	18, 119
TRP1	IV	N-(5'-phosphoribosyl) anthranilate isomerase	complementation in E. coli	44, 94
TRP5	VII	tryptophan synthetase	complementation in E. coli	34, 111
TYR1	II	prephenate dehydrogenase, tyrosine aminotransferase	complementation in yeast	46
URA3	V	orotidine-5-phosphate decarboxylase	complementation in E. coli	70

B. Other Well-Characterized Cloned DNA Sequences

Gene Product or Sequence Designation	Screening Method	References
actin	DNA-DNA hybridization (Dictyostelium probe)	26-28
glyceraldehyde-3-phosphate dehydrogenase	cDNA-DNA hybridization	10, 67, 114
histone H2A	mRNA-DNA hybridization/hybrid-selected mRNA translation	38
histone H2B	mRNA-DNA hybridization/hybrid-selected mRNA translation	38
ribosomal proteins	mRNA-DNA hybridization/hybrid-selected mRNA translation	37
rRNA[e]	rRNA-DNA hybridization	78, 80, 110, 113, 115-117, 120, 122-124

Table 1 (cont.)

Tabulation of Cloned Yeast Sequences[a]

B. Other Well-Characterized Cloned DNA Sequences (cont.)

Gene Product or Sequence Designation	Screening Method	References
$tRNA_3^{Arg}$[f]	tRNA-DNA hybridization	60, 109
$tRNA^{Asp}$	tRNA-DNA hybridization	60, 109
$tRNA_3^{Leu}$	tRNA-DNA hybridization	56, 57
$tRNA^{Met}$	tRNA-DNA hybridization	61
$tRNA^{Phe}$	tRNA-DNA hybridization	54
$tRNA_2^{Ser}$	tRNA-DNA hybridization	59, 119
$tRNA^{Trp}$	tRNA-DNA hybridization	55
δ (repetitive element)	DNA-DNA hybridization, transformation	25, 97–99, 107
Ty1 (repetitive element)	DNA-DNA hybridization, transformation	25, 97–99, 107
2μ plasmid	Direct examination of individual clones, DNA-DNA hybridization	85–88, 90–93

[a] In most cases, cloned sequences have been included only if there is published evidence as of 8/80 establishing the clone's identity.

[b] The most recent published genetic map of yeast is in ref. 1. This map is updated periodically in the abstracts of the biennial Cold Spring Harbor Meetings on the Molecular Biology of Yeast (last held 8/79; next scheduled 8/81). These abstracts may be ordered from the Cold Spring Harbor Laboratory.

[c] A survey of the gene products of yeast genetic loci may be found in ref. 1

[d] The cited references are limited to those that describe the cloning of the gene, the confirmation of its identity and the elucidation of the clone's physical structure (restriction mapping, sequencing, mapping of gene transcripts or other functions). Additional references may be found in the text to other types of experiments that involve the cloned gene.

[e] Although the rRNA gene cluster is not a mutationally defined genetic locus, it has been mapped to chromosome XII (52).

[f] Transfer RNA genes have been included only when the identity of the encoded tRNA has been established by sequencing. A large collection of tRNA gene clones, many of which were characterized by RNA-DNA hybridization to purified and partially-purified tRNAs, is described in ref. 109.

in which the circular chromosome forms are not rescued by mating to wild-type a cells that predominate in the culture, the event is lethal as a result of the deletion of many essential genes distal to MAT and HML. It proved possible to purify small quantities of intact DNA from the circular chromosome by equilibrium banding in cesium chloride-ethidium bromide gradients. Hybridization experiments showed that the large DNA circles, which are not observed in nonmutant control strains, contain sequences homologous to LEU2 and HIS4 clones, both of which derive from chromosome III. One of the important conclusions from this work is that the chromosomal DNA is evidently covalently continuous through the centromere.

FUTURE CHALLENGES

Rather than conclude with an overview of past accomplishments, it is more appropriate to look to the future, since we are discussing a field that is still in its infancy. Many of the problems that require further work are not particularly specific to yeast. Intensive efforts may be expected, for example, to develop regulated, template-dependent in vitro transcription and DNA replication systems both from yeast and from other organisms. If these efforts meet with success, a special advantage of yeast will be the ability to study the effects of mutations that alter not just the DNA template but other components of the system as well.

There are other opportunities, however, that arise more specifically out of the present state of yeast cloning experiments. One of these is the effort to use yeast as a host in which to study or exploit the expression of genes from higher organisms. The first attempt to achieve expression of mammalian genes in yeast, which involved a rabbit β-globin gene, was disappointing as no normal transcripts were produced (126). There is little doubt, however, that success could be achieved with a sufficient degree of manipulation of the vector and perhaps the exogenous gene. With no knowledge of what incompatibilities will prove most common between the genetic apparatus of yeast and the genes of higher organisms, it is premature to speculate on the ultimate promise of yeast as an all-purpose host for eukaryotic genes. However, many of the steps in yeast gene expression seem certain to be homologous with corresponding steps in higher organisms. Yeast may prove to offer an ideal system in which to subject these steps, in the case of specific exogenous genes, to genetic analysis. In terms of the commercial production of mammalian gene products, yeast may offer advantages over bacteria in its ability to carry out post-transcriptional and post-translational steps that do not occur in prokaryotes.

A second area of future challenges, which relates to the analysis of yeast genes themselves, concerns the development of better techniques for identifying the product and, more importantly, the function of the product of a cloned gene. This problem, of course, is also important to futher progress in a number of other systems, particularly in the case of viral genes and such cellular systems as heat shock genes, but yeast is the first eukaryote from which it has become possible to clone large numbers of genes whose products are essential to identified cellular processes, but about which no biochemical knowledge exists. Even the phenotypes of mutant alleles of these genes are often relatively nonspecific (e.g., the inability to mate, sporulate or initiate a new mitotic cell cycle). In principle, methods exist to use the cloned gene as the starting point for studies of the structure and function of the gene's products. For example, hybrid-selected message can be translated in vitro and antibodies raised against the translation products. These antibodies can then be used to assay for the protein during fractionation procedures or to study its subcellular localization by a method such as immunofluorescent microscopy.

Nonetheless, additional resources are badly needed. In E. coli, for example, minicell methodology allows specific labeling of copious quantities of the gene products encoded by plasmid clones. Spectacular overproduction of the products of cloned genes can also be engineered both on transducing phage and plasmids. Analogous methods could profitably be developed in yeast since the availability of preparative quantities of a gene's product will often be a prerequisite to functional studies.

Another area that appears promising is the prospect of carrying out detailed genetic analyses on genes that have been engineered in vitro and then reestablished in vivo by transformation. There has been much discussion of surrogate or reverse genetics in which cloned genes are mutagenized in vitro and then assessed for functionality in transcription systems or in intact cells. Important applications of this approach have involved efforts to define promoters for RNA polymerases II (127) and III (128,129) and to establish functions for intervening sequences (130,131). One such experiment in yeast led to the demonstration that an ochre allele of a tRNA^{Tyr} gene still exhibits a suppressor phenotype once its intervening sequence has been removed exactly (106).

The next level of sophistication in this area will be to subject the reintroduced, altered gene to a wide range of genetic selections in order to detect subtle phenotypic changes or to identify pathways by which the cell can circumvent the effects of the alteration. The experiments described earlier on unequal sister-strand crossing-over within the rDNA repeat (83,84) and a recent study on gene conversion between nonhomologous chromosomes (132) have some of this flavor. In these cases, novel chromosome structures were created by transformation and their genetic properties were analyzed. A similar approach to such in vitro constructions as gene fusions may open new avenues to the genetic analysis of gene expression and gene regulation.

A method has already been described for carrying out a key step in such a program: to substitute an altered gene for the wild-type copy without otherwise changing a strain's genotype (133). A his3 mutant gene containing an internal coding region deletion, which was generated in vitro, was substituted for a wild-type HIS3 gene in two steps. First, a normal integration event was selected between the wild-type chromosome and a URA3 plasmid containing the his3 deletion. This event produced the structure his3Δ-URA3-HIS3 by homologous recombination. A second recombination event, between the two copies of the duplicated sequence that lies between the deletion fusion point and the URA3 vector on one side and within the wild-type HIS3 gene on the other, produced a ura⁻his⁻ segregant that contained a perfect substitution of the altered gene for its wild-type allele.

A final example of a future challenge to yeast cloning technology, which now appears within reach, is the determination of the global physical structure of the yeast genome. With its small genome size and low amount of repetitive DNA, yeast is an ideal prospect for an analysis of the physical map of total nuclear DNA. With high capacity cloning vectors, one or two thousand clones would cover the entire genome and methods can be imagined for working out its restriction map. Yeast transformation would allow superposition of hundreds of genetically mapped genes on the physical map. Since tetrad analysis allows the genetic mapping of centromeres, these too could be localized. With a transformation assay for DNA replication origins and good techniques for mapping transcripts, the positions of a large number of functional sites could be specified. Given such an integrated genetic, physical and functional map, powerful new tools would become available for studying such phenomena as position effects, DNA replication, recombination, repetitive DNA, transposition and genetic instability, all of which are presently visible only when there is the chance involvement of a genetically marked site.

It is apparent that the coalescence of classical yeast genetics, cloning techniques and yeast transformation has opened the floodgates to a level of analysis in yeast molecular biology that only a few years ago seemed beyond reach for any eukaryotic cell. The extent to which the molecular details of yeast biology are ultimately elucidated no longer seems likely to be limited by technically insuperable problems. Instead, the degree of effort devoted to this endeavor will be determined by the extent to which the acquired information proves useful in understanding broader issues in biology and the pace at which still more complex organisms become accessible to a similar level of scrutiny.

REFERENCES

1. Sherman, F. and Lawrence, C.W. (1974) in Handbook of Genetics (King, R.C., ed.), Vol. I, pp. 359-393, Plenum Press, New York, NY.
2. Mortimer, R.K. and Hawthorne, D.C. (1969) in The Yeasts (Rose, A.H. and Harrison, J.S., eds.), Vol. I, pp. 385-460, Academic Press, New York, NY.
3. Fincham, J.R.S., Day, P.R. and Radford, A. (1979) Fungal Genetics, 4th Ed., University of California Press, Berkeley, CA.
4. Neve, R.L., West, R.W. and Rodriguez, R.L. (1979) Nature 277, 324-325.
5. Cohen, J.D., Eccleshall, T.R., Needleman, R.B., Federoff, H., Buchferer, B.A. and Marmur, J. (1980) Proc. Nat. Acad. Sci. U.S.A. 77, 1078-1082.
6. Ilgen, C., Farabaugh, P.J., Hinnen, A., Walsh, J.M. and Fink, G.R. (1978) in Genetic Engineering (Setlow, J.K. and Hollaender, A., eds.), Vol. 1, pp. 117-132, Plenum Press, New York, NY.
7. Bernard, H.U. and Helinski, D.R. (1979) in Genetic Engineering (Setlow, J.K. and Hollaender, A., eds.), Vol. 2, pp. 133-168, Plenum Press, New York, NY.
8. Williams, B.G. and Blattner, F.R. (1979) in Genetic Engineering (Setlow, J.K. and Hollaender, A., eds.), Vol. 2, pp. 201-281, Plenum Press, New York, NY.
9. Hohn, B. and Hinnen, A. (1979) in Genetic Engineering (Setlow, J.K. and Hollaender, A., eds.), Vol. 2, pp. 169-184, Plenum Press, New York, NY.
10. Holland, M.J. and Holland, J.P. (1979) J. Biol. Chem. 254, 5466-5474.
11. St. John, T. and Davis, R.W. (1979) Cell 16, 443-452.
12. Montgomery, D.L., Hall, B.D., Gillam, S. and Smith, M. (1978) Cell 14, 673-680.
13. Stewart, J.W. and Sherman, F. (1974) in Molecular and Environmental Aspects of Mutagenesis (Prakash, L., Sherman, F., Miller, M.W., Lawrence, C.W. and Taber, H.W., eds.), pp. 102-127, C.C. Thomas, Springfield, IL.
14. Szostak, J.W., Stiles, J.I., Tye, B.-K., Chiu, P., Sherman, F. and Wu, R. (1979) Methods Enzymol. 68, 419-428.
15. Mevarech, M., Noyes, B.E. and Agarwal, K.L. (1979) J. Biol. Chem. 254, 7472-7475.
16. Goodman, H.M., Olson, M.V. and Hall, B.D. (1977) Proc. Nat. Acad. Sci. U.S.A. 74, 5453-5457.
17. Kurjan, J., Hall, B.D., Gillam, S. and Smith, M. (1980) Cell 20, 701-709.
18. Olson, M.V., Page, G.S., Sentenac, A., Piper, P.W., Worthington, M., Weiss, R. and Hall, B.D. (1980) (in preparation).

19   Struhl, K. and Davis, R.W. (1977) Proc. Nat. Acad. Sci. U.S.A. 74, 5255-5259.
20   Struhl, K., Davis, R.W. and Fink, G.R. (1979) Nature 279, 78-79.
21   Montgomery, D.L., Leung, D.W., Smith, M., Shalit, P., Faye, G. and Hall, B.D. (1980) Proc. Nat. Acad. Sci. U.S.A. 77, 541-545.
22   Hicks, J., Strathern, J.N. and Klar, A.J.S. (1979) Nature 282, 478-483.
23   Strathern, J.N., Spatola, E., McGill, C. and Hicks, J.B. (1980) Proc. Nat. Acad. Sci. U.S.A. 77, 2839-2843.
24   Nasmyth, K.A. and Tatchell, K. (1980) Cell 19, 753-764.
25   Cameron, J.R., Loh, E.Y. and Davis, R.W. (1979) Cell 16, 739-751.
26   Gallwitz, D. and Seidel, R. (1980) Nucl. Acids Res. 8, 1043-1059.
27   Gallwitz, D. and Sures, I. (1980) Proc. Nat. Acad. Sci. U.S.A. 77, 2546-2550.
28   Ng, R. and Abelson, J. (1980) Proc. Nat. Acad. Sci. U.S.A. 77, 3912-3916.
29   Struhl, K., Cameron, J.R. and Davis, R.W. (1976) Proc. Nat. Acad. Sci. U.S.A. 73, 1471-1475.
30   Hicks, J. and Fink, G.R. (1977) Nature 269, 265-267.
31   Clarke, L. and Carbon, J. (1978) J. Mol. Biol. 120, 517-532.
32   Struhl, K., Stinchcomb, D.T. and Davis, R.W. (1980) J. Mol. Biol. 136, 291-307.
33   Struhl, K. and Davis, R.W. (1980) J. Mol. Biol. 136, 309-332.
34   Walz, A., Ratzkin, B. and Carbon, J. (1978) Proc. Nat. Acad. Sci. U.S.A. 75, 6172-6176.
35   Brennan, M.B. and Struhl, K. (1980) J. Mol. Biol. 136, 333-338.
36   Hitzeman, R.A., Chinault, A.C., Kingsman, A.J. and Carbon, J. (1979) ICN-UCLA Symp. Mol. Cell. Biol. 14, 57-68.
37   Woolford, J.L. Jr., Hereford, L.M. and Rosbash, M. (1979) Cell 18, 1247-1259.
38   Hereford, L., Fahrner, K., Woolford, J. Jr., Rosbash, M. and Kaback, D.B. (1979) Cell 18, 1261-1271.
39   Woolford, J.L. Jr., and Rosbash, M. (1979) Nucl. Acids Res. 6, 2483-2497.
40   Hinnen, A., Hicks, J.B. and Fink, G.R. (1978) Proc. Nat. Acad. Sci. U.S.A. 75, 1929-1933.
41   Clarke, L. and Carbon, J. (1979) Methods Enzymol. 68, 396-408.
42   Hinnen, A., Farabaugh, P., Ilgen, C. and Fink, G.R. (1979) ICN-UCLA Symp. Mol. Cell. Biol. 14, 43-50.
43   Beggs, J.D. (1978) Nature 275, 104-109.
44   Struhl, K., Stinchcomb, D.T., Scherer, S. and Davis, R.W. (1979) Proc. Nat. Acad. Sci. U.S.A. 76, 1035-1039.

45. Botstein, D., Falco, S.C., Stewart, S.E., Brennen, M., Scherer, S., Stinchcomb, D.T., Struhl, K. and Davis, R.W. (1979) Gene 8, 17-24.
46. Nasmyth, K.A. and Reed, S.I. (1980) Proc. Nat. Acad. Sci. U.S.A. 77, 2119-2123.
47. Broach, J.R., Strathern, J.N. and Hicks, J.B. (1979) Gene 8, 121-133.
48. Gerbaud, C., Fournier, P., Blanc, H., Aigle, M., Heslot, H. and Guerineau, M. (1979) Gene 5, 233-253.
49. Hsiao, C.-L. and Carbon, J. (1979) Proc. Nat. Acad. Sci. U.S.A. 76, 3829-3833.
50. Kingsman, A.J., Clarke, L., Mortimer, R.K. and Carbon, J. (1979) Gene 7, 141-152.
51. Storms, R.K., McNeil, J.B., Khandekar, G.A., Parker, J. and Friesen, J.D. (1979) J. Bacteriol. 140, 73-82.
52. Petes, T.D. (1979) Proc. Nat. Acad. Sci. U.S.A. 76, 410-414.
53. Olson, M.V., Loughney, K. and Hall, B.D. (1979) J. Mol. Biol. 132, 387-410.
54. Valenzuela, P., Venegas, A., Weinberg, F. Bishop, R. and Rutter, W.J. (1978) Proc. Nat. Acad. Sci. U.S.A. 75, 190-194.
55. Ogden, R.C., Beckmann, J.S., Abelson, J., Kang, H.S., Soll, D. and Schmidt, O. (1979) Cell 17, 399-406.
56. Venegas, A., Quiroga, M., Zaldivar, J., Rutter, W.J. and Valenzuela, P. (1979) J. Biol. Chem. 254, 12306-12309.
57. Johnson, J.D., Ogden, R., Johnson, P., Abelson, J., Dembeck, P. and Itakura, K. (1980) Proc. Nat. Acad. Sci. U.S.A. 77, 2564-2568.
58. Etcheverry, T., Colby, D. and Guthrie, C. (1979) Cell 18, 11-26.
59. Page, G.S. (1978) Ph.D. Dissertation, University of Washington, Seattle, WA.
60. Abelson, J.A. (1979) Annu. Rev. Biochem. 48, 1035-1069.
61. Olah, J. and Feldman, H. (1980) Nucl. Acids Res. 8, 1975-1986.
62. Garber, R.L. and Gage, L.P. (1979) Cell 18, 817-828.
63. Hagenbuchle, O., Larson, D., Hall, G.I. and Sprague, K.U. (1979) Cell 18, 1217-1229.
64. Hovemann, B., Sharp, S., Yamada, H. and Soll, D. (1980) Cell 19, 889-895.
65. Muller, F. and Clarkson, S.G. (1980) Cell 19, 345-353.
66. Stinchcomb, D.T., Thomas, M., Kelly, T., Selker, E. and Davis, R.W. (1980) Proc. Nat. Acad. Sci. U.S.A. 77, 4559-4563.
67. Holland, J.P. and Holland, M.J. (1980) J. Biol. Chem. 255, 2596-2605.
68. Smith, M., Leung, D.W., Gillam, S., Astell, C.R., Montgomery, D.L. and Hall, B.D. (1979) Cell 16, 753-761.

69 Herskowitz, I., Blair, L., Forbes, D., Hicks, J., Kassir, Y., Kushner, P., Rine, J., Sprague, G. and Strathern, J. (1979) in The Molecular Genetics of Development (Loomis, W. and Leighton, T., eds.), Academic Press, New York, NY (in press).
70 Bach, M.-L., Lacroute, F. and Botstein, D. (1979) Proc. Nat. Acad. Sci. U.S.A. 76, 386-390.
71 Zitomer, R.S., Montgomery, D., Nichols, D.L. and Hall, B.D. (1979) Proc. Nat. Acad. Sci. U.S.A. 76, 3627-3631.
72 Hopper, A.K., Banks, F. and Evangelidis, V. (1978) Cell 14, 211-219.
73 Knapp, G., Beckmann, J.S., Johnson, P.F., Fuhrman, S.A. and Abelson, J. (1978) Cell 14, 221-236.
74 O'Farrell, P.Z., Cordell, B., Valenzuela, P., Rutter, W.J. and Goodman, H.M. (1978) Nature 274, 438-445.
75 DeRobertis, E.M. and Olson, M.V. (1979) Nature 278, 137-143.
76 Koski, R.A., Clarkson, S.G., Kurjan, J., Hall, B.D. and Smith, M. (1980) Cell 22, 415-425.
77 Melton, D.A., DeRobertis, E.M. and Cortese, R. (1980) Nature 284, 143-148.
78 Philippsen, P., Thomas, M., Kramer, R.A. and Davis, R.W. (1978) J. Mol. Biol. 123, 387-404.
79 Klemenz, R. and Geiduschek, E.P. (1980) Nucl. Acids. Res. 8, 2679-2689.
80 Petes, T.D., Hereford, L.M. and Skryabin, K.G. (1978) J. Bacteriol. 134, 295-305.
81 Petes, T.D. and Botstein, D. (1977) Proc. Nat. Acad. Sci. U.S.A. 74, 5091-5095.
82 Petes, T.D. (1979) J. Bacteriol. 138, 185-192.
83 Szostak, J.W. and Wu, R. (1980) Nature 284, 426-430.
84 Petes, T.D. (1980) Cell 19, 765-774.
85 Beggs, J.D., Guerineau, M. and Atkins, J.F. (1976) Mol. Gen. Genet. 148, 287-294.
86 Hollenberg, C.P., Degelmann, A., Kustermann-Kuhn, B. and Royer, H.D. (1976) Proc. Nat. Acad. Sci. U.S.A. 73, 2072-2076.
87 Cameron, J.R., Philippsen, P. and Davis, R.W. (1977) Nucl. Acids. Res. 4, 1429-1448.
88 Gubbins, E.J., Newlon, C.S., Kann, M.D. and Donelson, J.E. (1977) Gene 1, 185-207.
89 Livingston, D.M. and Klein, H.L. (1977) J. Bacteriol. 129, 472-481.
90 Broach, J.R., Atkins, J.F., McGill, C. and Chow, L. (1979) Cell 16, 827-839.
91 Broach, J.R. and Hicks, J.B. (1980) Cell 21, 501-508.
92 Hollenberg, C.P. (1978) Mol. Gen. Genet. 162, 23-34.
93 Hartley, J.L. and Donelson, J.E. (1980) Nature 286, 860-865.
94 Stinchcomb, D.T., Struhl, K. and Davis, R.W. (1979) Nature 282, 39-43.

95   Beach, D., Piper, M. and Shall, S. (1980) Nature 284, 185-187.
96   Lauer, G.D., Roberts, T.M. and Klotz, L.C. (1977) J. Mol. Biol. 114, 507-526.
97   Roeder, G.S., Farabaugh, P.J., Chaleff, D.T. and Fink, G.R. (1980) Science 209, 1375-1380.
98   Farabaugh, P.J. and Fink, G.R. (1980) Nature 286, 352-356.
99   Gafner, J. and Philippsen, P. (1980) Nature 286, 414-418.
100 Chinault, A.C. and Carbon, J. (1979) Gene 5, 111-126.
101 Clarke, L. and Carbon, J. (1980) Proc. Nat. Acad. Sci. U.S.A. 77, 2173-2177.
102 Shalit, P., Loughney, K., Olson, M.V. and Hall, B.D. (1980) (manuscript in preparation).
103 Strathern, J.N., Newlon, C.S., Herskowitz, I. and Hicks, J.B. (1979) Cell 18, 309-319.
104 Rothstein, R. (1979) Cell 17, 185-190.
105 Philippsen, P., Cameron, J.R. and Davis, R.W. (personal communication).
106 Wallace, R.B., Johnson, P.F., Tanaka, S., Schold, M., Itakura, K. and Abelson, J. (1980) Science 209, 1396-1400.
107 Roeder, G.S. and Fink, G.R. (1980) Cell 21, 239-249.
108 Chaleff, D.T. and Fink, G.R. (1980) Cell 21, 227-237.
109 Beckmann, J.S., Johnson, P.F. and Abelson, J. (1977) Science 196, 205-208.
110 Bell, G.I., DeGennaro, L.J., Gelfand, D.H., Bishop, R.J., Valenzuela, P. and Rutter, W.J. (1977) J. Biol. Chem. 252, 8118-8125.
111 Carbon, J., Ratzkin, B., Clarke, L. and Richardson, D. (1977) Brookhaven Symp. Biol. 29, 277-296.
112 Citron, B.A., Feiss, M. and Donelson, J.E. (1979) Gene 6, 251-264.
113 Ferguson, J. and Davis, R.W. (1978) J. Mol. Biol. 123, 417-430.
114 Holland, J.P. and Holland, M.J. (1979) J. Biol. Chem. 254, 9839-9845.
115 Kramer, R.A., Cameron, J.R. and Davis, R.W. (1976) Cell 8, 227-232.
116 Kramer, R.A., Philippsen, P. and Davis, R.W. (1978) J. Mol. Biol. 123, 405-416.
117 Maxam, A.M., Tizard, R., Skryabin, K.G. and Gilbert, W. (1977) Nature 267, 643-645.
118 Olson, M.V., Hall, B.D., Cameron, J.R. and Davis, R.W. (1979) J. Mol. Biol. 127, 285-295.
119 Olson, M.V., Page, G.S., Sentenac, A., Loughney, K., Kurjan, J., Benditt, J. and Hall, B.D. (1980) in Transfer RNA: Biological Aspects (Soll, D., Abelson, J.N. and Schimmel, P.R., eds.), pp. 267-279, Cold Spring Harbor Laboratory, Cold Spring Harbor, NY.
120 Philippsen, P., Kramer, R.A. and Davis, R.W. (1978) J. Mol. Biol. 123, 371-386.

121 Ratzkin, B. and Carbon, J. (1977) Proc. Nat. Acad. Sci. U.S.A. 74, 487-491.
122 Skryabin, K.G., Maxam, A.M., Petes, T.D. and Hereford, L. (1978) J. Bacteriol. 134, 306-309.
123 Valenzuela, P., Bell, G.I., Masiarz, F.R., DeGennaro, L.J., and Rutter, W.J. (1977) Nature 267, 641-643.
124 Valenzuela, P., Bell, G.I., Venegas, A., Sewell, E.T., Masiarz, F.R., DeGennaro, L.J., Weinberg, F. and Rutter, W.J. (1977) J. Biol. Chem. 252, 8126-8135.
125 Davis, R.W., Struhl, K. and St. John, T. (1979) J. Supramolecular Structure Suppl. 3, 34.
126 Beggs, J.D., van den Berg, J., van Vyen, A. and Weissman, C. (1980) Nature 283, 835-840.
127 Corden, J., Wasylyk, B., Buchwalder, A., Sassone-Corsi, P., Kedinger, C. and Chambon, P. (1980) Science 209, 1406-1414.
128 Bogenhagen, D.F., Sakonju, S. and Brown, D.D. (1980) Cell 19, 27-35.
129 Sakonju, S., Bogenhagen, D.F. and Brown, D.D. (1980) Cell 19, 13-25.
130 Hamer, D.H. and Leder, P. (1979) Cell 18, 1299-1302.
131 Gruss, P. and Khoury, G. (1980) Nature 286, 634-637.
132 Scherer, S. and Davis, R.W. (1980) Science 209, 1380-1384.
133 Scherer, S. and Davis, R.W. (1979) Proc. Nat. Acad. Sci. U.S.A. 76, 4951-4955.
134 Williamson, V.M., Bennetzen, J., Young, E.T., Nasmyth, K. and Hall, B.D. (1980) Nature 283, 214-216.
135 Losson, R. and Lacroute, F. (1979) Proc. Nat. Acad. Sci. U.S.A. 76, 5134-5137.
136 Boss, J.M., Darrow, M.D. and Zitomer, R.S. (1980) J. Biol. Chem. 255, 8623-8628.

CLONING RETROVIRUSES: RETROVIRUS CLONING?

William L. McClements and George F. Vande Woude

National Institutes of Health
National Cancer Institute
Laboratory of Molecular Virology
Bethesda, Maryland 20205

INTRODUCTION

The retroviridae are viruses that possess the unique ability to transcribe their genomic RNA into a DNA copy through a process called reverse transcription (1,2) (Figure 1). In infected cells, the DNA intermediates (unintegrated provirus) become stably associated with the host chromosome via an as yet uncharacterized integration process to form the provirus (3,4). These viruses, some of which produce malignant disease in host animals (see below) can be horizontally or vertically transmitted and have been studied intensively (3,5,6). Until very recently, rigorous studies of the structure of the genomes of these viruses were hampered by lack of sufficient quantities of intact viral RNA and restricted to those few viruses produced in abundance in tissue culture. With the advent of molecular cloning, these limitations are eliminated and all of the molecular biological techniques developed for analyzing DNA can now be easily applied to cloned proviral DNA. These applications include: 1) S1 nuclease analyses of hybrids between in vivo transcribed mRNA and proviral DNA (7) for identifying splice sites in cellular transcribed RNA species; 2) in vitro transcription (8,9) from proviral DNA for identifying promoter sequences; 3) hybrid-arrest protein synthesis to assign polypeptide products to specific regions of the genome (10); 4) electron microscopy heteroduplex (11) and R-loop analysis (12) for analyzing gross structural features of the provirus; 5) gene transfer analysis (13) for analyzing the biological activity of the viral genome in direct DNA transfection assays (14); 6) site-specific mutagenesis (15) for determining properties of specific regions of the provirus, and 7) DNA

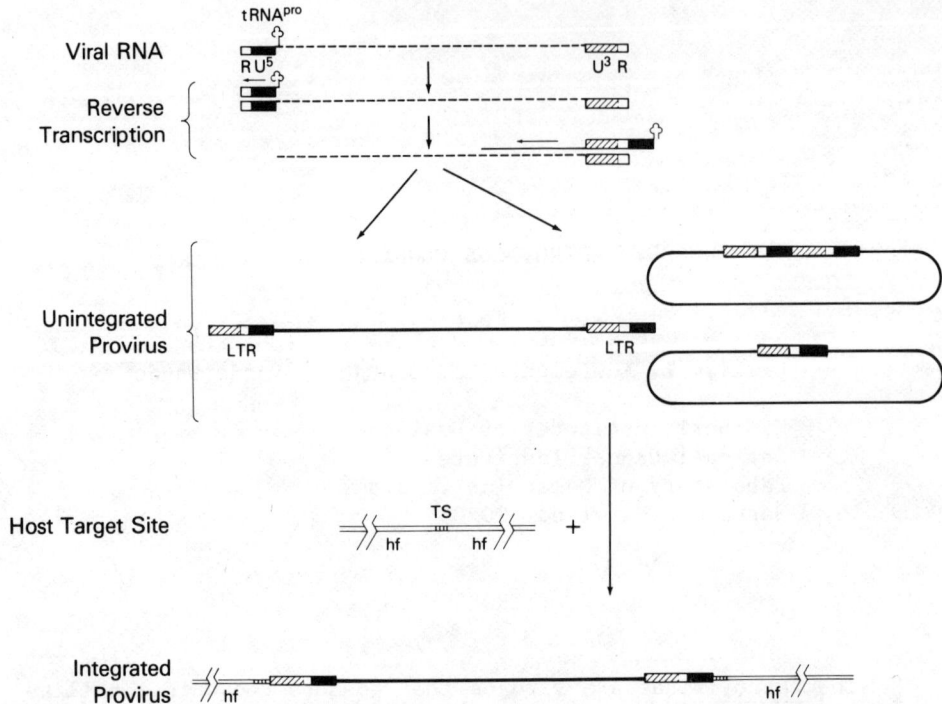

Figure 1. Retrovirus reverse transcription and integration. The dashed line represents the single-stranded RNA genome; light solid line, minus strand DNA; heavy solid line, double-stranded proviral DNA and heavy open line, double-stranded host DNA. In the viral RNA genome, the unique 5' sequences (U^5) are shown as a solid rectangle; the unique 3' sequences (U^3) as a cross-hatched rectangle; the terminally redundant sequences (R) as open rectangles. The t-RNA^{pro} primer binding site is identified by the clover-leaf structure. The long terminal repeat (LTR) of the provirus is composed of U^3, R and U^5 sequences and designated by the symbols described above. The retroviral target site (TS) in the host is represented by four vertical bars and the host sequences flanking the target site are labeled hf. The target site sequence, TS, is duplicated during integration as shown in the drawing representing integrated provirus.

sequencing for determination of actual nucleic acid structure (16,17). Application of molecular cloning techniques to retroviruses provides a means for studying the structures of both integrated and unintegrated proviruses and thereby provides an indirect means for studying the mechanism of integration itself. These analyses gave the first direct evidence that retroviruses resemble prokaryote and eukaryote transposable elements (18-21). One of the most powerful applications of recombinant DNA technology to retroviruses has been to provide a means for studying and identifying the essential genetic elements that contribute to oncogenic transformation (22-25). Some avian and mammalian retroviruses rapidly produce a variety of neoplasias in host animals (3,5,6). Most of these viruses are defective and lack a full complement of retrovirus replication genes, but they possess specific sequences that are homologous to sequences in the host genome. These sequences are presumably transduced from the normal genetic information of the host and there is overwhelming evidence correlating their expression in the provirus with the transforming phenotype. These sequences are highly conserved unique sequences in avian and mammalian cell DNA (3,5,6). They can be cloned from the normal host chromosome and tested for their ability to transform cells before and after being genetically engineered to contain portions of the viral genome (22,24). Experiments of this type have provided the first characterization of how a normal cell sequence can be activated to transform cells. In this chapter, after reviewing the structure of the retrovirus genome, we will describe the advantages and disadvantages of strategies that can be used to clone retroviruses and discuss the potential of using retroviruses as cloning vehicles.

THE RETROVIRUS AND PROVIRUS GENOMES

A generalized structure of a retrovirus plus strand RNA genome is depicted in Figure 1. The virion contains a low proportion of nucleic acid (about 1%). Nuclease sensitivity of the RNA and the difficulty in obtaining sufficient quantities of virus make them prime candidates for molecular cloning. Viral RNA genomes can be from 5,000 to 10,000 nucleotides in length. Two copies are found per virion (26) and there is evidence to indicate that they are noncovalently dimerized at their 5' ends (27). In a replication competent retrovirus, there are three major structural genes, gag, pol and env, oriented 5' to 3' (28). The gag (group specific antigen) gene product is synthesized as a polyprotein and subsequently cleaved to smaller virion structural polypeptides (29,30). While literature abounds on the properties of these polypeptides, little is known about how they function in viral replication. The pol gene product (reverse transcriptase) is responsible for synthesis of the provirus (1-3); pol is synthesized from genomic viral RNA as part of a gag-pol read-through product (31,32) which is cleaved to yield active virion reverse transcriptase. The env (envelope) gene product is also synthesized as a polyprotein and subsequently cleaved to smaller structural virion envelope polypeptides (33,34). Virion RNA can be translated in vitro to yield a gag and gag-pol readthrough product (35,36). The env gene product is translated from a subgenomic message containing leader sequences from the 5' end of the genome spliced to the body of the env gene mRNA (37).

The in vitro synthesis of a DNA copy of the retrovirus genome by reverse transcriptase is primed by a tRNA hybridized to a primer binding site located near the 5' end of the viral genome (38). The first major product is a short minus strand DNA copy of the 5' end of the genome containing unique 5' sequences (U^5) and the terminally-repeated sequence (R) also found at the 3' end of the genome (see Figure 1). The R sequences allow translocation of the short minus strand copy to the 3' end of viral RNA for the completion of minus strand DNA synthesis (39-45). Plus strand DNA synthesis begins in the 3' unique sequence (U^3) region of the genome to generate a short DNA product that includes U^3, R and U^5 sequences (46-48) (Figure 1). These short plus strand sequences form or contribute to the formation of the provirus long terminal repeat (LTR) (18-20,46-48), but the actual steps in this process have not yet been elucidated.

Three major unintegrated proviral intermediates have been identified in newly infected cells (49-52) (Figure 1). Two forms of closed circular provirus have been identified that contain either one or two copies of LTR. A linear form with two LTR elements is also found (49). This form of the provirus is coextensive with the integrated provirus found in the host chromosome (14,53,54) (Figure 1). It is not known whether one, all or any of these major forms are intermediates to provirus integration. The integrated provirus is coextensive with the genomic viral RNA but is longer and contains 3' sequences at its 5' end (in the form of U^3) and 5' sequences at its 3' end (in the form of U^5). These sequences are in turn bracketed by host cellular sequences (Figure 1). It is clear from numerous studies that retroviruses integrate at many sites in the host chromosome but at only a single site in the provirus (14,18,20,54-58). Because of the unique biology of retroviruses it is possible to clone (a) in vitro reverse transcribed complementary DNA (cDNA), (b) cellular derived unintegrated proviral DNA, and (c) integrated proviral DNA. We will discuss the advantages and disadvantages of each and then describe the principles we have used for cloning integrated forms of the retroviruses.

CLONING cDNA COPIES OF RETROVIRUS GENOMES

The cDNA cloning of retroviruses can be achieved in several ways. It has been demonstrated that full length, infectious DNA can be synthesized in an endogenous reaction (59). From this reaction (depicted in Figure 1), an entire provirus or its incomplete intermediate could be cloned (19). Alternatively, purified viral RNA can be reverse transcribed using an oligo(dT) primer hybridized to the poly(A) tail of the viral RNA (60,61). However, the synthesis of complete copies with either the endogenous or the oligo(dT) primed mRNA reaction is limited. When only a portion of the viral genome is required for use as a hybridization probe or if the DNA is to be used for gross structural analysis or in hybrid-arrest translation studies, then cloning a cDNA copy is the method of choice. This approach is advantageous because the product to be cloned is easily obtained, is precisely what is desired and requires a minimal amount of screening. Furthermore, a cDNA copy of messenger RNA provides the ultimate means of identifying RNA (viral) processing sites relative to the primary gene product. In studies involving a more rigorous analysis such as infectivity or DNA sequencing, the cloning of a cDNA copy is not satisfactory for the following reasons. First, most retrovirus preparations can contain a high percentage of defective particles, some of which can be due to errors in the genomic RNA. The endogenous reverse transcriptase reaction and the subsequent molecular cloning will certainly not discriminate these molecules. Another source of errors can be introduced by infidelity during cDNA reverse transcription (62). These problems, plus inefficient completion of full length cDNA copies, can lead to complicated screening and characterization of isolated clones.

Reverse transcribed cDNA copies of portions of the avian Rous sarcoma virus (RSV) (60,61) and Moloney murine leukemia virus (M-MuLV) (19) have been cloned in the plasmid pBR322. The M-MuLV clone was derived from the endogenous reaction (tRNA primer-dependent) while the RSV clone was obtained from an oligo(dT)-primed cDNA copy of the 3' portion of the virus-specific 21S RNA. These clones have provided us with new information about retroviruses. The M-MuLV clone provided additional evidence for the translocation of U^5 sequences to the 3' end of genomic RNA during minus strand reverse transcription: the sequences in this cloned DNA fragment extend beyond the U^3 region of the genome and into the env structural gene revealing an open reading frame capable of coding for a previously unidentified peptide (19). A small C-terminal peptide was chemically synthesized based on this sequence and used to raise antibody. The antibody subsequently served to demonstrate that this polypeptide is expressed in M-MuLV infected cells (63). This is truly a remarkable demonstration of the combined power of current technologies. The nucleotide sequence of the RSV cDNA clone revealed that the 21S mRNA did not terminate at the LTR stop signal, but continued beyond the 3' end of the LTR and into the host sequence. Thus, LTR transcription stop signals are not always used (60).

CLONING UNINTEGRATED PROVIRUS

Unintegrated proviral DNA can be readily obtained from cells early after infection by several extraction procedures. These include the Hirt extraction (64) and a urea extraction-differential sedimentation-hydroxylapatite (HAP) chromatography (65,66). Circular proviral DNA can be further purified by density gradient equilibrium centrifugation in the presence of intercalating dyes such as ethidium bromide or propidium di-iodide (67,68). Size fractionation of the DNA on sucrose gradients or by gel electrophoresis provides additional purification. Preparative gel electrophoresis is particularly useful because it takes advantage of the different forms of proviral DNA. Linear, small circle (one LTR) and large circle (two LTRs) proviral DNA can be identified (in gels) in a manner similar to that shown in Figure 2d by Southern blot analysis (69) with a suitable viral cDNA probe (58, 70). This enriched DNA can be cloned directly or treated with the appropriate restriction endonucleases to alter its electrophoretic mobility for a second round of purification by electrophoresis. For example, circular DNA can be converted to linear DNA or subgenomic fragments can be generated from any form of the provirus. In all cases, the second electrophoretic separation should serve to enrich for viral-specific fragments relative to contaminating cellular DNA fragments. Obviously, these procedures can be combined to yield enriched DNA but a balance must be struck between potential losses of material and the need to screen large numbers of recombinants by plaque blotting (71) or colony hybridization techniques (72). For example, a combination of the urea-sedimentation-HAP chromatography and gel electrophoresis procedures was used to prepare feline leukemia virus (FeLV) unintegrated proviral DNA for cloning (73). One in 300 λ phage recombinant clones screened contained the desired (FeLV) sequences derived from linear and circular proviral DNA. Additional enrichment could have increased the relative proportion of FeLV related clones, but the plaque blot screening (71) was far more expedient for identifying recombinants than attempting additional proviral DNA enrichment procedures.

A complete copy of the retroviral genome can be obtained from linear or circular provirus (Figure 1) by using a restriction enzyme that cuts once in the LTR; this copy will contain only one LTR element, part of which will be at each end of the DNA (73). This strategy was used to clone FeLV, feline sarcoma virus (FeSV) and RSV (68,73). Another approach uses single restriction sites in the circular provirus to clone a permuted form of the retrovirus genome. The Harvey sarcoma virus genome was cloned in this fashion and has been shown to be biologically active (74). Clones obtained this way have at least one copy of the LTR intact. Cloned unintegrated M-MuLV proviral DNA has been extremely useful for "freezing" cellular intermediates of retrovirus replication and for demonstrating that certain intermediates are reminiscent of bacterial transposons (21). However, cloning unintegrated proviruses has some of the same disadvantages as cloning cDNA copies of retrovirus genomes; that is, a percentage of any infecting virus population can be defective in some gene function(s). The activity of a clone cannot be ascertained until there is a considerable investment of the investigator's time. FeSV cloned in this fashion (73) was biologically inactive (C. Sherr, unpublished data) and it was subsequently shown that transformation-defective components were present in the infecting virus population (75). Another liability arises if the genome is cloned permuted and additional manipulations are necessary in order for the proviral DNA to be made infectious. However, if the physical (restriction) map of the provirus has been determined, cloning from unintegrated proviral DNA is an expedient method for obtaining genomic or subgenomic portions of the viral genome.

CLONING INTEGRATED PROVIRUS

For integrated proviruses there are at least three approaches for cloning genomic DNA: 1) libraries of randomly cleaved genomic DNA (76); 2) libraries prepared with DNA cleaved completely with an enzyme that does or does not cut within the provirus (77,78), and 3) genomic DNA restriction fragments that have been enriched for the desired provirus-containing sequence (14,77,79,80). These methods require prior knowledge of the number of integrated copies per cell genome. The latter two require prior knowledge of the proviral restriction sites as well as the size of the fragment into which the provirus is integrated so that a suitable cloning vehicle will be employed. The first two methods require extensive screening of genomic libraries, but may be unavoidable when cloning integrated proviruses that cannot be obtained intact after cleaving with restriction enzymes that are compatible with available vectors. The libraries can be constructed from partially- or fully-cleaved digests of genomic DNA in the conventional manner used by Maniatis et al. (76). To find a single-copy DNA sequence 10^6 to 10^7 plaque-forming units must be screened and it is still possible that recombinants containing provirus sequences will contain less than a complete provirus. Integrated avian myeloblastosis virus (AMV) has been cloned from cell genomic DNA in this manner (78). It is also possible to clone subgenomic portions of the provirus from genomic DNA with conventional enrichment procedures (79) (Figure 2).

Although more time consuming, there are many advantages to cloning integrated proviruses. First, by choosing a productively infected cell culture, possessing a single copy of the integrated provirus (14,18,58,78), the cloned integrated provirus will

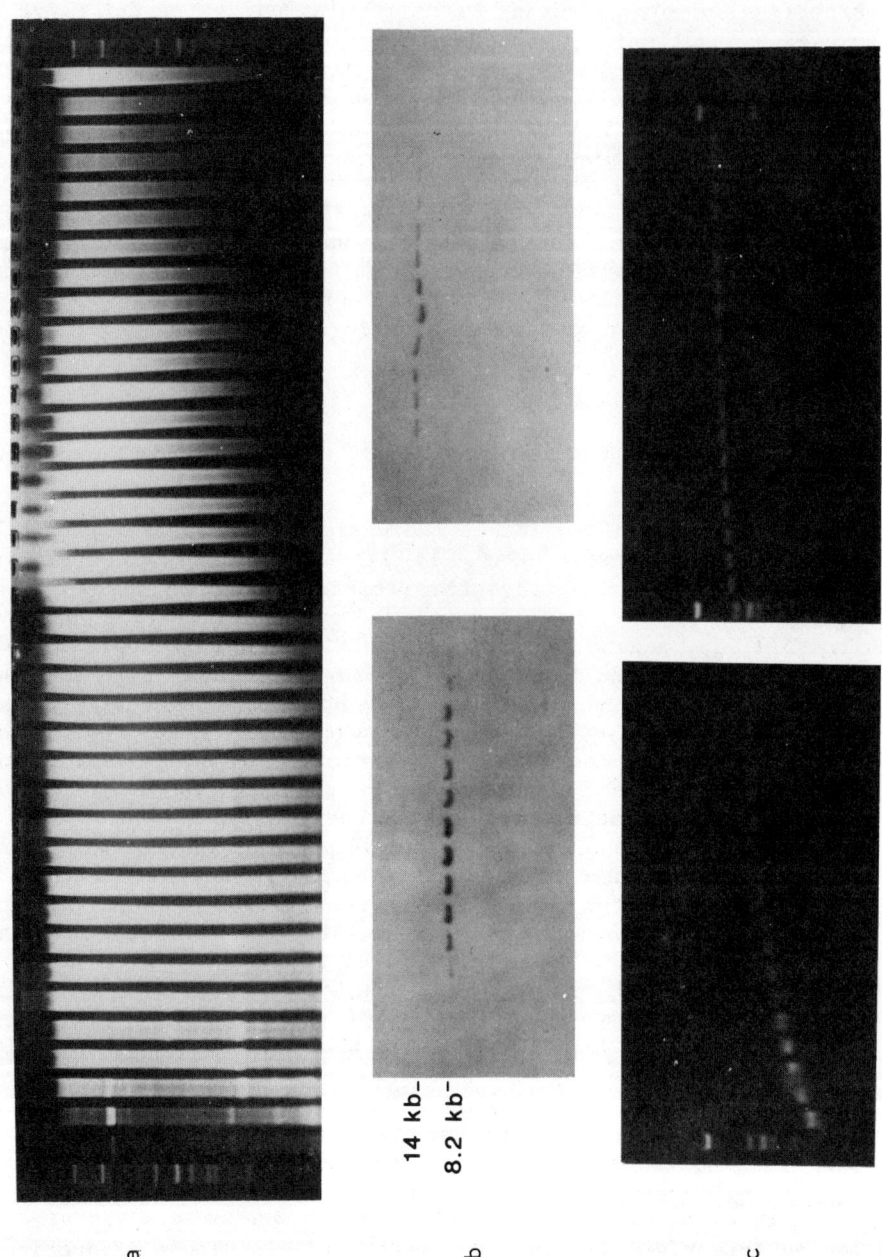

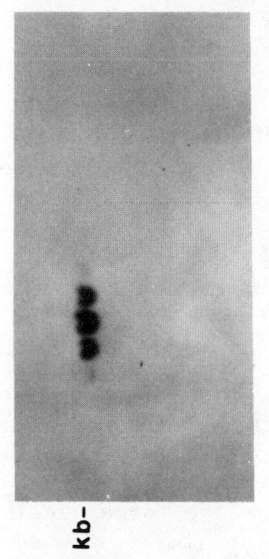

8.2 kb-

d

Figure 2. Purification of integrated retroviral DNA by RPC-5 chromatography and gel electrophoresis (11,14,58,86). Briefly, high molecular weight DNA extracted from cells transformed with MSV was digested to completion with EcoRI. The DNA (10 to 100 mg) was then fractionated by RPC-5 chromatography. DNA from individual fractions was subjected to agarose gel electrophoresis. The ethidium bromide stained analytical gel (a) shows the broad size distribution of DNA fragments in each RPC-5 fraction. The gel was analyzed by the Southern blot hybridization technique (69) using a cloned V-mos sequence as probe. Autoradiograms of the blots (b) indicate two species of DNA are homologous with the V-mos probe. The ~14 kilobase pair DNA fragment eluting from RPC-5 has been previously identified as the endogenous C-mos sequence (14,22,58). Thus, it is likely that the 8.2 kilobase DNA fragment contains the transforming sequences. Fractions containing the 8.2 kilobase DNA species were pooled, concentrated and subjected to preparative agarose gel electrophoresis. DNA eluted from the preparative gel was analyzed the same way as DNA from the RPC-5 chromatography in (a). The stained analytical gel (c) shows the narrow size distribution of DNA fragments in each fraction. An autoradiogram of the blotted gel (d) detects the DNA fragments of interest. Enrichment of a single-copy sequence is between 100- and 1,000-fold with this 2-step method.

likely be biologically active. The potential of cloning a defective provirus is less likely than with cloning cDNA and unintegrated proviruses as described in the previous sections. Second, the presence of the host flank sequences bracketing the integrated provirus allows examination of the host DNA sequences at the provirus integration site (14,18,20). These host sequences can serve as probes for the unoccupied target site in the genome of the uninfected host cell as well (14,81). This approach demonstrated that provirus integration resulted in a duplication of a few nucleotides in the host sequence at the target site (18,20). Thus, retrovirus integration resembles some aspects of bacterial transposition. Comparison of the nucleotide sequence of the integrated and unintegrated provirus may help elucidate the mechanism of viral integration.

The preceding information has been a general review of potential cloning methods; the following describes our procedures and practices for cloning integrated proviruses.

STRATEGY FOR CLONING INTEGRATED MSV, ITS UNOCCUPIED TARGET SITE AND THE NORMAL MOUSE SEQUENCE CONTAINING C-mos*

We have been specifically interested in identifying the molecular elements of the retroviral genome that are essential for its transforming phenotype. The Moloney sarcoma virus (MSV) has been especially useful in this regard since it contains a specific sequence, V-mos, believed to be responsible for transformation. This sequence is homologous to a portion of the normal mouse genome, C-mos (22,82-85). Our strategy was first to clone integrated MSV proviral DNA and characterize its structural and biological properties. We then used a portion of V-mos as probe to identify and clone a DNA fragment from the normal cell genome containing the cellular counterpart C-mos (22). Cloned C-mos was then tested for biological activity and portions of the provirus were recombined with C-mos to determine what molecular elements of the virus were sufficient to transform cells (22,24).

We chose to clone two different strains of MSV provirus from the DNA of transformed mink cells instead of mouse cells. This minimized potential background cross-hybridization of our probe (M-MuLV cDNA) with normal mouse DNA sequences including endogenous proviruses (14,58) and eliminated the requirement for extensive probe purification. Both MSV proviruses lacked EcoRI

---

*The new convention for the Moloney sarcoma virus-acquired sequence is V-mos in the virus genome and C-mos when part of the cellular genome. This convention supersedes the usage of src and sarc in Oskarsson et al. (22).

restriction sites, so cloned cellular EcoRI fragments should contain all of the proviruses as well as host flanking sequences.

DNA fragments carrying the proviruses were prepared by a two-dimensional enrichment procedure consisting of RPC-5 chromatography and preparative gel-electrophoresis (11,86). This procedure provided a 100- to 1,000-fold enrichment of the integrated proviral DNA. This approach was chosen because the RPC-5 chromatography (analyzed by the Southern transfer procedure (69) as shown in Figure 2a and 2b) provided a better resolution of the complex mixture of hybridizing species in the genomic DNA and, therefore, a greater degree of confidence that we were dealing with a single MSV provirus integration.

This enrichment procedure originally required large quantities of genomic DNA (about 50 mg) which created special problems for handling and extracting minimally sheared cellular DNA. We therefore developed a rapid extraction procedure using a detergent-lysis buffer at 60°C to purify nuclei directly from tissue culture monolayers, cell-packs or from frozen tissue specimens (ground with a mortar and pestle in liquid nitrogen) (58). Nuclei prepared in this manner have been stored at -70°C for over one year without indication of DNA deterioration. To reduce the viscosity of the DNA in the initial extraction without random shearing, nuclei were subjected to limited digestion with the same enzyme subsequently used for the complete cleavage of the DNA: in this case, EcoRI (58)). The viscosity of DNA at this stage is dramatically reduced, but the DNA is greater than 20 kilobase pairs in size and can be extracted by conventional procedures in small volumes.

DNA extracted from nuclei was digested (or redigested) to completion with the appropriate enzyme and, after phenol extraction and ethanol precipitation, was subjected to RPC-5 chromatography (14). Fractions from RPC-5 chromatography were analyzed by gel electrophoresis and Southern transfer hybridization (69) (typical examples are shown in Figure 2a and 2b, respectively, and are described elsewhere in detail (58)). The RPC-5 fraction(s) containing the desired fragment is subjected to preparative gel electrophoresis (11). Fractions eluted from a preparative gel were also analyzed by gel electrophoresis and Southern transfer hybridization (typical examples are shown in Figure 2c and 2d).

With this enrichment procedure, the desired recombinant occurs with a frequency of between 1/500 to 1/10,000 λ phage recombinants (22,58). With these procedures, we cloned two strains of integrated MSV from transformed mink cells (14,58) under highly restrictive P4 containment conditions. In a class III hood, it was simpler to start with extensively enriched proviral DNA than to screen large numbers of λ phage recombinants.

Comparison of the structural features of these clones demonstrated immediately that MSV integrates at different sites in the host chromosome, but the same site on the provirus (14,58). It

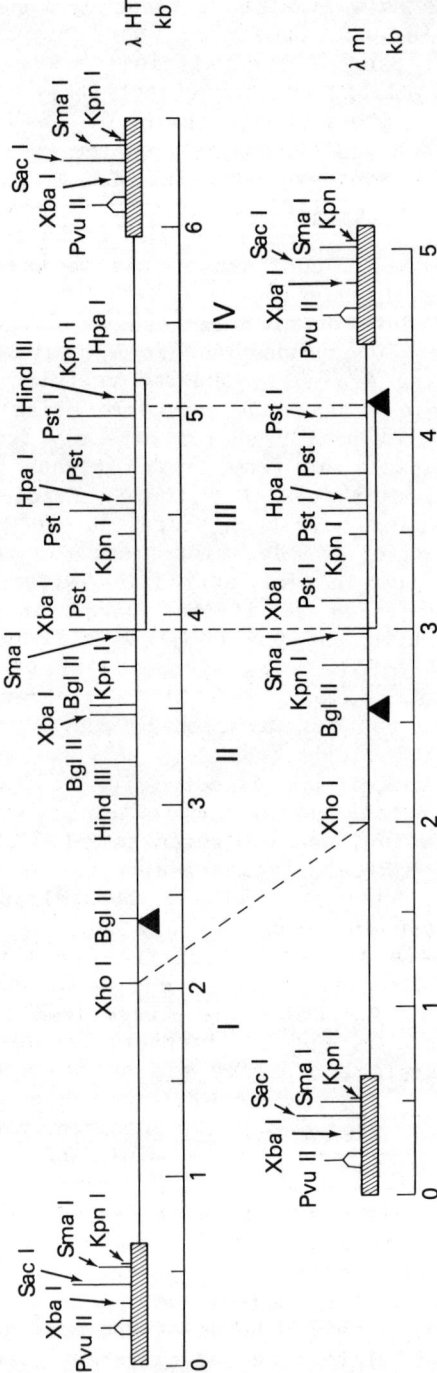

Figure 3. Physical maps of two strains of integrated MSV provirus. Physical maps of cloned m1 and HT1 MSV proviral DNA were developed from restriction endonuclease, heteroduplex, R-loop and hybridization data (14,21,58) with cloned DNA (λm1 and λHT1). The LTR elements are indicated by cross-hatched rectangles; the acquired sequences (V-mos) by the open rectangles. Deletions in the MSV genomes (compared to the parental M-MuLV) are indicated by the closed triangles. Domains designated I to IV denote regions of similarity and difference between the two strains of MSV. Sizes are given in kilobase pairs (kb).

also provided a means for direct comparison of the genomes of the two MSV strains, which were known to differ in certain biochemical and biological properties (87). A comparison of their physical maps is shown in Figure 3 with their genomes divided into four domains; obvious similarities exist in domains I and III, but the viruses differ in domains II and IV. This comparison has been described in detail (14). R-loop analysis and hybridization with M-MuLV RNA has shown that domain III contains the acquired V-mos sequence. We used V-mos to identify (14) and clone (22) the normal mouse genomic DNA fragment containing C-mos. V-mos was shown to be indistinguishable from C-mos by heteroduplex and restriction analysis (22).

We have used the cloned provirus to identify and clone an unoccupied MSV integration target site from normal mink cells (81). A heteroduplex between this DNA fragment and the fragment containing integrated MSV provirus is shown in Figure 4. It is a dramatic demonstration of a provirus inserted into the host chromosome. DNA sequencing of several host viral junctions and unoccupied target sites has revealed that retrovirus LTR elements resemble bacterial transposable elements (18-21) both structurally and functionally. These transposon-like properties of the retroviral LTR include putative transcription control features and allow some speculation about the potential use of retroviral sequences as cloning vehicles in eukaryotic cells.

RETROVIRUSES AS CLONING VECTORS?

The primary reason for considering retroviruses as potential cloning vehicles is that all of the rapid neoplastic transforming viruses appear to have transduced genomic host cell DNA (3,5,6). Comparison of some of these transduced sequences and their cellular homologs indicates that the cellular sequence may contain introns or intervening sequences (C. Sherr, personal communication). One possible explanation for the absence of introns in the virus sequence is that these sequences are transduced into the provirus as mRNA. The expression of the transduced cellular sequence is viral mediated, presumably controlled by the LTR. Gene products occur either as fusion products with a portion of the viral gag gene, or are independently expressed from subgenomic viral messages (see ref. 6 for review). If these speculations are correct, retroviruses are naturally-occurring mRNA cloning vehicles and are selected by virtue of their ability to cause acute oncogenic transformation.

From our own experiments, we know that cells transformed with a recombinant DNA containing the 5' LTR and gag gene sequences together with either V-mos (23) or C-mos DNA, but lacking the 3' end of the viral genome, can be rescued by superinfection with a competent helper virus (D. Blair et al., unpublished

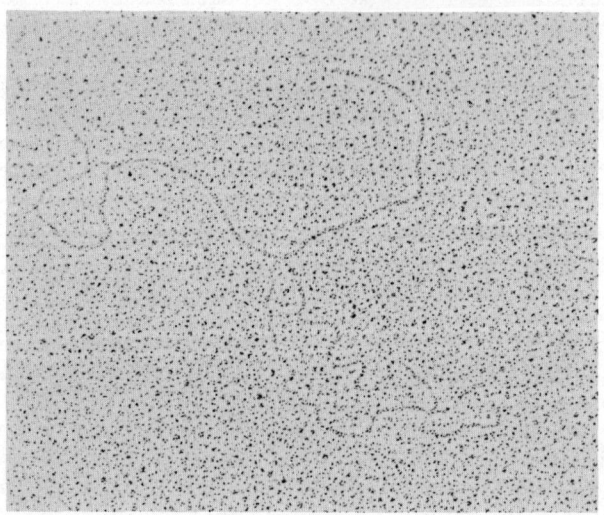

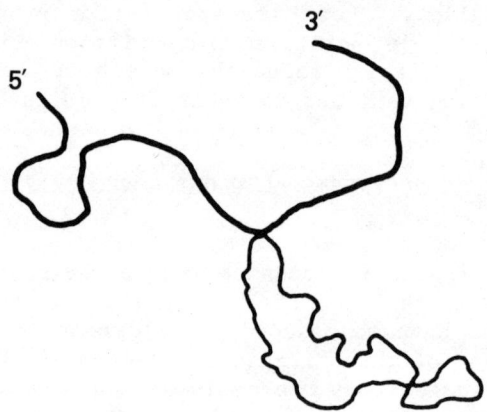

Figure 4. Heteroduplex between an integrated MSV provirus and its unoccupied target site. The cloned DNA fragment containing integrated MSV provirus (14,58) also contains flanking host DNA (see Figure 1). Host flank was used as a probe to detect the identical sequences in normal uninfected cell DNA (14,58). This normal DNA contains an unoccupied target site for MSV (see Figure 1) and was cloned by conventional methods (14,81). Physical maps of the sequence bearing the unoccupied target site were identical to maps of the host sequences flanking the integrated MSV (14,58, 81). Before heteroduplex formation, the recombinant λ-phage DNAs were digested with EcoRI to eliminate participation of λ DNA in the heteroduplex. The 5' and 3' duplex regions are 3.3 kilobases and 2.3 kilobases long, respectively; the single-stranded loop (proviral DNA) is 6.7 kilobases long.

data). The 3' portion of the rescued transforming virus genome is likely to have been derived from the helper virus. The C-mos transforming potential was activated by ligating the LTR-gag gene DNA of MSV 5' to C-mos (22). This DNA recombinant transformed cells even though it lacked MSV (or M-MuLV) sequences 3' to C-mos. The rescue of a virus containing C-mos from cells transformed with the LTR:gag:C-mos recombinant DNA is the first evidence that a genetically engineered retrovirus can be used to clone cellular sequences. This experiment could have been performed with any cell DNA fragment for which a selectable marker exists, for example thymidine kinase (13). We have shown that the LTR alone is capable of activating the transforming potential of C-mos even when located some distance from its 5' end (24). A retrovirus vector containing 5' LTR plus gag may also be able to transduce genes at some distance from their 5' ends permitting the generation of a retrovirus gene library by covalently linking partially digested cellular DNA to a retrovirus vector.

Acknowledgments: The authors are indebted to T.Gordon Wood and Marianne K. Oskarsson for helpful discussion and critical review and to Kathleen Barry for assistance in preparing this manuscript.

REFERENCES

1 Temin, H. and Baltimore, D. (1972) Adv. Virus Res. 17, 129-186.
2 Temin, H. (1976) Science 192, 1075-1080.
3 Bishop, J.M. (1978) Annu. Rev. Biochem. 47, 35-88.
4 Weinberg, R.A. (1980) Annu. Rev. Biochem. 49, 197-226.
5 Fischinger, P.J. (1979) in Molecular Biology of RNA Tumor Viruses (Stephenson, J.R., ed.), pp. 163-198, Academic Press, New York, NY.
6 Duesberg, P.H. (1979) Cold Spring Harbor Symp. Quant. Biol. 44, 13-29.
7 Berk, A.J. and Sharp, P.A. (1978) Proc. Nat. Acad. Sci. U.S.A. 75, 1274-1278.
8 Weil, P.A., Luse, D.S., Segall, J. and Roeder, R.G. (1979) Cell 18, 469-484.
9 Manley, J.L., Sharp, P.A. and Gefter, M.L. (1979) Proc. Nat. Acad. Sci. U.S.A. 76, 160-164.
10 Paterson, B.M., Roberts, B.E. and Kuff, E.L. (1977) Proc. Nat. Acad. Sci. U.S.A. 74, 4370-4374.
11 Tiemeier, D.C., Tilghman, S.M., Polsky, F.I., Seidman, J.G., Leder, A., Edgell, M.H. and Leder, P. (1978) Cell 14, 237-245.

12. Thomas, M., White, R.L. and Davis, R.W. (1976) Proc. Nat. Acad. Sci. U.S.A. 73, 2294-2298.
13. Wigler, M., Silverstein, S., Lee, L.-S., Pellicer, A., Cheng, Y.-C. and Axel, R. (1977) Cell 11, 223-232.
14. Vande Woude, G.F., Oskarsson, M., McClements, W.L., Enquist, L., Blair, D.G., Fischinger, P.J., Maizel, J.V. and Sullivan, M. (1979) Cold Spring Harbor Symp. Quant. Biol. 44, 735-745.
15. Shortle, D. and Nathans, D. (1978) Proc. Nat. Acad. Sci. U.S.A. 75, 2170-2174.
16. Maxam, A.H. and Gilbert, W. (1977) Proc. Nat. Acad. Sci. U.S.A. 74, 560-564.
17. Maat, J. and Smith, A.J.H. (1978) Nucl. Acid Res. 5, 4537-4545.
18. Shimotohno, K., Mizutani, S. and Temin, H. (1980) Nature 285, 550-554.
19. Sutcliffe, J.G., Shinnick, T.M., Verma, I.M. and Lerner, R.A. (1980) Proc. Nat. Acad. Sci. U.S.A. 77, 3302-3306.
20. Dhar, R., McClements, W.L., Enquist, L.W. and Vande Woude, G.F. (1980) Proc. Nat. Acad. Sci. U.S.A. 77, 3937-3941.
21. Shoemaker, C., Goff, S., Gilboa, E., Paskind, M., Mitra, S.W. and Baltimore, D. (1980) Proc. Nat. Acad. Sci. U.S.A. 77, 3932-3936.
22. Oskarsson, M., McClements, W.L., Blair, D.G., Maizel, J.V. and Vande Woude, G.F. (1980) Science 207, 1222-1224.
23. Blair, D.G., McClements, W.L., Oskarsson, M., Fischinger, P.J. and Vande Woude, G.F. (1980) Proc. Nat. Acad. Sci. U.S.A. 77, 3504-3508.
24. McClements, W.L., Dhar, R., Blair, D.G., Enquist, L.W., Oskarsson, M.K. and Vande Woude, G.F. (1980) Cold Spring Harbor Symp. Quant. Biol. 45 (in press).
25. Chang, E.H., Maryak, J.M., Wei, C.-M., Shih, T.Y., Shober, R., Cheung, H.L., Ellis, R.W., Hager, G.L., Scolnick, E.M. and Lowy, D.R. (1980) J. Virol. 35, 76-92.
26. Beemon, K.L., Faras, A.J., Haase, A.T., Duesberg, P.H. and Maisel, J.E. (1976) J. Virol. 17, 525-537.
27. Bender, W. and Davidson, N. (1976) Cell 7, 595-607.
28. Baltimore, D. (1974) Cold Spring Harbor Symp. Quant. Biol. 39, 1187-1200.
29. Vogt, V.M., Eisenman, R. and Diggelmann, H. (1975) J. Mol. Biol. 96, 471-493.
30. Arlinghaus, R.B., Naso, R.B., Jamjoom, G.A., Arcement, L.J. and Karshin, W.L. (1976) in Animal Virology, ICN-UCLA Symp. Mol. Cell. Biol. (Baltimore, D., Huang, A. and Fox, C.F., eds.), Vol. 4, pp. 689-715.
31. Oppermann, H., Bishop, J.M., Varmus, H.E. and Levintow, L. (1977) Cell 12, 993-1005.
32. Jamjoom, G.A., Naso, R.B. and Arlinghaus, R.B. (1977) Virology 78, 11-34.

33 Naso, R.B., Arcement, L.J., Karshin, W.L., Jamjoom, G.A. and Arlinghaus, R.B. (1976) Proc. Nat. Acad. Sci. U.S.A. 73, 2326-2330.
34 Moelling, K. and Hayami, M. (1977) J. Virol. 22, 598-607.
35 Kerr, I.M., Olshevsky, U., Lodish, H.F. and Baltimore, D. (1976) J. Virol. 18, 627-635.
36 Purchio, A.F., Erikson, E. and Erikson, R.L. (1977) Proc. Nat. Acad. Sci. U.S.A. 74, 4661-4665.
37 Rothenberg, E., Donoghue, D.J. and Baltimore, D. (1978) Cell 13, 435-451.
38 Taylor, J.M. (1977) Biochim. Biophys. Acta 473, 57-71.
39 Haseltine, W.A., Kleid, D.G., Panet, A., Rothenberg, E. and Baltimore, D. (1976) J. Mol. Biol. 106, 109-132.
40 Haseltine, W.A., Maxam, A.M. and Gilbert, W. (1977) Proc. Nat. Acad. Sci. U.S.A. 74, 989-993.
41 Schwartz, D.E., Zamecnik, P.C. and Weith, H.L. (1977) Proc. Nat. Acad. Sci. U.S.A. 74, 994-998.
42 Shine, J., Czernilofsky, A.P., Friedrich, R., Bishop, J.M. and Goodman, H.M. (1977) Proc. Nat. Acad. Sci. U.S.A. 74, 1473-1477.
43 Coffin, J.M. and Haseltine, W.A. (1977) Proc. Nat. Acad. Sci. U.S.A. 74, 1908-1912.
44 Stoll, E., Billeter, M.A., Palmenberg, A. and Weissmann, C. (1977) Cell 12, 57-72.
45 Coffin, J.M., Hagerman, T.C., Maxam, A.M. and Haseltine, W.A. (1978) Cell 13, 761-773.
46 Varmus, H.E., Heasley, S., Kung, H.J., Oppermann, H., Smith, V.C., Bishop, J.M. and Shank, P.R. (1978) J. Mol. Biol. 120, 55-82.
47 Gilboa, E., Mitra, S.W., Goff, S. and Baltimore, D. (1979) Cell 18, 93-100.
48 Dina, D. and Benz, E.W. (1980) J. Virol. 33, 377-389.
49 Shank, P.R., Hughes, S.H., Kung, H., Majors, J.E., Quintrell, N., Guntaka, R.V., Bishop, J.M. and Varmus, H.E. (1978) Cell 15, 1383-1395.
50 Gilboa, E., Goff, S., Shields, A., Yoshimura, F., Mitra, S. and Baltimore, D. (1979) Cell 16, 863-874.
51 Shank, P.R., Cohen, J.C., Varmus, H.E., Yamamoto, K.R. and Ringold, G.M. (1978) Proc. Nat. Acad. Sci. U.S.A. 75, 2112-2116.
52 Shank, P.R. and Varmus, H.E. (1978) J. Virol. 25, 104-114.
53 Hsu, T.W., Sabran, J.L., Mark, G.E., Guntaka, R.V. and Taylor, J.M. (1978) J. Virol. 28, 810-818.
54 Hughes, S.H., Shank, P.R., Spector, D., Kung, H.-J., Bishop, J.M., Varmus, H.E., Vogt, P.K. and Breitman, M.L. (1978) Cell 15, 1397-1410.
55 Keshet, E. and Temin, H.M. (1978) Proc. Nat. Acad. Sci. U.S.A. 75, 3372-3376.
56 Steffen, D. and Weinberg, R.A. (1978) Cell 15, 1003-1010.

57 Sabran, T.L., Hsu, T.E., Yeater, C., Kaji, A., Mason, W.S. and Taylor, J.M. (1979) J. Virol. 29, 170–178.
58 Vande Woude, G.F., Oskarsson, M.K., Enquist, L., Nomura, S., Sullivan, M. and Fischinger, P.J. (1979) Proc. Nat. Acad. Sci. U.S.A. 76, 4464–4468.
59 Rothenberg, E., Smotkin, D., Baltimore, D. and Weinberg, R.A. (1977) Nature 269, 122–126.
60 Yamamoto, T., Jay, G. and Pastan, I. (1980) Proc. Nat. Acad. Sci. U.S.A. 77, 176–180.
61 Yamamoto, T., Tyagi, J.S., Fagan, J.B., Jay, G., de Crombrugghe, B. and Pastan, I. (1980) J. Virol. 35, 436–443.
62 Andersson, P., Goldfarb, M.P. and Weinberg, R.A. (1979) Cell 16, 63–75.
63 Sutcliffe, J.G., Shinnick, T.M., Green, N., Lui, F.-T., Niman, H.L. and Lerner, R.A. (1980) Nature (in press).
64 Hirt, B. (1967) J. Mol. Biol. 26, 365–369.
65 Shoyab, M. and Sen, A. (1978) J. Biol. Chem. 253, 6654–6656.
66 Sherr, C.J., Fedele, L.A., Donner, L. and Turek, L.P. (1979) J. Virol. 32, 860–875.
67 Tronick, S.R., Robbins, K.C., Canaani, E., Devare, S.G., Andersen, P.R. and Aaronson, S.A. (1979) Proc. Nat. Acad. Sci. U.S.A. 76, 6314–6318.
68 Ju, G., Boone, L. and Skalka, A.M. (1980) J. Virol. 33, 1026–1033.
69 Southern, E.M. (1975) J. Mol. Biol. 38, 503–517.
70 Benveniste, R.E. and Scolnick, E.M. (1973) Virology 51, 370–382.
71 Benton, D. and Davis, R.W. (1977) Science 196, 180–182.
72 Grunstein, M. and Hogness, D.S. (1975) Proc. Nat. Acad. Sci. U.S.A. 72, 3961–3965.
73 Sherr, C.J., Fedele, L.A., Oskarsson, M., Maizel, J.V. and Vande Woude, G.F. (1980) J. Virol. 34, 200–212.
74 Hager, G.L., Chang, E.H., Chan, H.W., Garon, C.F., Israel, M.A., Martin, M.A., Scolnick, E.M. and Lowy, D.R. (1979) J. Virol. 31, 795–809.
75 Donner, L., Turek, L.P., Ruscetti, S.K., Fedele, L.A. and Sherr, C.J. (1980) J. Virol. 35, 129–140.
76 Maniatis, T.R., Hardison, R.C., Lacy, E., Lauer, J., O'Connell, C., Quon, D., Sim, G.K. and Efstratiadis, A. (1978) Cell 15, 687–701.
77 Donehower, L.A., Andre, J., Berard, D.S., Wolford, R.G. and Hager, G.L. (1980) Cold Spring Symp. Quant. Biol. 44, 1153–1159.
78 Souza, L.M, Komaromy, M.C. and Baluda, M.A. (1980) Proc. Nat. Acad. Sci. U.S.A. 77, 3004–3008.
79 Skalka, A., DeBona, P., Hishinuma, F. and McClements, W. (1979) Cold Spring Harbor Symp. Quant. Biol. 44, 1097–1104.
80 Lowy, D.R., Rands, K., Chattopadhyay, S.K. and Hager, G.L. (1979) Cold Spring Harbor Symp. Quant. Biol. 44, 1143–1151.

81 McClements, W.L., Enquist, L.W., Oskarsson, M., Sullivan, M. and Vande Woude, G.F. (1980) J. Virol. 35, 488-497.
82 Scolnick, E.M., Howk, R.S., Anisowicz, A., Peebles, P.T., Scher, C.D. and Parks, W.P. (1975) Proc. Nat. Acad. Sci. U.S.A. 72, 4650-4654.
83 Dina, D., Beemon, K. and Duesberg, P. (1976) Cell 9, 299-309.
84 Frankel, A.E. and Fischinger, P.J. (1976) Proc. Nat. Acad. Sci. U.S.A. 73, 3705-3709.
85 Frankel, A.E. and Fischinger, P.J. (1977) J. Virol. 21, 153-160.
86 Tilghman, S.M., Tiemeier, D.C., Polsky, F., Edgell, M.H., Seidman, J.G., Leder, A., Enquist, L.W., Norman, B. and Leder, P. (1977) Proc. Nat. Acad. Sci. U.S.A. 74, 4406-4410.
87 Robey, W.G., Oskarsson, M.K., Vande Woude, G.F., Naso, R.B., Arlinghaus, R.B., Hapaala, D.K. and Fischinger, P.J. (1977) Cell 10, 79-89.

REPEATED DNA SEQUENCES IN DROSOPHILA

Michael W. Young

The Rockefeller University

New York, New York 10021

In this chapter we will present a short review of the properties of repeated DNA sequences in Drosophila. Since Drosophila melanogaster has been the most extensively characterized species within this genus, most of our review will be focused on the structure of its genome. The chromosomal DNA of D. melanogaster can be broadly separated into three components by reassociation kinetics. Approximately two-thirds of this DNA, about 110,000 kilobases, is nonrepetitious; the remainder is reiterated and can be further subdivided into sequences that are either highly repetitive, with an average reiteration frequency of about 24,000, or moderately repetitive, with a reiteration frequency usually falling in a range of 10 to 100 copies per haploid genome. Highly repeated and moderately repeated DNA sequences each compose about one-sixth of the D. melanogaster DNA (1-3).

Much of the highly repeated DNA of D. melanogaster can be separated from moderately repeated and single-copy DNA by centrifugation in CsCl equilibrium gradients because the structures of these DNA sequences are quite simple; short (5 to 378 base pairs) nucleotide sequences are repeated, tandemly, tens of thousands of times per genome (4,5). Such simple sequence DNAs are unlikely to code for proteins and there is no evidence for their transcription. Inasmuch as the highly repeated DNAs are predominantly confined to heterochromatic chromosomal regions, some of which include the centromeres, several of the more popular hypotheses concerning the function of these sequences involve them in the mitotic and meiotic pairing of homologous chromosomes. Nevertheless, substantial deletions of the heterochromatic regions of Drosophila chromosomes often cause no disruption of chromosome pairing. In fact, autosomal pairing appears to be strongly dependent on euchromatic rather than heterochromatic homologies

(6,7). Deletions of X chromosomal heterochromatin do, however, reduce X chromosome recombination, and such reductions seem to be directly proportional to the amount of heterochromatin that is removed (8). Thus, highly repeated DNA sequences may be important for proper meiotic recombination.

Four major highly repetitive DNA species can be separated by density gradient centrifugation. These have buoyant densities of 1.672 g/cc, 1.686 g/cc, 1.688 g/cc and 1.705 g/cc. Each species composes about 4% of the D. melanogaster chromosomal DNA, and the primary sequence of each repeating unit has been determined (9-11). Two related DNA sequences can be detected in the 1.672 species, a pentamer (5' AATAT) and a derivative septamer (5' AATATAT). The 1.686 species can also be subdivided, but about 90% of this DNA has the sequence (5' AATAACATAG). The 1.705 tandem repeat is composed of the pentamer (5' AAGAG) although longer (septamer and decamer) related sequences have been identified (see also ref. 5). The 1.688 species is considerably more complex than the three highly repeated DNA sequences reviewed thus far. It consists of a 378 base-pair repeat that appears to have been generated, to a large extent, from an earlier multiplication of the sequence (5' TTTCC); nearly 50% of the 1.688 repeating unit is composed of this sequence or single base change derivatives of it (5).

The chromosomal distribution of each of these highly repeated DNA sequences has been established by in situ hybridization to metaphase chromosomes (12,13). The results of their localization can be summarized as follows: 1) each repeat species can be found as part of a long tandem array on more than one chromosome, and the same DNA sequence can often be mapped to several heterochromatic locations on a single chromosome; 2) quantitative analyses of the distribution of each DNA species make it clear that few moderately repetitive or nonrepetitive DNA sequences are located in the heterochromatic regions of D. melanogaster chromosomes; 3) the chromosomal arrangement of each highly repeated DNA sequence appears to be species-specific, as variations in the chromosomal location of the sequences are not seen when different strains of D. melanogaster are examined.

Highly repetitive DNA sequences have also been characterized in a sibling species of D. melanogaster, D. simulans, and eight major repeating units have been defined (13). These generally differ in buoyant density from those identified in D. melanogaster, and the chromosomal locations of four of these have been mapped by in situ hybridization. As in the case of the D. melanogaster DNAs, each of these highly repeated sequences has a species-specific distribution that includes heterochromatic regions on more than one chromosome (13). Attempts to cross hybridize the major D. simulans and D. melanogaster highly repeated DNAs indicate that most of these sequences are nonhomologous, as might be expected from their differing buoyant densities; only the 1.672 sequence forms a major repeat sequence in both species.

However, two of the major repeat sequences from D. simulans have been found as minor species in D. melanogaster, and the 1.705 repeat of D. melanogaster is represented at a low level in the D. simulans genome (the 1.705 sequence is approximately 5 times less abundant in D. simulans DNA than in conspecific DNA (13)).

The chromosomal distributions of both the 1.672 and the 1.705 sequences are homologous in D. simulans and D. melanogaster. For example, the 1.705 sequence occupies three Y chromosomal positions in D. melanogaster that correspond to the three Y chromosomal positions carrying the 1.705 sequence in D. simulans (13). Similarly, the two major repeat sequences from D. simulans that are present as minor repeats in D. melanogaster occupy, to a large extent, homologous chromosomal positions in both species (13). Apparently changes in the chromosomal distribution of these highly repeated DNAs have not accompanied their interspecific gain or loss.

ORGANIZATION OF MODERATELY REPEATED DNA
IN D. melanogaster CHROMOSOMES

Most of the moderately repetitive DNA of D. melanogaster is composed of families of chromosomally dispersed sequences. Among randomly selected, cloned segments of moderately repeated DNA, 75% contain sequences that can be found at 10 to 100 scattered chromosomal positions (14,15). Members of many of these repeat families are large, having a rather uniform size distribution over a range of 0.15 kilobases to 13 kilobases, and they are usually embedded in nonrepetitive DNA (2,16). These long segments of repeated DNA, interspersed with even longer segments of single-copy DNA, shape the molecular topography of the Drosophila genome. This topography has been called long-period interspersion (17).

Recently, a second arrangement for some segments of dispersed moderately repeated DNA has been described (18). This organization is called clustered and scrambled and consists of homologous and nonhomologous short repetitive DNA elements associated end-to-end to form longer segments of repeated DNA (see Figure 1). Short repeats that cross hybridize can be found in such clusters at scattered chromosomal locations, but the arrangement of the repeated elements within each cluster tends to vary from one chromosomal position to another. Wensink et al. (18) have estimated that most of the short repetitive elements making up these clusters are no more than 500 base pairs long.

By weight, short repetitive DNA segments like those composing permuted clusters (<500 base pairs) could make up nearly a third of the dispersed sequences. This conclusion is drawn in part from the size distribution of heteroduplexes formed by reassociating 2 to 3 kilobase long fragments of total, moderately

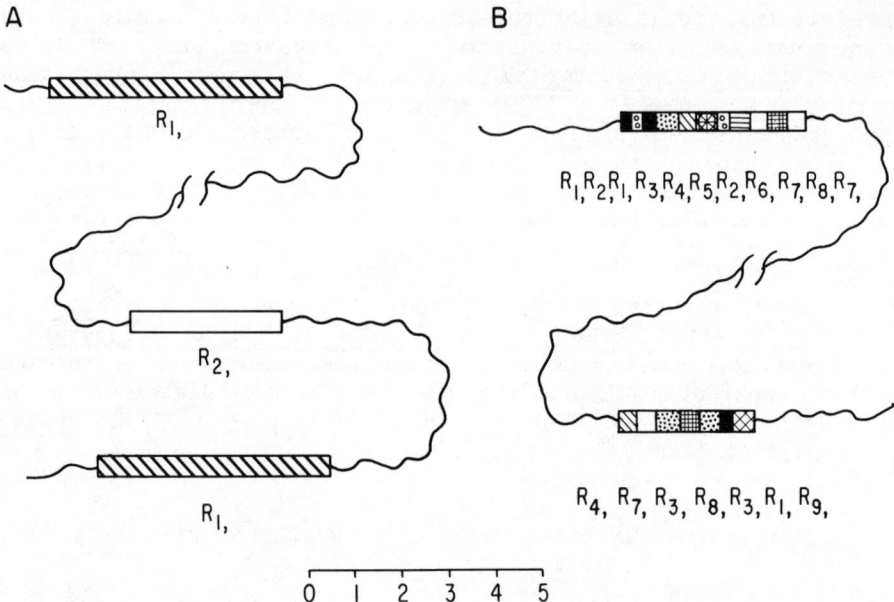

Figure 1. Schematic representation of two topographies of dispersed repeated DNA segments in D. melanogaster chromosomes. A. Long-period interspersion; blocks R1 and R2 portray members of two moderately repeated sequence families interspersed with non-repetitive DNA (wavy lines). B. Clustered and scrambled repeats; blocks R1 to R9 depict members of nine moderately repeated sequence families associated in a permuted fashion at different chromosomal positions. Wavy lines represent nonrepetitive DNA. Scale at bottom is in kilobase pairs and refers to both A and B.

repeated D. melanogaster DNA, which are then trimmed with S1 nuclease (16). Partially cross-hybridizing clusters of repeated DNA will reassociate to form either molecules with double-stranded regions separated by single-stranded segments of varying lengths, or multiply-nucleated structures in which several strands have reannealed to form a tangle. The S1 nuclease-resistant portions of both kinds of molecules should correspond to the repetitive elements making up the clusters. Approximately 15% of this S1-resistant DNA is shorter than 0.5 kilobases and could therefore contribute to the formation of short repeat clusters (16).

Another 5 to 10% of the total moderately repeated DNA in D. melanogaster may be composed of clusters of short repeat elements if these sequences tend to be found in foldback DNA. In the S1 nuclease analyses reviewed above, relatively long strands of DNA were reassociated. If strands containing clusters carry two or more copies of the same short repeat sequence, they may form intramolecular hybrids exhibiting the reassociation kinetics usually observed for highly repeated DNA. Sequences in and around the duplexed portions of these foldback molecules would have been eliminated, along with highly repeated DNA, from the material characterized by Crain et al. (16). From the observations of Schmid et al. (19), it can be concluded that about 7% of all moderately repeated DNA could have been lost in this fashion; 6.5% of all 8 kilobase long segments of randomly sheared total D. melanogaster DNA will foldback, and 19% of the sequences in and around the duplexed portions of these molecules are moderately repetitive.

How many short repeat clusters do we expect to find in the Drosophila genome? If we assume that most clusters are as large as those characterized by Wensink et al. (18) (four clusters had a number average length of 14.5 kilobases), then there can be no more than 400 to 500. This can be compared to roughly 3,000 long repeat segments interspersed with single-copy DNA that would form most of the remaining middle repetitive DNA (a basis for calculating the number of long repeat segments has been described in ref. 14).

DISPERSED REPEATS AND NOMADIC DNA

Another general property of families of dispersed repeated DNA is that they appear to occupy no fixed chromosomal locations. For this reason we have referred to these sequences as nomadic. The conclusion that dispersed repeated DNA is nomadic comes from two kinds of observations: different chromosomal positions are occupied by families of dispersed repeated DNA sequences in different D. melanogaster strains, and new chromosomal arrangements of these sequences can evolve within a strain following prolonged laboratory culture (14,15). In this review, we will describe only those observations related to differences between strains. Figure 2 illustrates the variable arrangements of several dispersed repeated DNA segments in two D. melanogaster strains. In each case, a randomly cloned segment of D. melanogaster DNA (Dm segment) carrying moderately repeated DNA is hybridized in situ, with salivary gland polytene chromosomes from strains g-1 and g-X11. For example, Figures 2a and 2b show the dispersed arrangement of the Dm 2074 sequence within chromosome arm 3R (see ref. 21 for nomenclature). The repeated sequence can be found at two positions in this chromosome arm in strain g-X11 (Figure 2a) and hybridizes to three positions in the homologous arm from strain g-1 (Figure 2b). The arrangement of the repeated sequence within the g-X11 chromosome differs substantially from that observed in the g-1 chromosome. Analogously, Figures 2c and

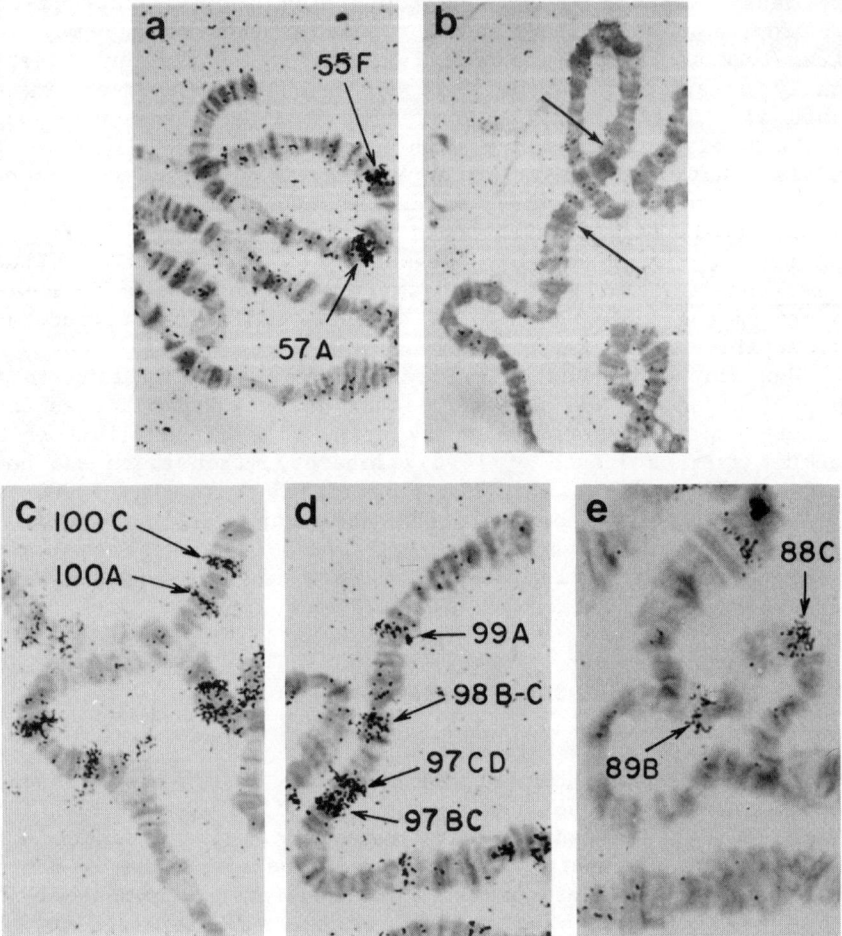

Figure 2. Localization of nomadic Dm segments in the polytene chromosomes of two strains of D. melanogaster. Each autoradiograph was obtained by hybridizing polytene chromosomes in situ with ^{3}H-cRNA transcribed from the indicated cDm plasmid. (For a description of this procedure, see ref. 20.) cDm plasmids were formed by joining mechanically sheared D. melanogaster embryo DNA to ColEl. Each Dm segment treated here was randomly chosen from a library of approximately 17,000 cDm plasmids (14). A description of the D. melanogaster strains g-1 and g-X11 has been presented elsewhere (14). cDm 2074 sequences were hybridized with salivary gland chromosomes from strain g-X11 (a) and strain g-1 (b). cDm 2015 sequences were hybridized with chromosomes from g-X11 (c) or g-1 (d). Asynapsed chromosomes (e) from a g-1/g-X11 heterozygous larva were hybridized with cDm 1142, copia, sequences. Number and letter designations refer to the Bridges (21) divisions of the D. melanogaster polytene chromosomes.

2d show differences in the hybridization patterns of Dm 2015 sequences in these strains. Figure 2e illustrates such differences in another way. The chromosomes shown in Figure 2e were derived from a heterozygous larva formed by crossing flies from the g-1 and g-X11 strains. Usually, homologous chromosomes are paired in salivary gland nuclei, but in the center of this figure homologs are partially asynapsed. When the nomadic Dm segment, Dm 1142, is hybridized to these chromosomes, only one of the two homologs is labeled in the interval in which pairing has been interrupted. The two positions of hybridization are 88C and 89B. By comparing these results with our chromosome maps for Dm 1142 in the parental strains g-1 and g-X11 (data not shown), it was determined that the labeled chromosome was derived from the g-X11 parent.

NOMADIC DNA IN Drosophila simulans

Nomadic DNA segments isolated from the D. melanogaster genome are usually not well represented in the D. simulans genome (15). Figure 3a shows a nomadic Dm segment, Dm 2088, which has been hybridized with equal amounts of total genome DNA from the two species. For each species, the genomic DNA has been digested with EcoRI, separated by gel electrophoresis and transferred to a nitrocellulose filter. Thus, the filter shown in Figure 3a contains two DNA spectra: D. simulans on the left and D. melanogaster on the right. While a few nomadic DNA segments from D. melanogaster can hybridize intensely to the genomic DNAs of both species (e.g., see Figure 3c), most nomadic sequences, like Dm 2088, hybridize much more efficiently to conspecific DNA than to that of D. simulans. The hybridization potentials of 15 randomly cloned nomadic Dm segments have been tested in this fashion and each hybridization has been analyzed quantitatively (15). The results of these analyses can be summarized as follows. 1) Approximately one-third of the Dm segments are at least 20 times more abundant in D. melanogaster DNA than in that of D. simulans. Some of these sequences may be altogether missing from the DNA of the latter species (e.g., Figures 3a and 3b). 2) Another one-third of these Dm segments are represented 5 to 10 times less often in D. simulans DNA. 3) Only about a fifth of these Dm segments are equally represented in the genomic DNAs of both species (e.g., Figure 3c). Nomadic Dm segments that are more abundant in D. simulans DNA than in conspecific DNA have not been found; this has led to the suggestion that the D. simulans genome harbors less nomadic DNA than does that of D. melanogaster (15).

EXTRACHROMOSOMAL MIDDLE REPETITIVE DNA

A heterogeneous population of small covalently-closed circular DNAs has been isolated from D. melanogaster nuclei (24).

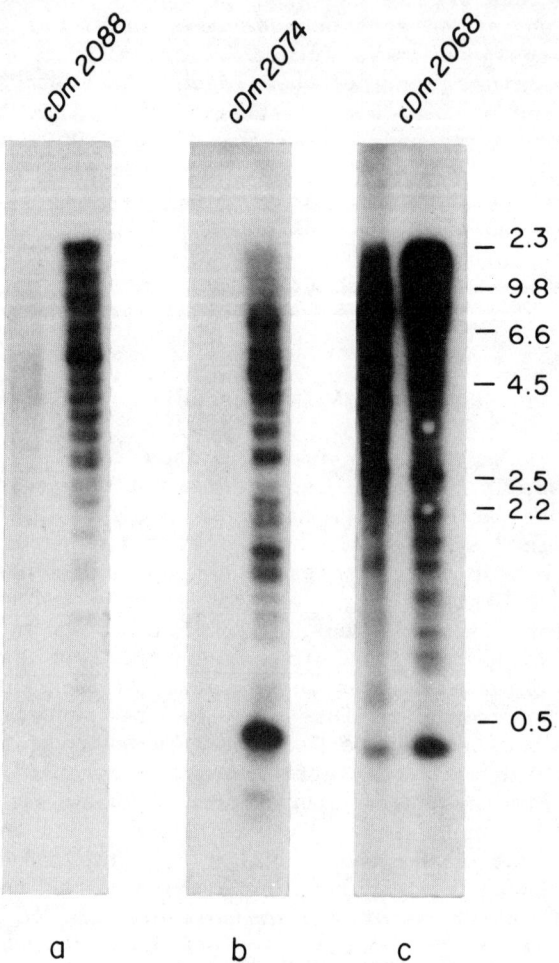

Figure 3. Restriction fragments homologous to nomadic Dm segments in D. simulans and D. melanogaster. Equal amounts of D. simulans (left) and D. melanogaster (right) total genome adult DNAs were digested with a restriction enzyme (EcoRI), separated by electrophoresis in a 0.7% agarose gel and transferred to a nitrocellulose filter with Southern's blotting technique (22) to form each panel. (a) The autoradiograph obtained when these DNA fragments were hybridized with Dm 2088. (b) Formed by hybridizing the DNA spectra with Dm 2074 sequences. (c) Results from the hybridization of Dm 2068 sequences. The lengths (kilobases) of HindIII fragments of bacteriophage λ are presented on the right (from ref. 23).

These circles have been recovered from two sources: eggs at the blastema stage and cultured cells (Schneider's line 2, see ref. 25). The circles have a buoyant density in CsCl gradients of 1.703 g/cc, which overlaps that of the chromosomal main band DNA of D. melanogaster. In cultured cells, circles can make up as much as 0.03% of the total cellular DNA. The fact that the circles are covalently closed has been demonstrated by CsCl/ethidium bromide gradient analysis.

The size distribution of the circles from cultured cells is broad, with a range of 0.25 to 10.0 kilobases. Their number average length is about 3.3 kilobases. It has been estimated that at least 11 distinct size classes of circular molecules can be found in cultured cell nuclei; this degree of size heterogeneity can be found in cloned sublines of Schneider's line 2 cells (24). These results support the conclusion that molecules of many different sizes coexist in the same cell nucleus. Digestion of small circular DNAs with EcoRI also indicates that homologous DNA sequences are not often found in molecules of different sizes (24).

Small circular DNAs contain sequences homologous to chromosomal DNA. Reassociation of circular DNA with a vast excess of total genomic DNA has shown that the homologous chromosomal sequences are moderately repetitive; no significant contribution of highly repetitive or nonrepetitive DNA sequences has been detected (26). The circles are also homologous to poly(A)$^+$ and poly(A)$^-$ RNAs expressed in cultured cells. It is interesting to note that these RNAs are predominantly nuclear and do not include ribosomal or transfer RNA sequences (27). Later in this chapter we will discuss RNAs with similar properties that are homologous to nomadic DNA segments.

NOMADIC DNA SEGMENTS AND ABUNDANT Poly(A)$^+$ RNAS

Most nomadic DNA segments appear to be homologous to abundant poly(A)$^+$ RNAs expressed in cultured D. melanogaster cells (15). For this reason, members of nomadic gene families are often isolated by a general procedure designed to obtain cloned segments of Drosophila DNA coding for the abundant mRNAs expressed by these cells (28). In this procedure, a random set of cloned Dm segments carrying chromosomal sequences from every part of the genome is screened by the colony hybridization method of Grunstein and Hogness (29). In its earliest applications, ^{32}P-labeled cytoplasmic polyadenylated RNA prepared from Eschalier's Kco line of D. melanogaster cells was used as the hybridization probe (28).

Colonies hybridizing with this RNA define two classes of poly(A)$^+$ RNA sequences. The first class is composed of extremely abundant RNAs, termed copia. Copia RNA accounts for at least 3%

of the total cytoplasmic poly(A)$^+$ RNA expressed in cultured cells. The second class of poly(A)$^+$ RNA is somewhat less abundant and probably includes several hundred RNA species. Each of these poly(A)$^+$ RNAs makes up between 0.1% and 0.6% of the total cytoplasmic poly(A)$^+$ RNA. Collectively, sequences in the second class compose nearly 60% of the cytoplasmic polyadenlyated RNA (see also ref. 30). The remaining poly(A)$^+$ RNAs (about 35% of the total) are present at concentrations less than 0.1%, and are not usually recognized in these colony hybridization experiments (28,31). Copia RNA, apparently the most abundant poly(A)$^+$ RNA expressed by cultured cells, is homologous to members of a nomadic gene family. Several nomadic gene families that code for RNAs comprising the second abundancy class of polyadenlyated sequences have also been identified. In the following section, we will review the properties of copia and of 412, a family coding for RNAs found in the second class. These genes have been especially well studied and may serve as models for much of the moderately repeated DNA in D. melanogaster.

NOMADIC GENE FAMILIES COPIA AND 412

The gene families copia and 412 are each composed of 20 to 30 chromosomally dispersed members. The nomadic character of both families has been demonstrated by hybridization of these sequences to salivary gland polytene chromosomes in situ (31,32; Young and Hogness, unpublished data) (see also Figure 2e). Figure 4 shows a restriction map of Dm 1142 which includes a copia sequence derived from a region near the tip of chromosome arm 3L. The chromosomal location of this sequence was determined by hybridizing restriction fragments surrounding the gene to polytene chromosomes in situ (Young and Hogness, unpublished data). The RNA homologous sequence is 5 kilobases long and has been mapped by hybridization of ^{32}P-labeled polyadenylated RNA from Kco cells to restriction fragments of Dm 1142 with Southern's blotting method (22). Only the contiguous restriction fragments marked as the RNA region in Figure 4 show detectable homology with the cultured cell probe (31; Young and Hogness, unpublished data).

The copia homologous regions of 16 additional Dm segments have been mapped (31,33; Young and Hogness, unpublished data). Most of the restriction fragments derived from the internal portions of these genes are equal in length to those found in the Dm 1142 RNA region, but there are noteworthy exceptions (see below). In contrast to these results, restriction fragments containing the ends of the genes usually vary in size among the Dm segments. This result is expected given the scattered chromosomal arrangement of these genes; most genes are embedded in differ-

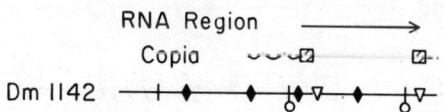
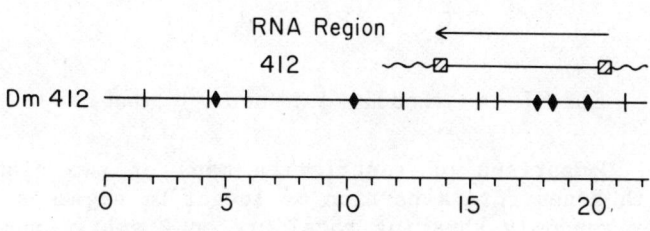

Figure 4. Restriction maps of Dm 1142 and Dm 412. The positions of the nomadic sequences of copia and 412 are indicated by horizontal lines bordered by hatched blocks. Hatched blocks represent direct terminal repeats. Wavy lines depict sequences flanking the nomadic gene in each Dm segment. The arrows indicate the extents of RNA homologies and the directions of transcription. Dm 1142 and Dm 412 were cloned in ColE1, but only the D. melanogaster portions of these plasmids are represented. An asterisk indicates that HhaI sites are given only for that portion of Dm 1142 that includes or is adjacent to the RNA homologous region. The scale at the bottom of the figure is in kilobases.

ent DNA sequences at these dispersed chromosomal locations (31; Young and Hogness, unpublished data).

As we have indicated, the primary sequence of most cloned copia genes appears to be well conserved, but some variants have been recognized. Four of these variants have been analyzed in detail. Dm 1142 is one of these and this gene differs from the canonical copia sequence by a deletion of about 200 base pairs.

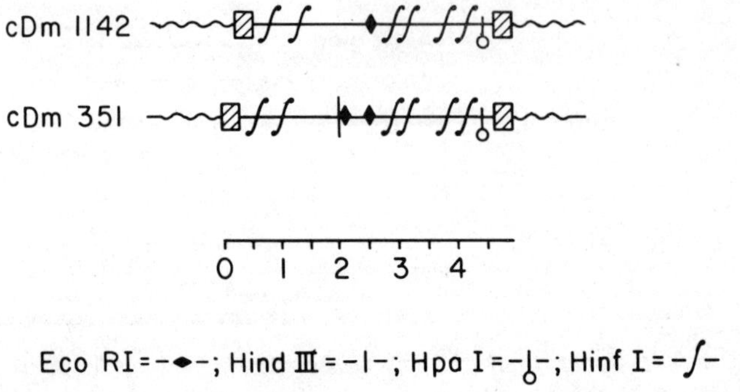

cDm 1142 ~~~▨ƒƒ　♦ƒƒ ƒƒ↓▨~~~

cDm 351 ~~~▨ƒƒ ▶♦ƒƒ ƒƒ↓▨~~~

```
├──┼──┼──┼──┼──┤
0   1   2   3   4
```

Eco RI = -●-; Hind III = -|-; Hpa I = -↓-; Hinf I = -ƒ-

Figure 5. Comparison of restriction maps of two cloned copia genes. Both genes form a portion of longer Dm segments that were produced by randomly shearing total Oregon R embryo nuclear DNA, which was then inserted into ColE1 using the poly(dA)/poly(dT) method of Wensink et al. (20). Dm 1142 and Dm 351 are homologous only for the copia sequence. The Dm 1142 copia map can be derived from that of Dm 351 by a simple deletion of approximately 0.2 kilobases. The DNA segment that is missing from the 1142 gene includes one HindIII and one EcoRI site that can be found in Dm 351. The scale at the bottom of the figure is in kilobases.

The sequence that is missing maps to a region near the center of the gene and is present in most of the remaining cloned genes studied by Young and Hogness (unpublished data). Figure 5 compares the structure of the more prevalent gene sequence, exemplified by Dm 351, and that of Dm 1142. Interestingly, Dm 1122, a second variant, also appears to carry a deletion for at least a portion of the same 200 base-pair region, but differs from both Dm 351 and Dm 1142 by the addition of a short sequence (probably less than 100 base pairs) in another region of the gene (Young and Hogness, unpublished data). Two more variants have been isolated by Carlson and Brutlag (33). Dm 139.9 is a truncated gene; it is only 3.7 kilobases long and includes the 5' terminus of the gene so that a little over 1 kilobase of DNA is missing from its 3' end. The deletion apparently includes the 3' terminal repeat sequence. Dm 139, a full length gene bordered by 1.688 satellite DNA, differs from the canonical sequence by the addition of a HinfI restriction endonuclease site that maps within 0.5 kilobases of the 5' end of the gene. No obvious addition or deletion of DNA accompanies this sequence variation.

The four variant genes reviewed were recognized following the characterization of only 17 cloned genes and whole genome analyses of restriction fragments homologous to copia support the conclusion that a number of variants are included in the family (15; Young and Hogness, unpublished data). Thus, copia is a polymorphic sequence. Variations of the canonical sequence are easily isolated and could make up nearly a quarter of all the members of this nomadic gene family.

When the homologies among copia genes were being mapped, it was observed that restriction fragments carrying the termini of each cloned gene cross hybridize. It was further established by electron microscope analysis that the sequence responsible for this cross hybridization was a direct rather than inverted repeat just less than 300 base pairs long (31; Young and Hogness, unpublished data) (see also Figures 4 and 5). Recently, Levis et al. (34) and Dunsmuir et al. (35) have determined the nucleotide sequence of this end repeat DNA and several of their findings are of particular interest. 1) A 5 base-pair sequence present at the chromosomal site at which copia is inserted is found duplicated at either end of the gene following insertion. 2) Comparable to the insertion of certain transposable elements in bacterial systems (for a review, see ref. 36), the 5 base-pair target sequence that undergoes this duplication is variable; copia generates different 5 base-pair duplications at different sites of insertion. 3) There is no detectable sequence homology between chromosomal sites of copia integration. 4) The direct repeats associated with a single copia gene are identical to each other but the end repeats can vary from gene to gene. Thus, it seems likely that one of the two end repeat sequences serves as a template for the synthesis of the other, perhaps during the process of copia rearrangement.

Members of the nomadic gene family 412 have not been as well defined as copia. However, 412 and copia share several properties. The primary sequence of 412 genes, which are 7.5 kilobases long, is well conserved and terminally redundant (Figure 4). The conserved sequence is more or less coextensive with the sequence defined by RNA homology, and the repeats at either end of this region are direct and about 500 base pairs long. Members of both of these nomadic gene families are often surrounded by nonrepetitious DNA (31). Thus, both nomadic gene families appear to be composed of nonpermuted long repetitive elements of the sort originally observed by Manning et al. (2).

COPIA AND 412 TRANSCRIPTION PRODUCTS

Copia and 412 were isolated because of their homologies to poly(A)$^+$ cultured cell RNAs. However, sequences homologous to

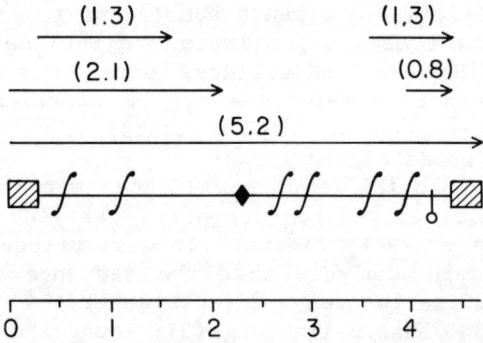

Figure 6. Overlapping RNAs homologous to copia genes. The restriction map shown is for the gene from Dm 1142 (see Figure 5). Arrows refer to DNA regions showing poly(A)$^+$ RNA homology and probable direction of transcription; as much as 4% of the total cellular copia RNA may not be transcribed in the indicated direction so that a portion of the RNA comprising any of the four size classes could be transcribed in the opposite direction. Numbers in parentheses indicate sizes (kilobases) of copia homologous poly(A)$^+$ RNAs hybridizing to each gene interval. A description of the procedures used to isolate and map copia poly(A)$^+$ RNAs can be found elsewhere (15; Schwartz and Young, in preparation). The scale at the bottom of the figure is in kilobases.

these genes are also found in poly(A)$^-$ RNA. Again, copia RNA has been better defined than that of 412 and about half of the total cellular copia RNA produced by Eschalier's Kco cells is polyadenylated. The poly(A) associated with this RNA appears to be terminal and heterogeneous ranging from about 50 to greater than 170 bases long. The size distribution of these poly(A) segments is comparable to that found for total polysomal poly(A)$^+$ RNA extracted from cultured cells (15; Schwartz and Young, in preparation).

Copia and 412 sequences are found in several size classes of poly(A)$^+$ RNA expressed in cultured cells. The 412 genes are homologous to at least three classes of poly(A)$^+$ RNA, these being about 7.5, 2.2 and 1.6 kilobases long (Schwartz and Young, in preparation), and four RNAs that are 5.2, 2.1, 1.3 and 0.8 kilobases long are homologous to copia (15; Schwartz and Young, in preparation). Copia RNAs have been mapped to overlapping regions of the gene sequence as shown in Figure 6, and each of the RNAs is likely to be transcribed in the same direction. The direction of transcription has been established by cloning copia in a single-stranded bacteriophage. By inserting copia in both orientations with respect to bacteriophage f1 replicative form DNA, which is double-stranded, two classes of phages have been produced. Each class of phages carries a different strand of the gene and more than 96% of all copia RNA produced by cultured Drosophila cells hybridizes with only one of these two classes (Schwartz and Young, in preparation).

Copia and 412 poly(A)$^+$ and poly(A)$^-$ RNAs are somewhat unusual in that both are enriched in nuclei (15; Schwartz and Young, in preparation). Cell fractionation studies have indicated that although total cellular copia RNA has a half life of about 4 hrs, 75% of the copia RNA produced by cultured cells over a period of 4.5 hours maps to nuclei. In these experiments, less than 35% of the total poly(A)$^+$ RNA is nuclear. More than 85% of the total cellular 412 RNA is associated with nuclei in these experiments.

Cytoplasmic copia and 412 RNAs have been further analyzed. When displayed on polyribosome gradients, over 90% of the RNA homologous to 412 is localized in a region falling between monosomes and trisomes (15; Schwartz and Young, in preparation). This is a region in which actively translating mRNAs are predominantly short (<300 bases) and code for correspondingly small proteins (37-39). Copia cytoplasmic RNAs are more broadly distributed in these gradients and as much as one-third of this RNA may be associated with polyribosomes (15; Schwartz and Young, in preparation). However, the modal distribution of copia RNA is centered about the region of these polyribosome gradients where 412 sequences have accumulated. Thus, most of the RNAs abundantly expressed by members of these two nomadic gene families are likely to be, at best, inefficiently translated in cultured cells.

NOMADIC GENES IN OTHER EUKARYOTIC SYSTEMS

A family of about 35 genes in Saccharomyces cerevisiae has been described that has many properties comparable to those of copia and 412. Most members of the yeast Ty1 family are 5.6 kilobases long and include direct terminal repeats of about 300 base pairs. The genes are also nomadic, occupying different chromosomal positions in different yeast strains; changes in the chromosomal arrangement of these genes have been detected after culturing a cloned cell line for as little as a few hundred generations (40). Like copia, the insertion of Ty1 at new chromosomal positions generates a 5 base-pair duplication of target DNA; the same 5 base-pair host sequence is found at each end of the gene following integration and different 5 base-pair repeats flank genes at different chromosomal positions (41,42).

Movement of Ty1 sequences into and out of yeast chromosomes has been rigorously demonstrated (43,44). Insertions of members of the Ty1 family at the His$^+$ locus lead to the formation of unstable mutations and reversion is often accompanied by chromosome rearrangement. Reversion to His$^+$ has also been shown to be associated with loss of the nomadic DNA segment. At least some of the His$^+$ revertants arise by recombination across the direct terminal repeats leaving a single copy of the end repeat sequence at the His locus (41). Thus, such excisions are not due to a simple reversal of the insertion process.

Like the nomadic gene families of Drosophila, the yeast gene family is also transcribed to form abundant poly(A)$^+$ RNAs. RNA homologous to Ty1 composes about 10% of the total yeast poly(A)$^+$ RNA so that only the ribosomal RNAs and double-stranded killer RNAs are more abundantly expressed in this organism (40).

Many of the properties of copia and 412 are also shared by certain integrated retrovirus DNAs. Examples include avian sarcoma, Moloney murine leukemia and spleen necrosis proviruses (45-49). Each provirus sequence can occupy various host chromosomal positions; retrovirus DNA integrates at different chromosomal locations in different infected cells. The proviruses are also terminally redundant, and the repeats are direct for each provirus, ranging in size from about 0.3 kilobases for avian sarcoma provirus to about 0.6 kilobases for spleen necrosis provirus. Again, integration of viral DNA is accompanied by an oligonucleotide duplication of host chromosomal DNA.

RNA plays a central role in the reproduction of each of these retroviruses. The proviral DNA sequence is transferred to the chromosomes of newly infected cells through reverse transcription of an RNA intermediate (50) and we can imagine that this mode of replication is shared by certain nomadic DNA segments in other eukaryotic sytstems; most of the RNA expressed by copia and 412 is confined to cell nuclei and each of these genes produces a major transcript that probably carries all of the sequence information found in the nomadic DNA segment (15; Schwartz and Young, in preparation). It may also be important that the end repeats of each provirus are made homosequential, presumably by replication through an RNA intermediate. The transcription of retrovirus RNA is initiated within one proviral end repeat sequence and is terminated in the other, so that only a portion of the terminal repeat sequence is present at either end of the transcription product (51,52). In the model of Gilboa et al. (53), it is assumed that one full length, end repeat DNA sequence is formed by reverse transcription, and this DNA, in turn, serves as a template for the synthesis of the second end repeat. The observation of Levis et al. (34) that end repeats from the same copia gene are identical while those of different copia genes may vary could be easily explained if a similar process governs the genomic rearrangement of copia. Earlier in this chapter, we reviewed the properties of circular DNA molecules that are likely to include extrachromosomal copies of D. melanogaster nomadic DNA segments. It should also be pointed out that upon infection, retrovirus RNA is copied to form nuclear, covalently-closed DNA circles and it has been proposed that these circles are integration precursors (47,52,54,55).

SUMMARY

Most of the repetitive DNA in D. melanogaster can be divided into two classes according to reiteration frequency and genomic arrangement. One half of this DNA has an average repetition frequency of 24,000 and consists, for the most part, of short sequence elements (5 to 10 nucleotide pairs long), tandemly repeated to form the heterochromatic regions of the chromosomes. Intraspecific variations in the arrangement and reiteration frequencies of these highly repeated DNA sequences have not been observed. In contrast to the tandem arrangement of these highly repeated DNAs, most of the moderately repeated DNA of D. melanogaster consists of sequences that are scattered about the genome. Many of these repeated DNA segments are large, having a number average length of 5.6 kilobases, and they are often interspersed with nonrepeated DNA. Usually these moderately repeated DNA sequences are reiterated 10 to 100 times per haploid genome. A further general property of dispersed repeated DNA segments in D. melanogaster is that they occupy no fixed chromosomal positions. For this reason such sequences are referred to as nomadic.

Members of two nomadic gene families, copia and 412, have been studied in detail and share a number of properties. 1) Like most of the dispersed sequences in Drosophila, the copia and 412 families are composed of long sequences of repeated DNA that are often interspersed with single-copy DNA sequences. 2) Genes from both families are usually composed of closely related sequences and they are terminally redundant. 3) The families are homologous to abundant poly(A)$^+$ RNAs. 4) The RNAs expressed by these genes appear to be enriched in nuclei. 5) Genes from both families are transcribed to form RNA molecules of several lengths, and for each gene family, an abundant RNA species is produced that contains all or most of the gene sequence. The architectures of such gene-length transcripts are remarkably similar to those of retrovirus RNAs.

REFERENCES

1. Schachat, F. and Hogness, D.S. (1974) Cold Spring Harbor Symp. Quant. Biol. 38, 371-381.
2. Manning, J.E., Schmid, C.W. and Davidson, N. (1975) Cell 4, 141-155.
3. Brutlag, D., Appels, R., Dennis, E.S. and Peacock, W.J. (1977) J. Mol. Biol. 112, 31-47.
4. Peacock, W.J., Brutlag, D.L., Goldring, E., Appels, R., Hinton, C.W. and Lindsley, D.L. (1974) Cold Spring Harbor Symp. Quant. Biol. 38, 405-416.
5. Brutlag, D., Carlson, M., Fry, K. and Hsieh, T.S. (1978) Cold Spring Harbor Symp. Quant. Biol. 42, 1137-1146.
6. Yamamoto, M. and Miklos, G.L.G. (1977) Chromosoma 60, 283-296.
7. Yamamoto, M. (1979) Chromosoma 72, 293-328.
8. Yamamoto, M. and Miklos, G.L.G. (1978) Chromosoma 66, 71-98.
9. Brutlag, D.L. and Peacock, W.J. (1975) in The Eukaryotic Chromosome (Peacock, W.J. and Brock, R.D., eds.), pp. 35-45, Australian National University Press, Canberra.
10. Endow, S.A., Polan, M.L. and Gall, J.G. (1975) J. Mol. Biol. 96, 665-692.
11. Carlson, M. and Brutlag, D. (1977) Cell 11, 371-381.
12. Peacock, W.J., Appels, R., Dunsmuir, P., Lohe, A.R. and Gerlach, W.L. (1977) in International Cell Biology 1976-1977 (Brinkley, B.K. and Porter, K.R., eds.), pp. 494-506, Rockefeller University Press, New York, NY.
13. Peacock, W.J., Lohe, A.R., Gerlach, W.L., Dunsmuir, P., Dennis, E.S. and Appels, R. (1978) Cold Spring Harbor Symp. Quant. Biol. 42, 1121-1135.
14. Young, M.W. (1979) Proc. Nat. Acad. Sci. U.S.A. 76, 6274-6278.
15. Young, M.W. and Schwartz, H.E. (1981) Cold Spring Harbor Symp. Quant. Biol. (in press).
16. Crain, W.R., Eden, F.C., Pearson, W.R., Davidson, E.H. and Britten, R.J. (1976) Chromosoma 56, 309-326.
17. Crain, W.R., Davidson, E.H. and Britten, R.J. (1976) Chromosoma 59, 1-12.
18. Wensink, P.C., Tabata, S. and Pachl, C. (1979) Cell 18, 1231-1246.
19. Schmid, C.W., Manning, J.E. and Davidson, N. (1975) Cell 5, 159-172.
20. Wensink, P.C., Finnegan, D.J., Donelson, J.E. and Hogness, D.S. (1974) Cell 3, 315-325.
21. Bridges, C.B. (1935) J. Heredity 26, 60-64.
22. Southern, E.M. (1975) J. Mol. Biol. 98, 503-517.
23. Murray, K. and Murray, N.E. (1975) J. Mol. Biol. 98, 551-564.
24. Stanfield, S. and Helinski, D.R. (1976) Cell 9, 333-345.

25 Schneider, I. (1972) J. Embryol. Exp. Morphol. 27, 353-365.
26 Stanfield, S.W. and Lengyel, J.A. (1979) Proc. Nat. Acad. Sci. U.S.A. 76, 6142-6146.
27 Stanfield, S.W. and Lengyel, J.A. (1980) Biochemistry 19, 3873-3877.
28 Young, M.W. and Hogness, D.S. (1977) in Eucaryotic Genetics System, ICN-UCLA Symp. Mol. Cell. Biol. (Wilcox, G. et al., eds.), Vol. 8, pp. 315-331, Academic Press, New York, NY.
29 Grunstein, M. and Hogness, D. (1975) Proc. Nat. Acad. Sci. U.S.A. 72, 3961-3965.
30 Levy W., B. and McCarthy, B. (1975) Biochemistry 14, 2440-2446.
31 Finnegan, D.J., Rubin, G.M., Young, M.W. and Hogness, D.S. (1978) Cold Spring Harbor Symp. Quant. Biol. 42, 1053-1063.
32 Strobel, E., Dunsmuir, P. and Rubin, G.M. (1979) Cell 17, 429-439.
33 Carlson, M. and Brutlag, D. (1978) Cell 15, 733-742.
34 Levis, R., Dunsmuir, P. and Rubin, G.M. (1980) Cell (in press).
35 Dunsmuir, P., Brorein, W.J. Jr., Simon, M.A. and Rubin, G.M. (1980) Cell (in press).
36 Calos, M.P. and Miller, J.H. (1980) Cell 20, 579-595.
37 Laird, C.D., Chooi, W.Y., Cohen, E.H., Dickson, E., Hutchinson, N. and Turner, S.H. (1974) Cold Spring Harbor Symp. Quant. Biol. 38, 311-327.
38 Mirault, M.E., Goldschmidt-Clermont, M., Moran, L., Arrigo, A.P. and Tissières, A. (1978) Cold Spring Harbor Symp. Quant. Biol. 42, 819-827.
39 Lindquist, S. (1980) J. Mol. Biol. 137, 151-158.
40 Cameron, J.R., Loh, E.Y. and Davis, R.W. (1979) Cell 16, 739-751.
41 Farabaugh, P.J. and Fink, G.R. (1980) Nature 286, 352-356.
42 Gafner, J. and Philippsen, P. (1980) Nature 286, 414-418.
43 Chaleff, D.T. and Fink, G.R. (1980) Cell 21, 227-237.
44 Roeder, G.S. and Fink, G.R. (1980) Cell 21, 239-249.
45 Hughes, S.H., Shank, P.R., Spector, D.H., Kung, H.J., Bishop, J.M., Varmus, H.E., Vogt, P.K. and Breitman, M.L. (1978) Cell 15, 1397-1410.
46 Shimotohno, K., Mizutani, S and Temin, H.M. (1980) Nature 285, 550-554.
47 Shoemaker, C., Goff, S., Gilboa, E., Paskind, M., Mitra, S.W. and Baltimore, D. (1980) Proc. Nat. Acad. Sci. U.S.A. 77, 3932-3936.
48 Sutcliffe, J.G., Shinnick, T.M., Verma, I.M. and Lerner, R.A. (1980) Proc. Nat. Acad. Sci. U.S.A. 77, 3302-3306.
49 Van Beveren, C., Goddard, J.G., Berns, A. and Verma, I.M. (1980) Proc. Nat. Acad. Sci. U.S.A. 77, 3307-3311.
50 Temin, H.M. (1976) Science 192, 1075-1080.

51 Hsu, T.W., Sabran, J.L., Mark, G.E., Guntaka, R.V. and Taylor, J.M. (1978) J. Virol. 28, 810-818.
52 Shank, P.R., Cohen, J.C., Varmus, H.E., Yamamoto, K.R. and Ringold, G.M. (1978) Proc. Nat. Acad. Sci. U.S.A. 75, 2112-2116.
53 Gilboa, E., Mitra, S.W., Goff, S. and Baltimore, D. (1979) Cell 19, 93-100.
54 Shank, P.R., Hughes, S.H., Kung, H.J., Majors, J.E., Quintrell, N., Guntaka, R.V., Bishop, J.M. and Varmus, H.E. (1978) Cell 15, 1383-1395.
55 Yoshimura, F. and Weinberg, R.A. (1979) Cell 16, 323-332.

MICROBIAL SURFACE ELEMENTS: THE CASE OF VARIANT SURFACE GLYCOPROTEIN (VSG) GENES OF AFRICAN TRYPANOSOMES

Kenneth B. Marcu

Biochemistry Department
State University of New York at Stony Brook
Stony Brook, New York 11794

and

Richard O. Williams

ILRAD
P.O. Box 30709
Nairobi, Kenya

INTRODUCTION

Trypanosoma brucei and its related species (T. congolense, T. equiperdum and T. vivax) are hemoparasitic, flagellated protozoa that are the causative pathogenic agents of African sleeping sickness in humans (T. brucei rhodesiense and T. brucei gambiense) and in domestic animals. T. brucei is transmitted by the tsetse fly (genus Glossina), thereby largely restricting these diseases to the African continent. During its life cycle in the fly and subsequently in the mammalian host, T. brucei normally appears to be pleomorphic in nature exhibiting several distinctive morphological forms (1). Slender bloodstream forms are present in the initial stages of an infection while nondividing, stumpy forms accumulate at later times. The stumpy forms are ingested by the fly and are transformed into procyclic forms (the only stage lacking a defined surface coat) in the midgut of the fly (2,3). The procyclic forms eventually migrate into the salivary glands where they differentiate into a heterogeneous population of metacyclic forms which are transmitted back to the host to complete the life cycle. Evolution has imparted intricate mechanisms to the trypanosomes for maintaining chronic infections

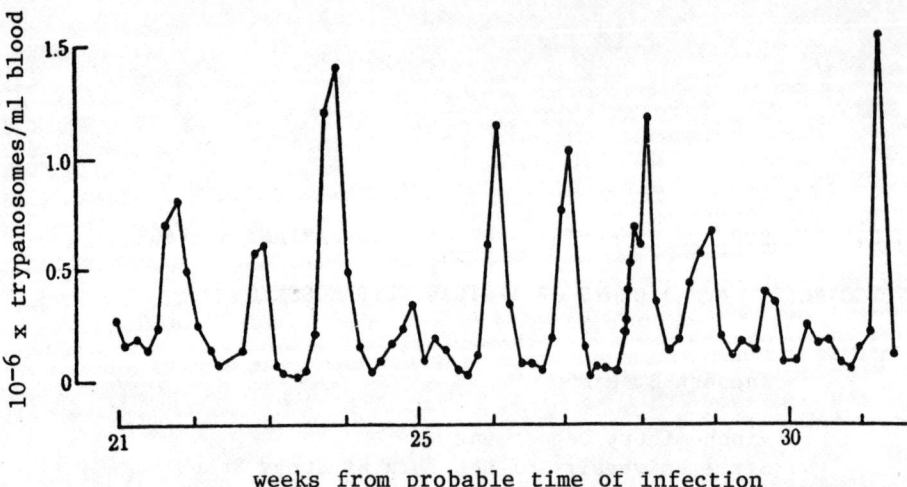

Figure 1. Fluctuation of parasitemia in a case of human trypanosomiasis (T. gambiense). (Redrawn from ref. 8).

in their mammalian hosts, thereby allowing for cyclical transmission via the tsetse fly. This unique phenomenon appears to be manifested by the ability of infectious trypanosomes to temporally alter this surface antigen structure thereby evading the host's immune response (1,4-7). The net result is that T. brucei is a brilliant parasite.

Antigenic variation in trypanosomiasis is manifested through the sequential expression of a series of antigenically distinct, homogeneous polypeptides on the parasite membrane surface (6). Ideally, in a normal host, the initial population of parasites derived from one trypanosome uniformly expresses a single, highly immunogenic membrane surface polypeptide (6). If the host is not killed by the initial parasitemia, the level of parasites will begin to decline as antibodies appear against the first membrane antigen. After a substantial decrease in blood parasites, subsequent parasitemias rise and fall in waves with each new parasite population carrying a new immunologically distinct membrane surface antigen (7). A series of periodic parasitemias originally observed in 1910 in a case of human sleeping sickness is depicted in Figure 1 (8). In reality, the contents of all these peaks are probably quite heterogeneous with respect to the biochemical characteristics of the major surface antigen. This would appear to be dependent on the previous environment and/or history of the particular infective trypanosome population (1,6,9): (a) Has it arisen directly from the fly? (b) Was it significantly attenuated by syringe passage in laboratory animals? The former material is more inclined to behave in a polymorphic fashion even though a particular variant may predominate the initial population. However, a single trypanosome derived from extensively syringe-passaged material tends to yield a clone of trypanosomes expressing one surface antigen followed by the usual sequential variations (see below) (1,6,7,9,10).

Sequentially arising populations of trypanosomes originating from a clone have the ability at infrequent intervals in normal infections to re-express antigens that have appeared earlier in the same series of variants (7). These results suggest that the genes coding for the variant antigens are stably inherited and do not undergo irreversible alteration or loss in progeny cells. Contrary to original proposals (4), the sequence of specific antigens that appear in a normal infection is not fixed or absolute (1,7,10). With any given starting antigen, there is a predominance for the expression of certain antigens. Even so, new antigens can occasionally appear while others may not be expressed at all (7). Frequently after the fourth or fifth parasitemia, one cannot predict what antigens will appear next and the parasitemias are often heterogeneous with a number of antigen types being expressed at the same time (1,7,10). The potential antigen repertoire of a single trypanosome is unknown. The largest number of antigenic variants derived from a clone is 101, reported for Trypanosoma equiperdum (11). Since surface antigen heterogeneity is found in the metacyclic trypanosomes of the tsetse fly (12) and variation has also recently been observed in long term tissue culture of bloodstream forms (13), the switch from one antigen to another may be under selective pressures from a hostile host immune system but is certainly not induced by interaction with antibody (1).

Vickerman (1,5,14) demonstrated that the variable antigens were located on the trypanosome cell surface. He showed by electron microscopy that trypanosomes possess an unusual membrane coat, 12 to 15 nm thick, external to the normal lipid bilayer (14). Using ferritin-labeled antibodies, these surface antigens were shown to be the cell coat seen by electron microscopy (5). Other workers demonstrated earlier that the variable antigens were glycoproteins (15,16).

The first definitive variable antigen characterization from cloned cells by Cross (6) showed that the antigens are homogeneous glycoproteins that constitute 95% of the protein on the membrane surface. Cross also demonstrated that a trypanosome population derived from a clone homogeneously expresses a single variable antigen and that each new wave of parasites expresses structurally different variable antigens. Following the purification of a series of antigens, Cross has shown that each possesses a unique isoelectric point (i.e., PI 5-9) and that the amino acid composition differs widely between different antigens (6). Subsequent to this work, the variable antigens have been referred to as "variant specific surface antigens" (VSSAs), "variant surface antigens" (VSAs) or, more descriptively accurate, as "variant surface glycoproteins"(VSGs). VSGs have a somewhat restricted molecular weight range of 56,000 to 67,000 (6). Amino acid sequence data are still quite limited. Five VSGs have been partially sequenced from the amino terminus and little if any homology is seen in the first 30 to 40 residues (17,18). The extent of glycosylation is variable as well, falling within a range of 7 to 17% of antigen mass (19). Mild periodate oxidation of the sugar moieties destroys precipitability of the antigens by concanavalin A but does not appear to affect precipitation by VSG-specific antisera (20). The carboxy terminus of six VSGs has recently been shown to be glycosylated with three of these precisely terminating in a glycosylated aspartic acid or asparagine residue (21).

Immunological cross-reactivity among VSGs falls into two classes. For the majority of variant antigens analyzed, there are a few minor antigenic determinants that do cross-react to different degrees with other unrelated VSGs (20,22). This immunological cross-reactivity can be reduced by an order of magnitude via mild periodate oxidation (2). It remains unclear whether the cross-reacting moieties are solely carbohydrate in nature or possibly represent carbohydrate bound to a short conserved block of amino acids (20). In any event, these antigenic determinants are only observed with purified VSGs and are probably normally buried in the surface membrane (20,22). The second class of immunologically cross-reactive VSGs constitutes a small set of very similar antigens. These VSGs have been called isotypes or isovariable antigen types (isoVATs) (23,24). Peptide mapping and fingerprint type analyses recently performed by VerVoort, Barbet and their collaborators (24,25) have indicated extensive regions of homologies between isoVATs isolated from different trypanosome subspecies, but the glycoproteins also appear to possess distinctive structural features. Therefore, these structurally similar VSGs would appear to be encoded by separate genes. A summary of distinctive properties of the VSGs is presented in Table 1.

For the remainder of this report we will concern ourselves solely with studies designed to assess the genetic basis for VSG gene expression and the extent of the VSG gene repertoire.

VSG mRNAs AND cDNA CLONING

VSG mRNAs have been shown to be polyadenylated and to possess a size range of 1.9 to 2.3 kilobases (26-31). Partial purification of VSG mRNAs by physical means and their subsequent translation in cell-free systems suggested early on that only one particular VSG mRNA was present in cloned trypanosomes and it appeared to be an abundant species of RNA (26,27). More recent studies have relied on polyribosome immunoprecipitation techniques to obtain highly purified VSG mRNAs (28,31,32). The only

Table 1

Some Properties of VSGs

1	Molecular weights of 56,000 to 67,000 with 7 to 17% of the VSG mass glycosylated.
2	Isoelectric points are distinct for different antigens (PI= 5-9).
3	Negligible amino acid sequence homologies exist.
4	Negligible immunological cross-reactivities are observed.
5	VSGs constitute about 10% of trypanosome dry weight and 95% of the membrane surface polypeptides.
6	Certain rare isoVATs share extensive structural similarities.
7	VSG mRNAs are abundant in the trypanosome cytoplasm.
8	Cloned trypanosomes only appear to possess one VSG on their surface and probably only one VSG mRNA in their cytoplasm.

other species of mRNA comparable in quantity to the specific VSG would appear to be the mRNAs encoding α- and β-tubulin (27,29,33). cDNA clones have been constructed and characterized for the VSG mRNAs of several genetically unrelated stocks of trypanosomes (i.e., Lump 227, S427 and EATRO 164) (29,30,33). cDNA clones have been identified initially by hybrid-arrested (34) and hybrid-selected in vitro translation (29,20,33). This procedure provides a highly definitive identification of the appropriate cDNA clone (34). A typical hybrid-arrested translation profile for the cDNA clone of the Iltat 1.2 antigen is shown in Figure 2. Only one mRNA species would appear to be diminished in translational capacity upon hybridization to its putative cDNA clone and this activity is neatly rescued by heating the hybrids, thereby confirming the identification. The protein band affected by these treatments has been identified as the Iltat 1.2 VSG by immunoprecipitation (29). Hoeijmakers and co-workers have performed RNA blotting (35) experiments with cloned cDNA probes of four antigenically unrelated VSGs (30,36). These studies have definitively demonstrated that transcription of the VSG genes would appear to be trypanosome clone-specific and these genes possess no detectable nucleotide sequence homology. In addition, a stock of procyclic trypanosomes (i.e., coatless forms) maintained in tissue culture have recently been shown to lack specific VSG mRNA sequences by Agabian and her colleagues (33). These results would collectively argue that controls at the translational level are not utilized by the trypanosome to alter VSG expression while factors responsible for determining the transcrip-

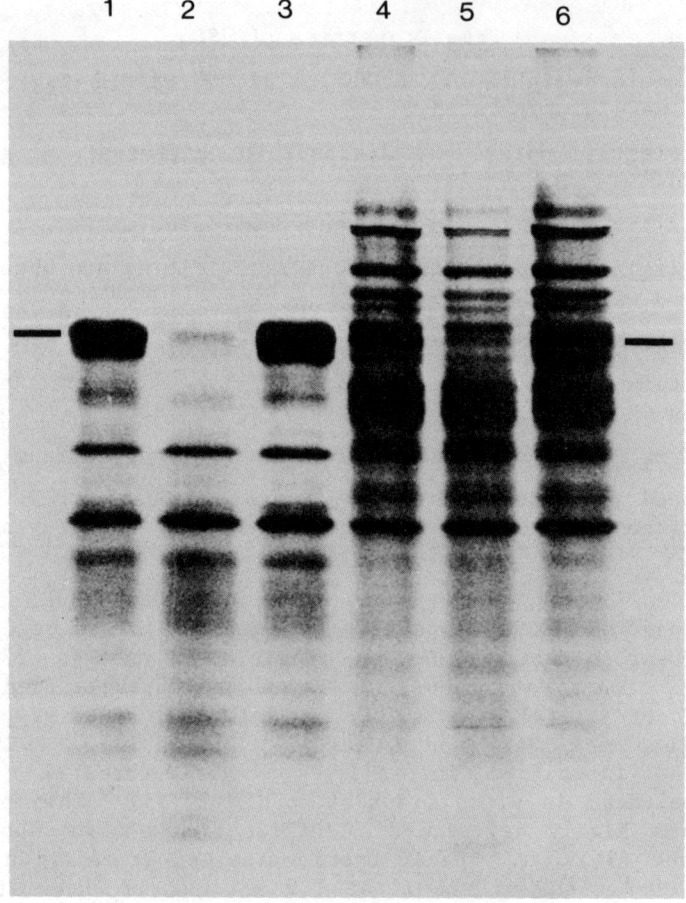

Figure 2. Hybrid-arrested translation identification of an Iltat 1.2 cDNA clone (pcB602). Translation products of partially purified Iltat 1.2 mRNA: lane 1, untreated mRNA control; lane 2, mRNA hybridized with pcB602; lane 3, mRNA-DNA hybrids were denatured by boiling. Translation products of total mRNA: lane 4, untreated; lane 5, hybridized with pcB602; lane 6, hybridized and then denatured. Hybridizations were performed according to Patterson et al. (34). Hybrids were denatured by boiling for 90 sec. Cell-free translations were performed in the rabbit reticulocyte system (56). Electrophoresis of ^{35}S-methionine-labeled cell-free products was performed on a 10% polyacrylamide SDS slab gel. The variable antigen band identified by immunoprecipitation is indicated by the horizontal bars. Control experiments with the plasmid vehicle pBR322 gave no inhibition of synthesis (data not shown).

TRYPANOSOME GLYCOPROTEIN GENES

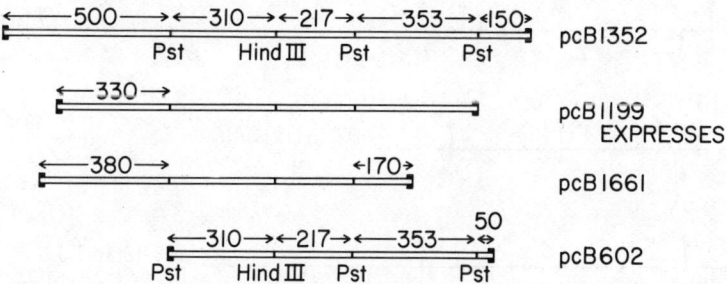

Figure 3. Comparative restriction enzyme maps of 4 overlapping Iltat 1.2 cDNA clones. Double-stranded cDNA was made essentially according to Wickens et al. (57) and inserted into the plasmid vector pBR322 by G-C tailing with terminal transferase (58). Transformed bacteria containing recombinant plasmids were screened by in situ hybridization (59). Iltat 1.2 cDNA clones were subsequently definitively identified by hybrid-arrested in vitro translation as shown in Figure 2.

tional competency of VSG genes would seem to be the major focus of VSG regulation.

Restriction enzyme maps of four cDNA clones of the Iltat (ILRAD Trypanosome Antigen Type) serodeme (originally derived from T. Brucei Lump Stock 227) are depicted in Figures 3 and 4 (see Figure 5 flow diagram for the sequential derivation of cloned trypanosomes of this serodeme). Four individually isolated cDNA clones of Iltat 1.2 were observed to possess overlapping sequences with the largest insert being 1.5 kilobases (see pcB1352 in Figure 3). The Iltat 1.2 VSG mRNA is about 2.0 kilobases in size. The complete DNA sequence of pcB1352 has revealed only one possible reading frame lacking termination codons and only appears to contain coding sequences (37). One of the Iltat 1.2 cDNA clones, pcB1199, has been observed to direct the synthesis of a portion of this particular VSG (i.e., Iltat 1.2) in bacterial cells as revealed by competitive radioimmune assays (RIAs) performed with purified, iodinated Iltat 1.2 and dilutions of bacterial extracts as competitor (38). This constitutes a hybrid protein since the cDNA insert is fused to the β-lactamase gene in this cloning protocol. The level of specific antigen produced is

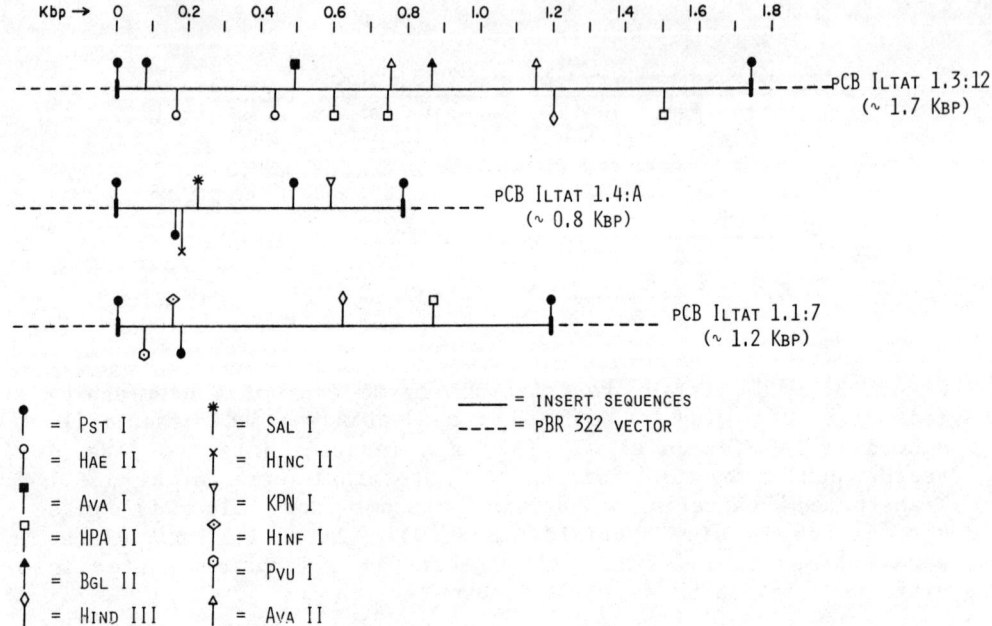

Figure 4. Comparative restriction enzyme maps of cDNA clones for Iltat 1.1, 1.3 and 1.4. These were prepared and characterized as described for the Iltat 1.2 cDNA clones in Figures 2 and 3.

quite low and can only be detected conveniently by the sensitive RIA procedure. This may be due to the low stability of such a hybrid protein or RNA species in E. coli or possibly inefficient usage of the β-lactamase promoter. cDNA clones have also been prepared and similarly identified for three other members of the Iltat serodeme (Iltat 1.1, 1.3 and 1.4) and their individual restriction maps and insert sizes are shown in Figure 4. Cross-hybridization studies performed with this antigenically unrelated, sequentially arising, group of VSGs (see Figure 5) confirms and extends the findings of Hoejimakers (36) demonstrating no detectable sequence homology among VSG mRNAs (39).

Boothroyd and colleagues (40) have recently determined the DNA sequence corresponding to the carboxy-terminus of a particular VSG (VSG 117 of trypanosome stock 427). These results indicate that information for an additional 23 carboxy-proximal amino acids is present in the VSG mRNA sequence. This hydrophobic stretch of amino acids would appear to be proteolytically removed from the mature VSG either prior to its export to the outer surface membrane or during the isolation of the VSG itself (21,40).

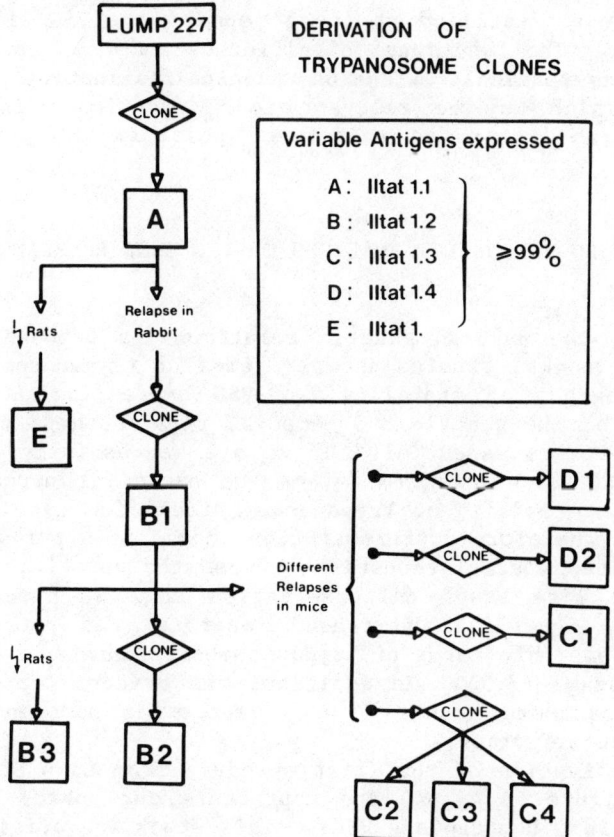

Figure 5. The history of the Iltat serodeme. Homogeneous, non-cross-reacting VSGs are indicated by the letters A-E. The word CLONE indicates passages through sublethally irradiated rats.

This particular VSG is then subsequently glycosylated at a terminal aspartic acid residue which is assumed to represent the site of cleavage. Therefore, this interesting hydrophobic extension may either function as a membrane anchor for the VSG or could possibly act as a signal sequence for VSG processing and secretion (i.e., comparable to amino-terminal leader type sequences) (40,41). However, a true amino-terminal leader sequence has also been identified at the 5' end of the VSG 117 cDNA sequence (42). The functional significance and time of removal of this carboxy-terminal extension remain fascinating unanswered questions which may be relevant to VSG antibody interactions (i.e., VSG shedding) and therefore, possibly VSG variation as well.

GENOMIC SOUTHERN BLOT ANALYSES OF VSG GENE REARRANGEMENTS

Due to the unknown genetic relationships between different laboratory stocks, strains and serodemes of trypanosomes and the variable degrees of stability for VSG expression, an accurate assessment of the genetic and temporal relationships of a series of cloned VSGs is essential. If we are successfully to use recombinant DNA techniques to define the basis for normal VSG expression, the capacity of trypanosome stocks for cyclical transmission and therefore differentiation should be a primary consideration. The interrelationships between the normal processes of trypanosome life stage differentiation and VSG production are fairly well known (1) and these properties can potentially be lost in monomorphic forms of trypanosomes attenuated by long term syringe passage (7,9). In addition, the criteria for purity of cloned trypanosome stocks (i.e., expressing homogeneous VSGs) must be quite rigorous.

The derivation of the Iltat serodeme from Lump Stock 227 is shown in Figure 5 (43,44) and represents our source of sequentially arising, homogeneous VSGs. This stock was originally isolated from an infected ox in 1964 in Kenya (originally designated UHEMBO/64/EATRO/795). The trypanosomes expressing the homogeneous VSGs (Iltat 1.1, 1.2, 1.3 and 1.4) are pleomorphic in appearance and will give rise to chronic infections in normal animals if initially passaged in low doses. Several of these antigens have been reisolated and recloned following cyclical transmission in the tsetse (45). Cloned trypanosome populations expressing specific VSGs are derived from infections of sublethally irradiated rats. These have been judged to be pure by a variety of criteria: (a) immune lysis tests (10); (b) indirect immunofluorescence microscopy (10,46), and (c) in some cases, by examination in a FACS apparatus with FITC conjugated antibody (resolution of one contaminant in 100,000 organisms) (47). Samples B1 and B2 represent independent isolates of cloned trypanosomes expressing the identical VSG (Iltat 1.2). Samples C2, C3 and C4 were all cloned from the same relapse infection of a single mouse and all express Iltat 1.3 (see Figure 5).

A summary of the data on the genomic context of the Iltat 1.2 gene in these sequentially arising variants is presented in Figure 6 and a restriction map of pcB1352 (an Iltat 1.2 cDNA clone) is shown for comparison. Williams, Young and Majiwa (29)

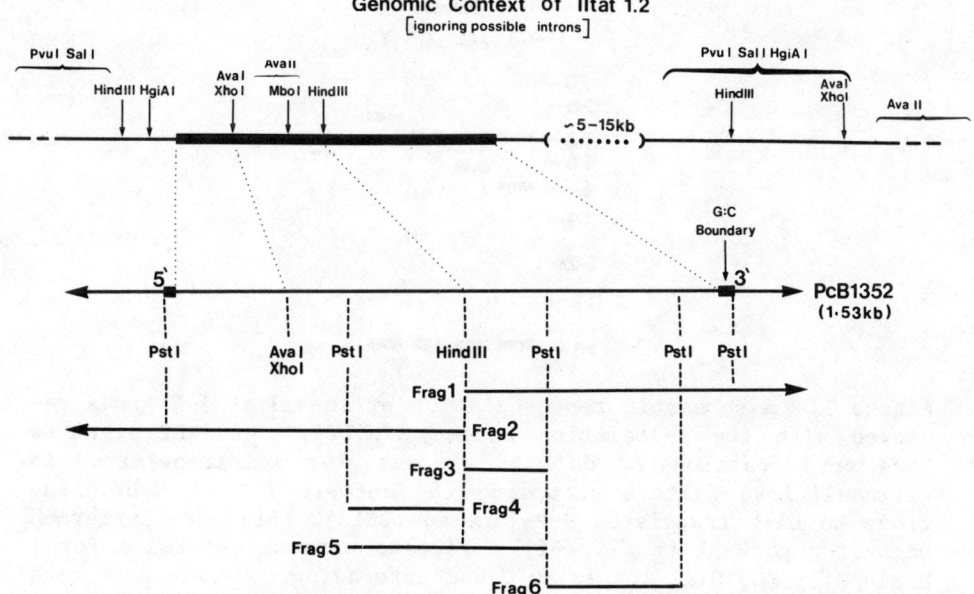

Figure 6. A summary of the restriction enzyme maps of the genomic Iltat 1.2 sequence in cloned variants of the Iltat serodeme. A restriction map of the Iltat 1.2 cDNA clone, pCB1352 and the location of appropriate restriction fragment probes are shown for comparison.

originally demonstrated that genomic rearrangements occur in the environment of the Iltat 1.2 gene in different cloned variants of the Iltat serodeme. Results with about 20 restriction enzymes have confirmed these rearrangements to be due to 3' flanking sequence length heterogeneity of this particular VSG gene (44). Restriction enzyme digestions performed with enzymes that cut the cDNA sequence yield three hybridizing bands in the genomes of cloned variants within the Iltat serodeme (Figures 7 to 9) but only two bands in two antigenically distinct clones derived from a genetically unrelated stock (i.e., S427) (9) as shown in Figure 7. When three bands are present, one appears constant in molecular weight in different variants while the size differences observed for the other bands remain constant with different restriction enzymes. When only two hybridizing bands are observed, neither band is constant in size in different variants but once again their relative size differences remain constant in different variants with all enzymes yielding two bands (see Figures 9 and 10). However, clones X and Y (S427 series) yield only one band with this group of enzymes (see Figure 10). Hybridization studies performed with 5' and 3' restriction fragment probes

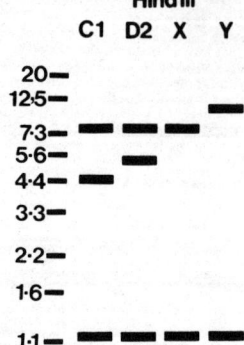

Figure 7. A schematic representation of the Iltat 1.2 genes revealed with the restriction enzyme HindIII. HindIII digested DNAs were fractionated on 0.6% agarose gels and transferred to nitrocellulose filters according to Southern (60). Hybridizations to nick translated ^{32}P-labeled pCB1352 (61) were performed according to Wahl et al. (62). Filters were washed twice for 1 hr in 0.1xSSC, 0.1% SDS at 60°C and autoradiography was performed at -80°C with an intensifier screen. Clones C1 and D2 are of the Iltat serodeme and clones X and Y were derived from trypanosome stock S427. Sizes of molecular weight markers in kilobases are indicated.

(prepared from the pcB1352 insert as shown in Figure 6) indicate that: (a) two Iltat 1.2 genes are present in all cloned variants of the Iltat serodeme, and (b) 3' flanking sequence length variation occurs for both gene copies (see Figure 11) (44). As shown in Figure 11, a low molecular weight band of high intensity exclusively hybridizes to 5' specific restriction fragments while two higher molecular weight bands of weaker intensity hybridize to 3' specific probes. Differential DNA methylation does not appear to be significant since isoschizomers like MboI and Sau3A appear to yield identical patterns (see Figure 12). The presence of single bands of high intensity may indicate that both genes are comigrating. This is evident for clones C2, C3 and C4 in Figure 12 and for clone D1 in Figure 10. Flanking sequence length variants occur in independent clones which express the same antigen (D1 vs. D2 and C1 vs. C2, C3, C4). B1 appears to be a pure

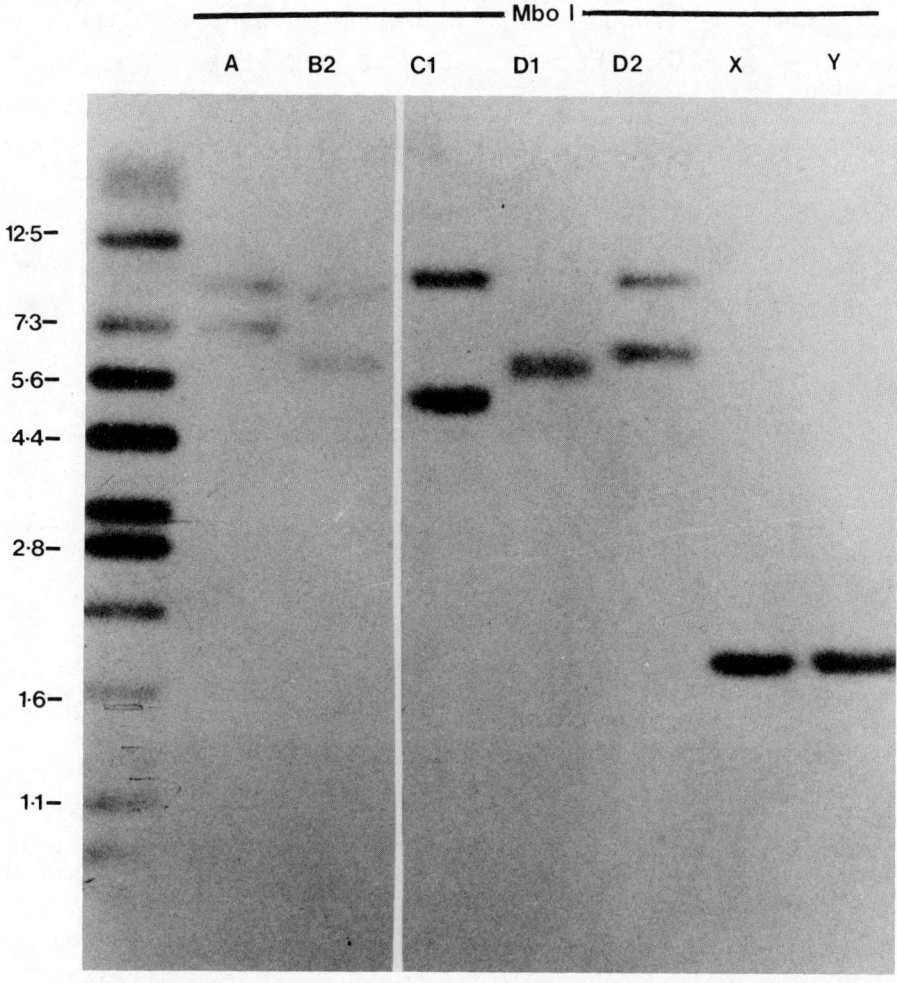

Figure 8. Results of a Southern transfer of MboI digested DNAs hybridized to the pcB1352 probe. The positions of radioactive marker bands are indicated adjacent to variant A. An additional hybridizing band of about 0.3 kilobase pairs has run off this particular gel but appears in Figure 12.

population with respect to VSG production but possesses several additional length variants of weaker intensity in comparison to clone B2 (Figure 10). In conclusion, comparisons performed between cloned variants that do or do not express the I1tat 1.2 antigen do not as yet reveal any correlation between DNA sequence

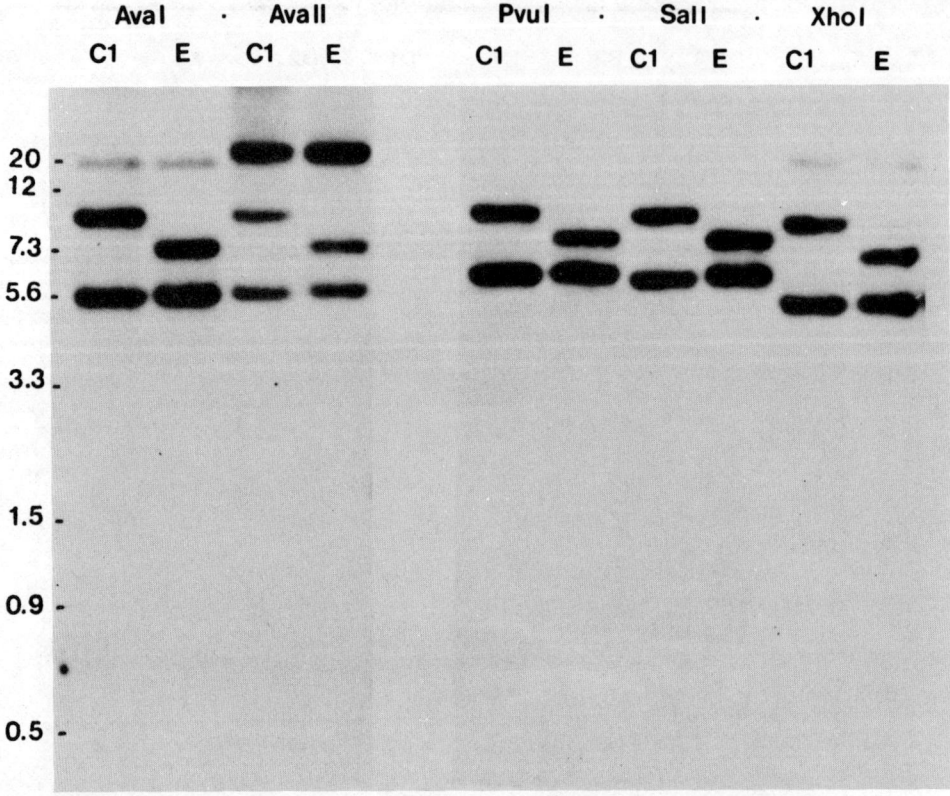

Figure 9. Results of a Southern transfer of AvaI, AvaII, PvuI, SalI and XhoI digested DNAs of variants C1 and E hybridized to the pcB1352 probe.

rearrangements and the expression of this particular VSG. Due to the lack of restriction sites, accurate mapping studies beyond a 2.1 kilobase limit on the 5' side of this gene will await the isolation of genomic clones.

Preliminary results obtained with cDNAs prepared for other VSGs of the Iltat serodeme are only somewhat analogous to the findings with Iltat 1.2. Hybridization studies performed with pcB Iltat 1.3:12 (see Figure 4 for restriction map) indicate the existence of 4 to 6 bands in different cell clones with 4 of these being variable in size in all clones. These results are comparable to the pcB1352 data for Iltat 1.2. However, pcB Iltat 1.1:7 and pcB Iltat 1.4:A hybridize to 4 and 3 bands, respectively, which so far appear to be constant in size in all cloned

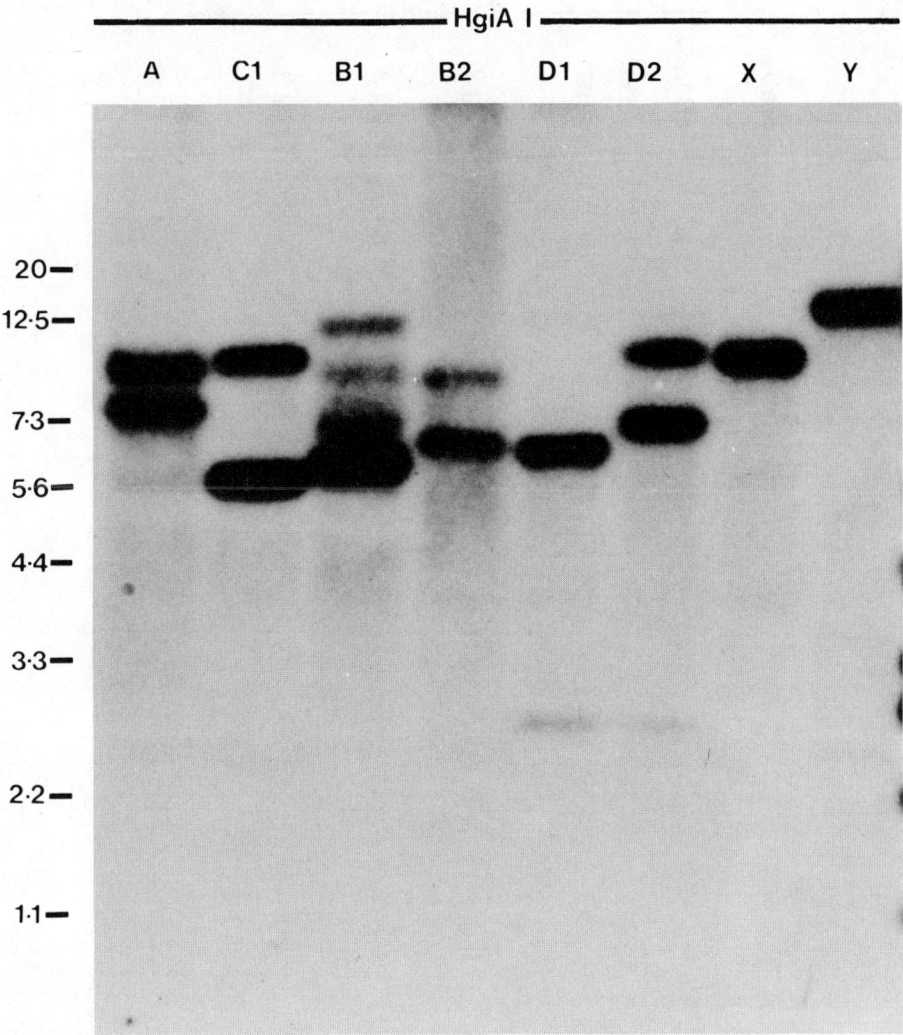

Figure 10. The pattern of Iltat 1.2 like sequences observed with Iltat serodeme variants A, C1, B1, B2, D1 and D2 and S427 derived clones X and Y with restriction enzyme HgiAI.

variants of the Iltat serodeme. These results were obtained by digestions with about 15 different restriction enzymes and the size range of the bands observed are 4 to 12 kilobase pairs. None of the genomic bands observed with these four noncross-hybridizing cDNA clones appears to comigrate, suggesting that these genes are a considerable distance apart in the genome.

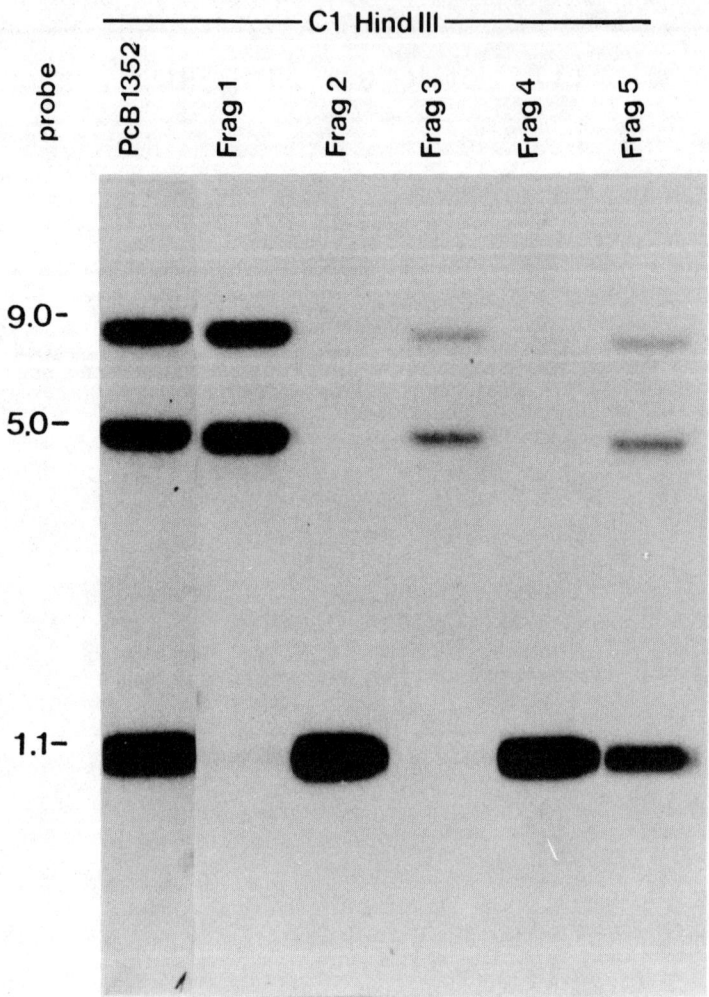

Figure 11. The identification of the genomic DNA fragments containing the 5' and 3' ends of the pcB1352 cDNA clones. Variant C DNA digested with HindIII was hybridized to whole pcB1352 and 5 restriction fragments specifying different portions of the pcB1352 insert (see Figure 6 for their locations).

Using a series of antigenically-unrelated VSGs derived from a monomorphic stock of trypanosomes (S427) originally characterized by Cross (6,9), Borst and his colleagues have suggested that the appearance of novel "expression linked copies" (ELC) of VSG genes are associated with the molecular events responsible for

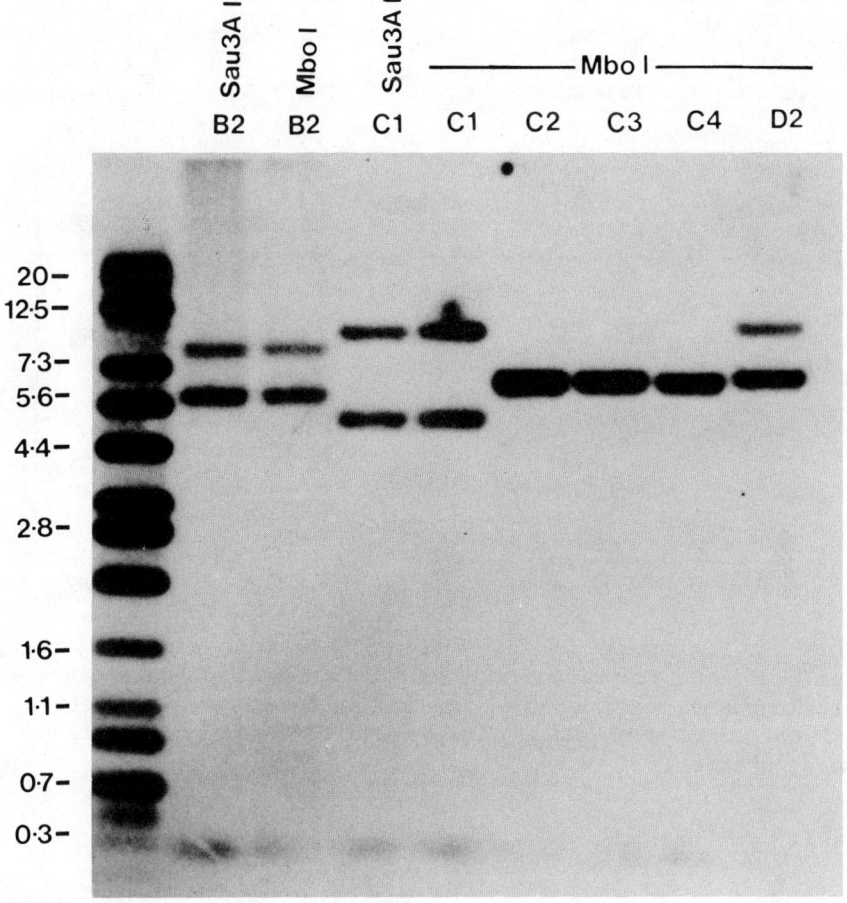

Figure 12. Results of a Southern transfer of Sau3A and MboI digested genomic DNAs of variants C1, C2, C3, C4, B2 and D2 hybridized to the pcB1352 probe.

the expression of VSGs (36,48). A schematic representation of these findings is shown in Figure 13. Results with two out of four nonhomologous VSG cDNA clones indicate that a new copy of the expressed VSG gene is present in the expressing clone of trypanosomes. This new copy is generally of higher molecular weight than several remaining homologous VSG gene copies. The unaltered homologous and presumably identical VSG genes are referred to here as basic library copies. Fine structure restriction enzyme mapping has recently confirmed that one of the ELCs (for VSG 118) appears to represent an intact VSG gene with a new 5' and 3'

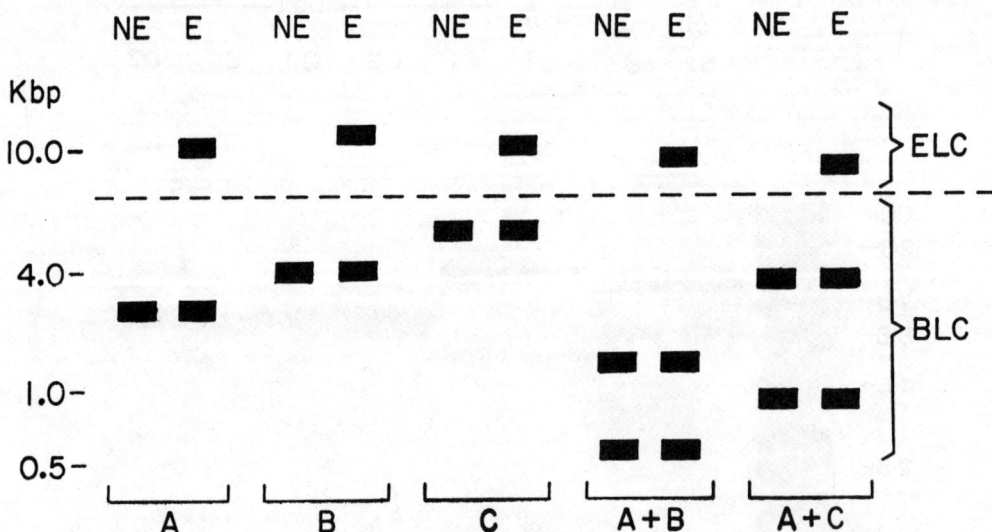

Figure 13. A schematic representation of ELCs and BLCs of VSG genes. NE and E represent a nonexpressing heterologous clone and a homologous expressing clone of trypanosomes, respectively. A, B, C and their combinations represent different restriction enzyme digestions.

flanking sequence environment compared to the basic copy of this gene. Other data of a more preliminary nature indicate that the presumptive ELCs for VSG 117 and VSG 121 (S427 stock) are not localized in similar DNA sequences (48). In addition, a VSG 121 cDNA clone has unexpectedly been observed to detect a new high molecular weight copy of this VSG gene in a clone of VSG 221 expressing trypanosomes. Due to the fact that the clone of trypanosomes expressing VSG 121 appeared prior to the 221 variant in a chronically infected rabbit (6), it is currently thought that this extra copy of the VSG 121 gene in 221 nuclear DNA represents an inactivated form of the previously expressed 121 gene (36, 48). However, since the cloned variants expressing noncrossreacting 117, 118, 121 and 221 VSG were derived by sampling a single chronically infected rabbit over a 43-day period, the periodicity of the parasitemias and the nature of the sequence of appearance of these variants actually remains unknown (6). Results with the 221 VSG cDNA clone are a bit more complicated to interpret due to the loss of specific bands in the 221-expressing clone and restriction pattern differences in heterologous clones (48). These authors have tentatively concluded that ELCs represent a pathway for the appearance of new VSGs (36,48). ELCs may arise by gene duplication followed by their insertion into a transcriptionally competent region of DNA or may appear following the insertion of a portable RNA polymerase promoter site (36). The latter possibility would appear less likely if ELC flanking sequences are nonhomologous. Agabian, Stuart and their colleagues (33) have also recently observed the appearance of an ELC for a VSG (referred to as Antigen D) of another unrelated trypanosome stock (ISTAT serodeme derived by Dr. Ken Stuart from EATRO 164), while another VSG (referred to as Antigen A) in the same series of antigenically distinct clones did not appear altered in genomic context. Preliminary results with Antigen D-expressing clones derived from relapses in Peromyscus as opposed to the rat indicate other alterations in the context of the Antigen D type genes (2 to 3 copies) complicating the assignment of the putative ELC (49). Recombinant DNA studies have recently been undertaken on a T. equiperdum serodeme (50) which consists of 101 antigenically distinct variants from a chronic rabbit infection (11) and it is hoped that this work will add much needed information on the capacity of a particular species of trypanosomes for VSG expression in this host.

Borst and his colleagues (48) have also observed that the number of genes detectable with several of their VSG cDNA clones can vary dramatically depending on the region of the cDNA used as a probe (i.e., 5' or 3' coding sequence) and the stringency of the hybridization protocol. Lower stringency filter washes performed with 118 cDNA blots produce a complex banding pattern which is only visible with a 3'-specific 118 probe. Results with a 3'-half probe for the 117 cDNA indicate the existence of 20 to 30 bands under standard conditions (3xSSC, 65°C) which are not visible with a 5'-half probe. In addition, the 3'-half probe 117 positive bands do not appear to be limited to the carboxy-terminal 29 amino acids or the 3'noncoding sequence of the mRNA. The 3'homologous 221 gene family is the most complex and probably represents at least 50 bands under low stringency conditions. Studies performed with other trypanosome strains (T. brucei 31, Lump 127 and EATRO 1125) and species (T. equiperdum) indicate that many of the multiple 3'-half positive bands are conserved with considerably less conservation of the 5'-halves. Borst et al. have concluded from these interesting studies that the rates of evolution of the 5' and 3' halves of the VSG genes may occur at different rates and these genes probably arose long before the appearance of different trypanosome species. However, it remains unclear whether all the 3'-half positive sequences in the

Table 2

Classifications of Iltat 1.3 Like Genomic Clones

Iltat 1.3 like genomic clones isolated from partial EcoRI Charon 4A genomic libraries (63,64) of the Iltat 1.2, 1.3 and 1.4 genomes. These comparisons are made on the basis of the sizes of EcoRI restriction fragments within the genomic clones and their hybridization responses to the Iltat 1.3 cDNA clone.

	Trypanosome DNA Sources			
	Iltat 1.3	Iltat 1.2	Iltat 1.4	Hybridization[a] Response
All like λG1	λG1	λG10	λG14	(+)
	λG2[b]	λG11		(+)
	λG6			(+)
Like λG12		λG12		(+)
All like λG17			λG17	(±)
	λG7[b]			(±)
			λG18[b]	(±)

[a] (+) = strong signal; (±) = weak signal

[b] Missing one EcoRI fragment.

genome actually possess more divergent 5' sequences. Are these 3' half sequences the vestiges or footprints of genes that have long since split the scene? More extensive studies of the appropriate genomic clones will presumably yield solid answers for these interesting findings.

PRELIMINARY FINDINGS WITH VSG GENOMIC CLONES

Borst and his colleagues (48) recently reported their preliminary findings on a cloned 8.2 kilobase genomic DNA fragment containing a VSG 117 basic copy gene. R-loop analysis indicates that no introns appear to be present within the size of the hybrid (i.e., about 1600 base pairs). This is about 600 nucleotides smaller than the predicted size of the 117 mRNA and a 3' single-stranded RNA tail of about 350 nucleotides was also observed. The 3' tail may represent a poly(A) stretch. Considering the overall size discrepancies, a more interesting possibility would be that this gene is not the real 117 gene but a highly homologous sequence with a dissimilar 3' terminus. This gene was isolated from a heterologous clone of trypanosomes expressing VSG 118. No other cDNA probes were observed to hybridize to this VSG 117-like gene (48).

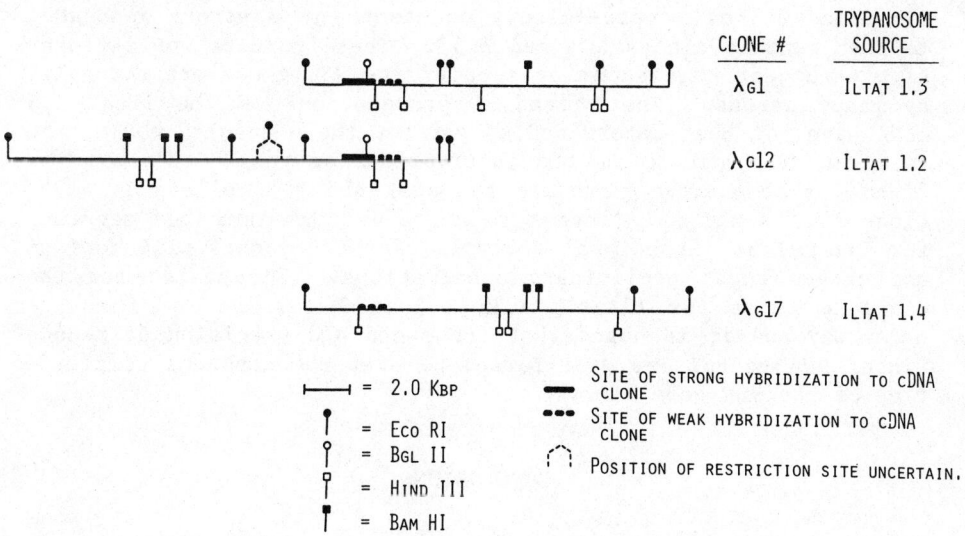

Figure 14. Comparative anatomy of three genomic clones containing Iltat 1.3 homologous sequences.

A description of the hybridization properties of a group of genomic clones isolated from partial EcoRI Charon 4A phage libraries of the Iltat 1.2, 1.3 and 1.4 genomes with an Iltat 1.3 cDNA clone (described in Figure 3 as Iltat 1.3:12) is presented in Table 2. Restriction enzyme maps of three of these Iltat 1.3 like genes are shown in Figure 14. Clones λG1 and λG12 were obtained from the genomes of different cloned variants but may in fact represent two large overlapping regions of DNA due to a conserved sequence of about 6 kilobases which hybridizes to the cDNA clone. If these cloned DNA fragments are indeed overlapping in the genome, they span a combined distance of about 30 kilobase pairs and our three remaining nonhomologous cDNA clones (see Figure 4) do not hybridize to these sequences. These preliminary results along with those of Borst (48) would collectively argue that at least some noncrossreacting VSG genes are not tightly linked in the trypanosome genome.

In addition, genomic clones λG1, λG12 and λG17 do not appear to contain a true Iltat 1.3 gene. In the case of λG1 and λG12, a strong hybridization response is observed over about 1.2 kilobases and several restriction sites appear to be in common with the Iltat 1.3 cDNA clone. However, the remainder of the cDNA sequence, which consists of about 500 base pairs, only weakly hybridizes (i.e., approximately one-tenth of a strong response) to both genomic clones λG1 and λG12. These results would tentatively suggest that at least some of the VSG genes are not split by many introns. The 5' and 3' orientations of the Iltat 1.3 cDNA have not been determined as yet but these results would seem to agree with and extend the observations of the VSG 5'-half vs. 3'-half probe blots performed by Borst and his colleagues (48). Clone λG17 would only appear to contain a low homology sequence for Iltat 1.3. Structural analyses of these clones will confirm and extend these preliminary observations. The search for the complete genes for Iltat's 1.1, 1.2, 1.3 and 1.4 is of course under way and it is hoped that the speed and precision of recombinant DNA techniques will persevere over the apparent complexities of the VSG gene story.

CONCLUSIONS

We have attempted to present an accurate account of the present state of the VSG gene story. The regulation of VSG expression would appear to reside at the transcriptional level. However, the appearance of "expression linked copies" (36,48) for some VSGs and the frequent occurrence of flanking DNA sequence rearrangements for other VSGs (29,44) would suggest that novel mechanisms possibly associated with gene mobility are at work here. Unfortunately, a clear picture of the absolute requirements for VSG expression has not emerged out of the apparent complexity of the VSG gene repertoire. Analogies to the casette model originally proposed for the expression of yeast mating types (51) have been suggested in regard to the presence of the ELC (36) but preliminary data on the structures of different ELCs indicate that this may be an overly simplistic interpretation of their significance (48). In another vein, a recent study on the organization of the α- and β-tubulin genes of T. brucei has demonstrated that the genomic context of this gene family is constant in different VSG variants (33). These important findings would suggest that the tendency for VSG gene rearrangements may be a property unique to this large gene family and not reflective of the trypanosome genome in general. Studies directed towards defining the initial transcripts of VSG genes (i.e., in heterogeneous nuclear RNA) would potentially open up other avenues for VSG regulation. Assuming that VSG gene flanking sequences are unique in structure (as our current data would suggest), the identification of high molecular weight initial transcription units of VSG genes may provide an assay for functional VSG genomic sequences and should be vigorously investigated in the future.

The appearance and loss of specific VSG genes appears to be quite common when the genomic content of VSGs expressed by genetically unrelated stocks or strains of trypanosomes is compared

(29,44,48). One explanation for some of these findings is that they may simply be due to variability in genome size. The ploidy of trypanosomes has often been disputed and has been stated to range from anywhere between haploid and tetraploid (36,48,52). However, the recent work of Tait (53), in which a number of specific enzyme polymorphisms were analyzed in a wild population, would suggest that the T. brucei genome is diploid with a high probability for genetic exchange. Our findings that all variants of the Iltat serodeme appear to possess two copies of the Iltat 1.2 gene would agree with these results. Further analyses of the type described by Tait (53) should be performed on cloned VSG variants that possess ELCs. In addition, an investigation of the genomic content and context of VSGs in procyclic forms (i.e., coatless, noninfectious forms) of trypanosomes would be most informative.

Our most recent findings now indicate that akin to the Iltat 1.2 and 1.3 genes, the Iltat 1.4 genes also appear to possess some type of 3' flanking sequence length variation (N.A. Penncavage, personal communication). Studies designed to assess the homology of this length variation region found adjacent to nonhomologous VSGs are currently under way.

If one considers the existence of a common 3' flanking sequence that contains numerous tandemly repetitious DNA sequences, such an intrinsically unstable region may promote VSG movement via some type of homologous recombination process. This may occasionally result in VSG movement to transcriptionally competent sites on a random or probabilistic basis. Such a model for VSG expression would satisfy the apparent lack of a true sequence for VSG appearance and also avoids the difficulties experienced with the current ELC controversy. Indeed, ELCs may occasionally arise via this genetic recombination phenomenon but may just as likely constitute inactive or active forms of a VSG gene. Interestingly (as a final note of comparison), small repetitive DNA sequences which undergo length variation in mammalian cells, and are also unstable during their propagation in a variety of cloning vehicles, have been implicated in a presumably novel "switch-recombination" process associated with the expression of heavy chain immunoglobulin constant region genes (64).

The direct benefits of determining the genetic basis of VSG expression for the treatment and/or prevention of African Trypanosomiasis remain unknown. We imagine, however, that the VSG represents a highly evolved form of camouflage for the parasite that has succeeded in adapting to the immune surveillance of the animal host by expanding the diversity of its genetic repertoire. The use of bacteria programmed with VSG cDNA clones as a possible source of large quantitites of VSGs for a host immunization regime would seem to be an ineffective approach considering the extent of VSG diversity. Predominant antigens of certain laboratory strains could be isolated with these techniques but there would still be no assurances for immunity of hosts continually subjected to wild trypanosome populations. This approach could be attempted with predominant metacyclic antigens that have a high probability of appearance in initial infections (1). However, sterile immunity would still not be attainable and the diversity of the metacyclic antigen may also present difficulties (1). A complete definition of the target of the trypanocidal activities of high density lipoproteins (isolated from the sera of different animal hosts) known to create host specific barriers against infection by particular species of trypanosomes (54,55) should also be pursued at the genetic level, if at all possible. The possibility of coordinate expression of other gene products associated with VSG production must also be considered and should be investigated throughout the differentiation process (i.e., procyclic to metacyclic to infective bloodstream forms to stumpy forms). Certainly, a thorough understanding of the molecular events associated with VSG production will be essential for the success of any of these scenarios. The trypanosome has had millions of years of evolution to become a smart parasite, but the genetic dissection of its complex life style has just seriously begun.

Acknowledgments: We would like to express our sincere gratitude to our colleagues and collaborators: J. Young, S. Shapiro, J. Doyle, P. Majiwa, J. Donelson, P. Englund, N. Penncavage and the staff of ILRAD, Nairobi for their contributions to various aspects of this work. We would also like to thank N. Agabian, K. Stuart, P. Borst, G.A.M. Cross, J.H.J. Hoeijmakers and J.C. Boothroyd for providing us with preprints of manuscripts prior to their publication. The understanding and good will of Dr. Jane Setlow in allowing us a generous extension of time for the assembly of the materials in this manuscript is also very much appreciated. KBM would also like to thank Lillian Geist for her undaunting patience in the preparation of this chapter.

REFERENCES

1. Vickerman, K. (1978) Nature 273, 613-617.
2. Seed, J.R. (1964) Parasitology 54, 593-596.
3. Barry, D. (1977) J. Protozool. 24, 5A.
4. Gray, A.R. (1965) J. Gen. Microbiol. 41, 195-214.
5. Vickerman, K. and Luckins, A.G. (1969) Nature 224, 1125-1126.
6. Cross, G.A.M. (1975) Parasitology 71, 398-417.
7. Doyle, J.J. (1977) in Immunity to Blood Parasites in Animals and May (Miller, L., Pino, J. and McKelvey, J., eds.), pp. 27-63, Plenum Press, New York, NY.
8. Ross, R. and Thomson, D. (1910) Proc. R. Soc. Lond. B82, 411-415.

9   Cross, G.A.M. and Manning, J.C. (1973) Parasitology 67, 315-331.
10 vanMeirvenne, N., Janssens, P.G., Magnus, E., Lumsden, W.H.R. and Herbert, W.J. (1975) Ann. Soc. Belge Méd. Trop. 55, 1-23.
11 Capbern, A., Giroud, C., Baltz, T. and Mattern, P. (1977) Exp. Parasitol. 42, 6-13.
12 LeRay, D., Barry, J.D. and Vickerman, K. (1978) Nature 273, 300-302.
13 Doyle, J.J., Hirumi, H., Hirumi, K., Lupton, E.N. and Cross, G.A.M. (1980) Parasitology 80, 359-369.
14 Vickerman, K. (1969) J. Cell Science 5, 163-194.
15 Allsopp, B.A., Njogu, A.R. and Humphryes, K.C. (1971) Exp. Parasitol. 29, 271-284.
16 Allsopp, B.A. and Njogu, A.R. (1974) Parasitology 69, 271-281.
17 Bridgen, P.J., Cross, G.A.M. and Bridgen, J. (1976) Nature 263, 613-614.
18 Johnson, J.G. and Cross, G.A.M. (1979) Biochem. J. 178, 689-697.
19 Cross, G.A.M. (1977) Am. J. Trop. Med. Hyg. 26, 240-244.
20 Cross, G.A.M. (1979) Nature 277, 310-312.
21 Holder, A.A. and Cross, G.A.M. (1980) Mol. Biochem. Parasitol. (in press).
22 Barbet, A.F. and McGuire, T.C. (1978) Proc. Nat. Acad. Sci. U.S.A. 75, 1989-1993.
23 vanMeirvenne, N., Magnus, E. and Vervoort, T. (1977) Ann. Soc. Belge Méd. Trop. 57, 409-423.
24 Vervoort, T. et al. (1980) submitted for publication.
25 Barbet, A.F. (personal communication).
26 Eggitt, M.J., Tappenden, L. and Brown, K.N. (1977) Parasitology 75, 133-141.
27 Williams, R.O., Marcu, K.B., Young, J.R., Rovis, L. and Williams, S.C. (1978) Nucl. Acids Res. 5, 3171-3182.
28 Lheureux, M., Lheureux, M., Vervoort, T., vanMeirvenne, M. and Steinert, M. (1979) Nucl. Acids Res. 7, 595-609.
29 Williams, R.O., Young, J.R. and Majiwa, P.A.O. (1979) Nature 282, 847-849.
30 Hoeijmakers, J.H.J., Borst, P., van den Burg, J., Weissmann, C. and Cross, G.A.M. (1980) Gene 8, 391-417.
31 Merritt, S.C. (1980) Mol. Biochem. Parasitol. 1, 151-166.
32 Shapiro, S.Z. and Young, J.R. (1980) J. Biol. Chem. (in press).
33 Agabian, N., Sibley, L., Milhausen, M. and Stuart, K. (1980) Am. J. Trop. Med. Hyg. (in press).
34 Paterson, B.M., Roberts, B.E. and Kuff, E.L. (1977) Proc. Nat. Acad. Sci. U.S.A. 74, 4370-4374.
35 Alwine, J.C., Kemp, D.J. and Stark, G.R. (1977) Proc. Nat. Acad. Sci. U.S.A. 74, 5350-5354.

36 Hoeijmakers, J.H.J., Frasch, A.C.C., Bernards, A., Borst, P. and Cross, G.A.M. (1980) Nature 284, 78–80.
37 Chen, K., Donelson, J.E. and Williams, R.O. (unpublished data).
38 Roelantz, G. (personal communication).
39 Young, J.R. (personal communication).
40 Boothroyd, J.C., Cross, G.A.M., Hoeijmakers, J.H.J. and Borst, P. (1980) Nature (in press).
41 Davis, B.D. and Tai, P.-C. (1980) Nature 283, 433–438.
42 Boothroyd, J.C. and Allen, G. (unpublished data).
43 Shapiro, S.Z. and Doyle, J.J. (unpublished data).
44 Williams, R.O., Young, J.R., Majiwa, P.A.O., Doyle, J. and Shapiro, S.Z. (1980) Am. J. Trop. Med. Hyg. (in press).
45 Shapiro, S.Z. (personal communication).
46 Doyle, J.J., Behin, R., Mauel, J. and Rowe, D.S. (1975) Ann. N.Y. Acad. Sci. 254, 315–325.
47 Doyle, J.J. (personal communication).
48 Borst, P., Frasch, A.C.C., Bernards, A., vand der Ploeg, L.H.T., Hoeijmakers, J.H.J., Annberg, A.C. and Cross, G.A.M. (1981) Cold Spring Harbor Symp. Quant. Biol. 45 (in press).
49 Agabian, N. (personal communication).
50 Longacre, S., Hibner, U., Baltz, T. and Eisen, H. (unpublished data).
51 Hicks, J., Strathern, J.N. and Klar, A.J.S. (1979) Nature 282, 478–483.
52 Borst, P., Fase-Fouler, F., Frasch, A.C.C., Hoeijmakers, J.H.J. and Weijers, P.J. (1980) Mol. Biochem. Parasitol. 1 (in press).
53 Tait, A. (1980) Nature 287, 536–538.
54 Targett, G.A.T. and Wilson, V.C.L.C. (1973) Intern. J. Parasitol. 3, 5–11.
55 Rifkin, M.R. (1978) Proc. Nat. Acad. Sci. U.S.A. 75, 3450–3454.
56 Pelham, H.R.B. and Jackson, R.J. (1976) Eur. J. Biochem. 67, 247–256.
57 Wickens, M.P., Buell, G.N. and Schimke, R.T. (1978) J. Biol. Chem. 253, 2483–2495.
58 Lobian, P.E. and Kaiser, A.D. (1973) J. Mol. Biol. 78, 453–471.
59 Grunstein, M. and Hogness, D. (1975) Proc. Nat. Acad. Sci. U.S.A. 3961–3965.
60 Southern, E.M. (1975) J. Mol. Biol. 98, 503–517.
61 Rigby, P.W.J., Diekmann, M., Rhodes, C. and Berg, P. (1977) J. Mol. Biol. 113, 237–251.
62 Wahl, C.M., Stern, M. and Stark, G.R. (1979) Proc. Nat. Acad. Sci. U.S.A. 76, 3683–3687.
63 Blattner, F.R., Blechl, A.E., Thompson, D.K., Faber, H.E., Richards, J.E., Slightom, J.L., Tucker, P.W. and Smithies, O. (1978) Science 202, 1279–1283.

64 Marcu, K.B., Banjeri, J., Penncavage, N.A., Lang, R. and Arnheim, N. (1980) Cell 22, 187-196.

MOUSE IMMUNOGLOBULIN GENES

P. Early and L. Hood

Division of Biology
California Institute of Technology
Pasadena, California 91125

During the past few years, rapid progress has been made in characterizing the genes encoding immunoglobulin molecules. Major aspects of the organization and rearrangements of immunoglobulin genes in antibody-producing (B) cells are now known. Almost all work in this field has used inbred strains of mice, particularly BALB/c, although very recent studies have yielded similar results for human immunoglobulin genes. As with much of eukaryotic molecular biology, the study of immunoglobulin genes has depended heavily on what may now be regarded as classical techniques of recombinant DNA, or genetic engineering.

The structure of a typical immunoglobulin protein is shown in Figure 1. It contains two identical light chains and two identical heavy chains. Each polypeptide chain contains a variable (V) and a constant (C) region. The antigen-binding sites are formed by a combination of one light chain V region and one heavy chain V region. Thus, the antibody molecule is divalent for antigen binding. As the name suggests, there are a large number of alternative variable regions in both the light and heavy chains of different immunoglobulin molecules. Many of the differences between V regions are concentrated in three hypervariable regions, which contain amino acid residues directly involved in antigen contact (Figure 1). Protein sequence analysis and X-ray crystallography have shown that immunoglobulin molecules are composed of several discrete structural domains, each containing about 110 amino acid residues. These domains each appear to have specific functions: for example, V regions form the antigen-binding domains. The presence of significant homologies between domains suggests that they evolved by duplication and divergence of a common ancestral homology unit encoding about 110 amino acid residues. Immunoglobulin molecules can be categorized

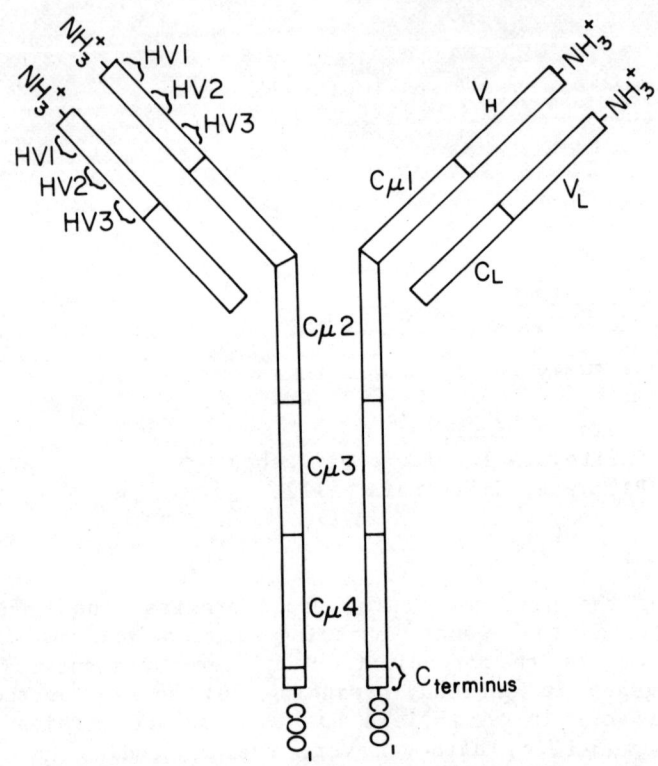

Figure 1. Structure of a typical immunoglobulin monomer from secreted IgM. The heavy (H) and light (L) chain polypeptides are held together by disulfide bonds and noncovalent interactions. The μ heavy chain contains four constant region domains ($C\mu$ 1-4) plus a short carboxy-terminal segment ($C_{terminus}$). The light chain contains only a single constant region domain. Antigen is bound by each pair of heavy and light chain variable regions (V_H and V_L). The V_H and V_L regions each contain three hypervariable regions (HV1-3) that form the walls of the antigen-binding site.

by class or subclass, depending on which of eight (in mice) possible heavy chain constant regions they contain (μ, δ, $\gamma 3$, $\gamma 1$, $\gamma 2b$, $\gamma 2a$, α, ϵ). An immunoglobulin with a μ heavy chain belongs to the IgM class, one with a $\gamma 1$ heavy chain to the IgG1 subclass, and so forth. An immunoglobulin molecule also contains one of two types of light chains, κ or λ, which have different constant regions and different sets of variable regions. In response to bound antigen, immunoglobulin molecules with different heavy chain constant regions perform different effector functions, such as complement fixation or the triggering of mast cells to release histamine and other pharmacologically active compounds.

The questions addressed by molecular biologists studying immunoglobulin genes have to a large extent been those formulated earlier when immunologists were limited to the techniques of serology, protein chemistry or classical genetics. Probably the chief stimulus to the study of immunoglobulin genes was the model proposed by Dreyer and Bennett in 1965 (1) to account for the presence of distinct V and C regions in immunoglobulin polypeptides. They hypothesized that immunoglobulins are encoded by separate V and C genes, with a single C gene being joined to one of many possible V genes in an antibody-producing cell. This model prompted an early series of experiments to enumerate V and C genes with immunoglobulin mRNA probes and eventually was confirmed by Tonegawa and his co-workers using cloned λ light chain genes (2).

Other intriguing phenomena of immunoglobulins have also been targets for molecular biologists. These include the origins of the vast repertoire of diversity in the V regions -- the immune system may be capable of synthesizing 10^7 or more different antibody molecules (3). Germline hypotheses proposed that this diversity is encoded by genes present in germline DNA, while somatic mutation hypotheses suggested that genes are generated anew in each individual during the course of B-cell differentiation (4).

Immunoglobulins are produced by lymphocytes of the B-cell lineage, derived from bone marrow (Figure 2). Immunoglobulin polypeptides are unusual in that they are produced from only one of two homologous chromosomes in any given B cell (5,6). This is the phenomenon of allelic exclusion, which is at least superficially similar to the expression of only one allele of an X chromosome-linked gene in a given mammalian cell.

Lymphocytes initially synthesize IgM molecules (primary immune response) but later they or their progeny can switch to the synthesis of another class of immunoglobulin molecules, IgG for example (secondary immune response) (Figure 2). The V domain (antigen specificity) and the light chain are unaltered by class switching but a new heavy chain C region replaces the C_μ region in the immunoglobulin molecule (7). As we will discuss subsequently, class switching is also mediated by DNA rearrangement.

Immunoglobulins are produced in two alternative forms by B cells: either as membrane-bound cell-surface receptors to be triggered by antigen or as secreted antibodies (Figure 2). Cell-surface immunoglobulins are predominantly IgM and IgD while serum-secreted antibodies are mostly IgG and IgM. Most or all classes of immunoglobulin exist in both forms with the carboxy-terminal portions of the heavy chains either attached to the hydrophobic environment of the cell membrane or free in the hydrophilic extracellular environment. Controversy arose from experi-

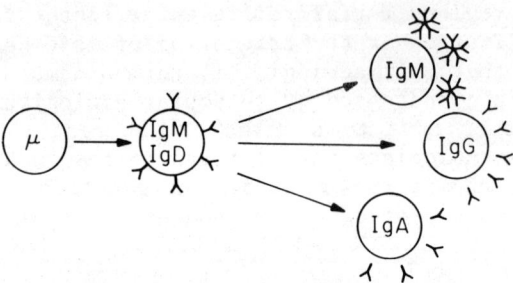

Figure 2. B-cell differentiation. Cells initially committed to immunoglobulin synthesis first express variable regions on membrane-bound IgM (and later IgD also, in most cases). After stimulation by antigen, these cells or their progeny can secrete immunoglobulins with the same variable region, and any one of the eight classes or subclasses of heavy chain. Secreted IgM molecules are pentamers.

ments to determine whether membrane-bound and secreted immunoglobulins differ in amino acid sequence or only in the carbohydrates or other secondary modification of the polypeptide chain (8,9). Recent experiments have shown that the heavy chains of secreted and membrane-bound IgM differ in the carboxy-terminal portions of their amino acid sequences (10,11). The two forms of μ heavy chain are translated from two different mRNAs that are both transcribed from the same gene as the result of developmentally regulated alternative pathways of RNA processing (10,12). This is the first example of such a mechanism of gene regulation in eukaryotic cells.

V GENE REARRANGEMENT

Heavy chains and κ and λ light chains are encoded by three unlinked families of genes as demonstrated by classical genetic studies in humans, mice and rabbits (13,14). More recently, the heavy chain genes have been localized to chromosome 12 and the κ light chain genes to chromosome 6 in mice (15-17). In the heavy chain gene family, markers for V and C regions have recombination frequencies of 0.4 to 7% suggesting that V and C genes may be separated by as much as 10^6 base pairs in germline DNA (18). Our current understanding of immunoglobulin gene organization in mice is depicted in Figure 3. The λ chain gene family contains two known V genes ($V_{\lambda I}$ and $V_{\lambda II}$) and two known C genes ($C_{\lambda I}$ and $C_{\lambda II}$). $V_{\lambda I}$ is expressed with $C_{\lambda I}$ and $V_{\lambda II}$ with $C_{\lambda II}$. The κ gene family contains multiple V genes,

Figure 3. Organization of immunoglobulin genes in BALB/c mice. Each gene family is located on a separate chromosome. The lambda gene family contains at least two units (λ_I and λ_{II}) organized as shown. The probable order of the heavy chain constant region genes is shown beneath the diagram of variable region gene segments. Each C_H gene contains multiple exons, as depicted for $C\mu$. Slash marks indicate linkage relationships that have not been firmly established. Distances are not to scale. (From ref. 56.) Recent work has shown that the C_ϵ gene is located between the $C\gamma 2a$ and $C\alpha$ genes (127).

any one of which can be expressed with the single C_K gene. The heavy chain gene family has both multiple V genes and multiple (eight) C_H genes.

The first immunoglobulin gene families to be studied were those encoding light chains. This was partly for technical reasons -- light chain mRNA was initially easier to isolate than heavy chain mRNA (19-21) -- but light chain gene structure has also proven to be simpler than that of the heavy chains. The usual sources of immunoglobulin mRNAs and rearranged immunoglobulin genes have been the many transplantable plasma cell tumors (myelomas) propagated in inbred BALB/c mice or NZB mice (22). These myeloma tumors also serve as a source of immunoglobulins for protein sequence studies, so it is possible to isolate mRNAs (and genes) encoding known immunoglobulin chains. The disadvantage of myeloma tumors is that they are generally subtetraploid and may contain chromosome abnormalities that can complicate the interpretation of some gene rearrangements. For this reason, some recent studies of immunoglobulin genes have used normal B cells obtained from mouse spleens or lymph nodes by the fluorescence-activated cell sorter (23; C. Nottenburg and I. Weissman, personal communication).

One important early result obtained by molecular biologists was the demonstration that a single species of mRNA encodes both the V and C regions of an immunoglobulin chain (24,25). This result eliminated the possibility that V and C regions are joined post-translationally and focused attention on the organization of V and C genes in germline (undifferentiated) and myeloma (differentiated) DNAs.

The first evidence for immunoglobulin gene rearrangements in myeloma tumors was found before the advent of the Southern blot technique by using preparative agarose gel electrophoresis to obtain size fractions of mouse embryo (assumed to be equivalent to germline) and myeloma DNAs that had been digested with BamHI or other restriction enzymes (26-28). These DNA fractions were assayed for hybridization to radiolabeled probes containing either V_K+C_K or only C_K sequences derived from κ mRNA purified from M321 myeloma tumors. In embryo DNA or other tissues not producing immunoglobulins, V_K and C_K genes were detected in two separate peaks in the DNA fractions, while in M321 myeloma DNA a single new peak containing both V_K and C_K genes was observed. A V_K gene unrelated to the M321 V_K gene was not rearranged in M321 DNA nor were $V_{\lambda I}$ or $C_{\lambda I}$ genes, indicating that DNA rearrangement is confined to the genes expressed in this myeloma tumor. Evidence for DNA rearraangement was also obtained for $V_{\lambda I}$ and $C_{\lambda I}$ genes in a λI-producing myeloma tumor (27,28). These data offered direct support for the Dreyer and Bennett hypothesis of V and C gene rearrangement during B-cell differentiation (1). However, elucidating the details of this gene rearrangement required use of recombinant DNA techniques.

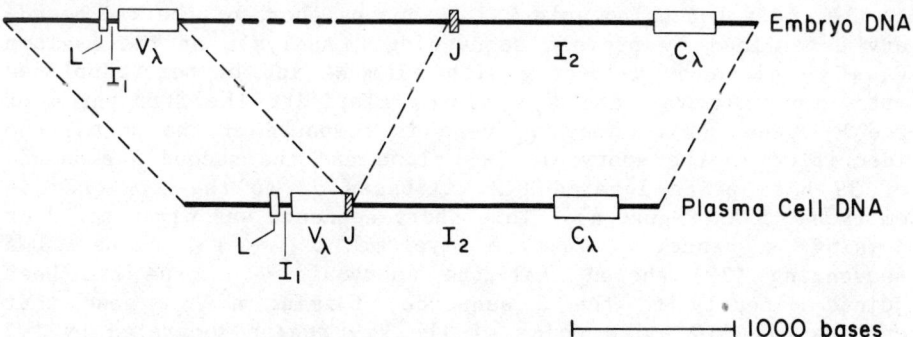

Figure 4. The organization and rearrangement of mouse λ_I light chain gene segments. The L exon encodes amino acids present in the precursor to mature light chains. I_1 and I_2 are introns. (redrawn from ref. 2.)

Immunoglobulin genes were first cloned using a modified λ phage vector and genomic DNA fractions enriched by size or by R-looping (29,30). The clones isolated included embryonic $V_{\lambda I}$ amd embryonic $C_{\lambda I}$ genes, and a clone from the λI-producing myeloma H2020, which contained both $V_{\lambda I}$ and $C_{\lambda I}$ genes. The structure of the myeloma $V_{\lambda I}+C_{\lambda I}$ clone confirmed the earlier conclusion (26-28) that V and C genes are brought closer together in the genome of a myeloma tumor expressing these genes (Figure 4). Unexpectedly, however, the $V_{\lambda I}$ and $C_{\lambda I}$ genes are not joined in the myeloma DNA, but remain 1.2 kilobases apart (30). This result was one of the first to show the need for RNA splicing in the synthesis of a normal eukaryotic cellular mRNA analogous to the RNA splicing seen previously in adenovirus transcripts (31,32). The transcription unit for a light chain gene may be as large as 10 kilobases, although the final mRNA contains only about 1000 nucleotides, excluding poly(A) (33-35,73).

The first DNA sequence of an embryonic $V_{\lambda II}$ gene (36) resulted in the discovery of an additional 93 base-pair intron in the N-terminal signal peptide at codon-4 preceding the mature $V_{\lambda II}$ sequence. The existence of a precursor N-terminal peptide in immunoglobulins had first been shown when <u>in vitro</u> translation of a purified κ mRNA in a rabbit reticulocyte system led to the synthesis of a polypeptide about 1500 daltons larger than authentic κ chain (37). Similar N-terminal peptides have since been found in the precursors of many other eukaryotic secretory proteins and are thought to provide a signal for secretion (38).

The embryonic $V_{\lambda II}$ gene includes the first two hypervariable regions but ends just after the third hypervariable region at the codon for amino acid 98, 14 codons short of the V/C boundary determined by protein sequencing. Analysis of the myeloma $V_{\lambda I}+C_{\lambda I}$ clone by R-looping with λI mRNA and by heteroduplexes with embryonic $V_{\lambda I}$ and $C_{\lambda I}$ clones clarified the structures of the V_λ genes (2). The $V_{\lambda I}$ gene is composed of two parts, one identified in the embryonic $V_{\lambda I}$ clone and the second a sequence of 39 base pairs located 1.2 kilobases 5' to the $C_{\lambda I}$ gene in embryonic DNA (Figure 4). This short sequence was named the J or joining sequence. In the myeloma $V_{\lambda I}+C_{\lambda I}$ clone, DNA sequencing (39) showed that the embryonic $V_{\lambda I}$ gene has been joined directly to the J sequence, forming a $V_{\lambda I}$ gene that encodes all 110 amino acids of the $V_{\lambda I}$ region separated by 1.2 kilobases from the $C_{\lambda I}$ gene. For clarity, in the remainder of this review embryonic sequences that can form parts of an expressed V gene will be referred to as gene segments. Thus, the embryonic $V_{\lambda I}$ and $J_{\lambda I}$ gene segments are joined in λI myeloma DNA to form the $V_{\lambda I}$ gene. The DNA between the $V_{\lambda I}$ and $J_{\lambda I}$ gene segments is apparently deleted by joining, since sequences originally adjacent to these gene segments are no longer present in the genome of a λI myeloma tumor (40).

The structure of the κ light chain gene family in mice has proven to be similar to that of the λ light chain genes but with larger numbers of V and J gene segments (40-45). Five J_κ gene segments are located 2.4 to 3.7 kilobases 5' to the C_κ gene (40,45) (Figure 3). One of these gene segments ($J_\kappa 3$) is not found in any known myeloma V_κ region and has accumulated mutations that apparently prevent it from functioning in RNA splicing (40,45) and perhaps in joining to a V_κ gene segment (46). The remaining four J_κ gene segments can account for amino acid residues 97 to 108 in all known V_κ regions, including some minor variants apparently due to somatic mutation.

The heavy chain genes are the most complex of the three immunoglobulin gene families in mice (Figure 3). There are at least eight C_H genes in a cluster that spans approximately 250 kilobases. The organization and function of the C_H genes will be considered in more detail later. The C_H gene at the 5' end of the cluster is C_μ, which encodes the C_H region of IgM, the first class of immunoglobulin molecules synthesized by a developing B lymphocyte. As in the κ light chain gene family, there are four functional J_H gene segments located about 8 kilobases 5' to the C_μ gene (46-48). The first sequences of heavy chain variable region genes and gene segments uncovered one fundamental difference between the heavy and light chain gene families (46): the presence of a third gene segment encoding part of the V_H region. In the heavy chain of a myeloma protein from the phosphorylcholine-binding group, S107, amino acid residues 1 to 101 of the V_H region are encoded by a germline V_H gene segment. Amino acid residues 107 to 123 (the end of the V_H region) are encoded by J_{H1}, the most 5' of the four germline J_H gene segments. The codons for amino acid residues 102 to 106 in the rearranged S107 V_H gene are not present in either of these two germline gene segments. The same is true for two other V_H genes related to S107 (46) and for two V_H genes from different groups (48,49). Comparisons of a large number of protein sequences suggest that this is the general case for heavy chains (50). On the basis of these observations and consideration of the probable mechanism outlined below for rearrangement of the immunoglobulin gene segments, it is likely that heavy chain variable regions are encoded by three germline gene segments: the V_H and J_H gene segments so far identified and a class of short D (diversity) gene segments that bridge the V_H and J_H gene segments to form complete V_H genes (46) (Figure 5). The small sizes of known D segments have made their identification in germline DNA difficult, but the discovery in a normal B cell of a rearranged D-J_H gene that lacks a V_H-D junction and therefore contains germline sequence 5' to the D gene segment should permit isolation of the complete germline D gene segment (P. Early, unpublished data).

The sequences of light chain V and J gene segments showed that two blocks of conserved nucleotides are always found 3' to V gene segments and as inverted complements 5' to J gene segments (40,45). The same blocks of conserved nucleotides are also found adjacent to V_H and J_H gene segments (46) (Table 1). It was initially suggested that these nucleotides might form a stem-loop structure, typical of inverted repeat sequences, to juxtapose V and J gene segments for joining (40,45). An alternative model for the DNA rearrangements in immunoglobulin variable regions was prompted by the observations that the same blocks of nucleotides are conserved in both light and heavy chain gene families and that the conserved nucleotides are always present as two blocks of 7 and 10 nucleotides separated by either one turn (11 ± 1 nucleotides) or two turns (22 ± 1 nucleotides) of the DNA helix (46) (Table 1).

This model, illustrated in Figure 5, hypothesizes that the combination of the same two blocks of conserved nucleotides with two different lengths of intervening spacer DNA results in two types of recognition sites for proteins mediating variable region gene rearrangement. Gene segments that can join to one another always have different types of recognition sites: V_κ gene segments, for example, have one-turn sites and J_κ gene segments have two-turn sites. This implies that two different kinds of proteins, one bound to a one-turn site and one bound to a two-turn site, must interact to join immunoglobulin gene segments. Although all three immunoglobulin gene families contain the same two types of recognition sites, separation of the gene families on different chromosomes presumably prevents joining between gene segments from different families.

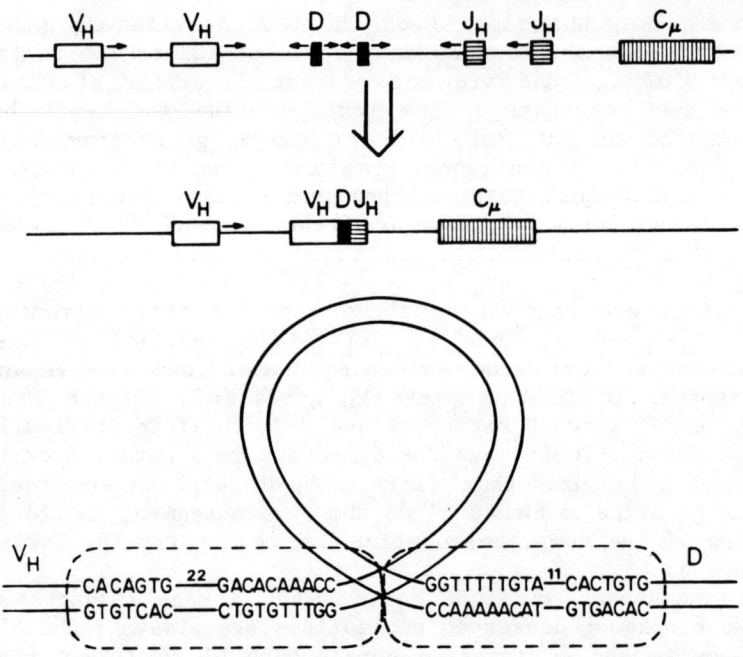

Figure 5. Organization of heavy chain gene segments and a model for DNA rearrangement. The upper diagram depicts possible germline D gene segments. Relative distances of the various gene segments from one another are undetermined. The short arrows indicate conserved noncoding sequences (Table 1) which may be involved in DNA rearrangement. In this model, DNA rearrangement joins V_H-D-J_H gene segments. Intervening DNA may be deleted or could undergo other types of rearrangement. The lower diagram shows paired V_H and D (alternatively D and J_H, or V_L and J_L) gene segments with 11 and 22 nucleotide spacers. Putative DNA-joining proteins might bind to the areas enclosed by dashed lines. The gene segments are represented as colinear to emphasize the symmetry of the conserved noncoding nucleotides. The actual structure may bring the ends of the two gene segments into close proximity. (From ref. 55.)

Both V_H and J_H gene segments are flanked by two-turn recognition sites. This suggests that the D nucleotides of rearranged V_H genes are contributed by D gene segments that are flanked on both sides by one-turn recognition sites. Germline sequence which flanks the 5' side of a D gene segment in a case

Table 1

Comparison of Recognition Nucleotides Adjacent to V and J Gene Segments

$V_{\kappa 41}$	CACAGTGATACAAATCATAACATAAACC	(11)
$V_{\kappa 2}$	CACAGTGATTCAAGCCATGACATAAACC	(11)
$V_{\kappa 3}$	CACAGTGATTCAAGCCATGACATAAACC	(11)
$V_{\kappa 21}$	CACAGTGCTCAGGGCTGAACAAAAACC	(10)
V_{H107}	CACAGTGAGAGGACGTCATTGTGAGCCCAGACACAAACC	(22)
$V_{\lambda I}$	CACAATGACATGTGTAGATGGGGAAGTAGATCAAGAACA	(22)
$V_{\lambda II}$	CACAATGACATGTGTAGATGGGGAAGTAGAACAAGAACA	(22)
$J_{\kappa 1}$	GGTTTTTGTAGAGAGGGGCATGTCATAGTCCTCACTGTG	(22)
$J_{\kappa 2}$	GGTTTTTGTAAAGGGGGCGCAGTGATATGAATCACTGTG	(23)
*$J_{\kappa 3}$	GGGTTTTGTGGAGGTAAAGTTAAAATAAATCACTGTA	(20)
$J_{\kappa 4}$	AGTTTTTGTATGGGGGTTGAGTGAAGGGACACCAGTGTG	(22)
$J_{\kappa 5}$	GGTTTTTGTACAGCCAGACAGTGGAGTACTACCACTGTG	(22)
J_{H1}	AGTTTTAGTATAGGAACAGAGGCAGAACAGAGACTGTG	(21)
J_{H2}	GGTTTTTGTACACCCACTAAAGGGGTCTATGATAGTGTG	(22)
J_{H3}	ATTTATTGTCAGGGGTCTAATCATTGTTGTCACAATGTG	(22)
$J_{\lambda I}$	GGTTTTTGCATGAGTCTATATCACAGTG	(11)

κ sequences from refs. 40 and 45; λ sequences from refs. 36 and 39; heavy chains from refs. 46 and 47.

*Possible nonfunctional J_κ gene segment not yet found in a rearranged gene.

The mRNA-sense strand is shown 3' to V gene segments and 5' to J gene segments. Right column lists the number of nucleotides between the underlined hepta- and decanucleotides. (From ref. 56.)

of D-J_H rearrangement shows the predicted one-turn recognition site (P. Early, unpublished data).

Recent studies of human V gene segments offer more support for the model of gene rearrangement outlined above (51,52). Human V_H and V_κ gene segments contain the same blocks of 7 and 10 conserved nucleotides seen in mouse V gene segments and the spacer lengths between these blocks (one-turn for V_κ, two-turns for V_H) are the same in human and mouse DNAs (51). The structures of the human and mouse V genes are identical: all V genes have an intron in the signal peptide at the fourth codon preceding the mature amino acid sequence; V_H gene segments end at the 5' boundary of the third hypervariable region, while V_κ gene segments end at the 3' boundary of the third hypervariable region (51). The sequences of genes in the V_{HIII} group also appear to be remarkably similar in humans and mice (51).

It is striking to notice the similarity of the conserved blocks of nucleotides adjacent to immunoglobulin gene segments (46) and a sequence present at the boundaries of the invertable phase variation gene in Salmonella (53). Perhaps a very ancient system for DNA rearrangement has been adopted by the immunoglobulin genes.

GENERATION OF V REGION DIVERSITY

An early, and to some extent continuing, controversy in immunology arose between theories of antibody diversity that relied exclusively on germline genes and those that invoked somatic mutational mechanisms. The discovery of V-J and V-D-J joining ended the dispute, since it became apparent that the process of joining separate gene segments to create a complete V gene is itself a source of antibody diversity. DNA joining amplifies diversity by allowing a variety of possible combinations between germline gene segments. In κ light chains, for example, a V_κ gene segment is able to combine with any one of four J_κ gene segments to form a complete V_κ gene.

In κ light chains, amino acid residue 96 is remarkable for its diversity (54). This residue lies at the V_κ/J_κ junction and the process of DNA rearrangement can generate variant codons at this position (40,45,54). By independently varying the sites of DNA joining on the V_κ and J_κ gene segments, the number of codons at the junction can be increased or decreased and hybrid codons (part V_κ, part J_κ) can be formed (Figure 6). The only restriction on variants is that the translational reading frame be preserved from V_κ to J_κ. Since the V_κ/J_κ junction occurs at the end of the third hypervariable region, diversity generated at this position by DNA rearrangement may alter the antigen-binding site and thereby contribute to the repertoire of antibody specificities (45).

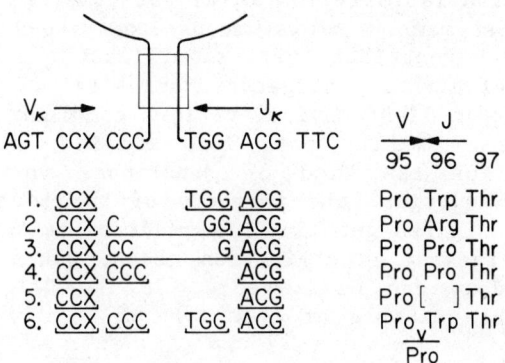

Figure 6. Junctional diversity in κ light chains. Joining of one V_κ gene segment to one J_κ gene segment at different points can generate the six different junctional variants seen in the $V_\kappa 21$ group. (From ref. 56.)

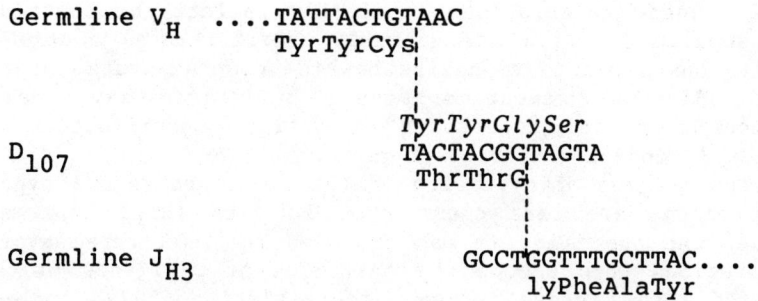

Figure 7. Alternative possible reading frames for a D gene segment. The protein sequence of the S107 D segment is shown in italics. V(D)J joining at the sites indicated by dotted lines can generate the A4 protein sequence from a germline A4-like V_H gene (S. Crews, unpublished data), the D_{107} gene segment and J_{H3} gene segment. (From ref. 56.)

The D segment in V_H genes encompasses almost the entire third hypervariable region (46,48,49), unlike light chains where the third hypervariable region is almost completely contained in the V_H gene segment. This means that V_H-D-J_H joining in heavy chains can contribute very significantly to the generation of a spectrum of diverse antigen-binding sites in immunoglobulins (46,50,55). As in light chains, various combinations of V, J and D can be made as well. Both V_H/D and D/J_H boundaries offer opportunities for the kind of junctional variation seen in κ chains. In addition, the presence of two boundaries for DNA joining allows a given germline D segment to be translated in any reading frame lacking a termination codon, since frameshifts at the V_H/D boundary can be corrected at the D/J_H boundary (55, 56). A possible example of this type of V_H diversity is shown in Figure 7.

These observations can be used to estimate the total antibody diversity arising from germline gene segments and DNA joining in third hypervariable region sequences. Two-hundred germline V_κ gene segments (57) and four J_κ gene segments can pair randomly with perhaps six alternative junctional variants (54,58) in each case for a total of 4800 V_κ regions. If heavy chains are assumed to be encoded by 200 V_H gene segments, 10 D gene segments and four J_H gene segments, six-fold junctional variation at two positions and three possible reading frames for D segments would generate nearly 90,000 V_H regions. If the light and heavy chains can be expressed in any possible pairwise combination by a B cell, more than 4×10^9 different antibodies are possible. These calculations completely neglect any contributions from somatic mutation and probably underestimate junctional diversity but serve to emphasize how the complex system of organization and rearrangement in immunoglobulin genes may generate an enormous potential repertoire of antigen specificities from a relatively modest number of genes.

The origins of diversity in the first and second hypervariable regions are less clear than for the third hypervariable region. At one time, it was proposed that all three hypervariable regions were encoded by batteries of minigenes which were inserted into V genes during differentiation (59). Indeed, the D gene segment encoding the third hypervariable region of heavy chains is such a minigene. Although much of the work on undifferentiated immunoglobulin genes has used early embryos or adult livers as a source of DNA, some studies have specifically used germline (sperm) DNA to study V_κ and V_H gene segments (46,60). The first and second hypervariable regions are already present in these truly germline V gene segments, as they are in V gene segments from embryos and liver. Without minigenes, other explanations must be sought for diversity in the first and second hypervariable regions.

It is still unclear how generally significant somatic mutation may be for the generation of antibody diversity. In the case of λ_I light chains, the prediction by hybridization kinetics (26,61) of one germline $V_{\lambda I}$ gene has been subsequently verified by Southern blots and cloning (2). However, more than six slightly different $V_{\lambda I}$ sequences are known from myeloma proteins (26). In the κ and heavy chain gene families, individual V gene segments can cross-hybridize with 6 to 10 related germline V gene segments, indicating a substantial germline contribution to V segment diversity (62-65). However, the number of different protein sequences in at least one group of closely related V_κ regions still appears to exceed the number of cross-hybridizing germline V_κ gene segments (66). Small somatic changes in V genes may be functionally significant events occurring in normal B cells or may merely be a peculiar feature of myeloma tumors that have undergone many cell generations since their initial transformation.

ALLELIC EXCLUSION

Allelic exclusion in immunoglobulin genes refers to the synthesis of an immunoglobulin chain from only one chromosome homolog in any given B cell (5,6,67). Since it would be unlikely for the DNA rearrangements described in the previous section to form the same V region on two chromosomes, allelic exclusion permits a B cell to synthesize a homogeneous type of immunoglobulin with a unique or limited range of antigen specificity.

Two general types of mechanisms can be suggested for allelic exclusion. One possibility is that variable region gene rearrangements have a fairly low probability of forming a functional V gene and that this generally occurs successfully on at most one homolog in any pair of chromosomes. This stochastic model of allelic exclusion predicts a relatively high frequency of abortively rearranged V genes (principally out-of-reading frame junctions) on unexpressed chromosomes. The other possibility is that a specific DNA rearrangement or other controlling event is responsible for turning off the unexpressed alleles. In this case, an unexpressed chromosome might contain a normally rearranged V gene or it might remain unrearranged.

At the present time, it is uncertain what mechanism is responsible for allelic exclusion within each immunoglobulin gene family and also what mechanism is responsible for the choice of one, but not both types of light chains. In many myeloma tumors, DNA rearrangements are seen on both chromosomes (sometimes more, since myelomas are often subtetraploid) of a pair containing a single immunoglobulin gene family (35,64,69). Aberrant rearrangement of a V_κ gene segment has been observed in the MPC11 myeloma (70-72). In this case, the V_κ gene segment has joined to a sequence in the J_κ-C_κ intron that resembles a recognition site for V/J joining (72). However, Southern blots of DNA from normal

B cells have shown that at least a substantial fraction of κ chain synthesizing cells have one J_κ-C_κ locus in the germline configuration (23). In contrast, both J_H loci are rearranged in virtually all normal B cells synthesizing IgM (C. Nottenburg and I. Weissman, personal communication). Detailed analysis of the J_H loci from these cells should determine whether half are abortively rearranged, as predicted by the stochastic model of allelic exclusion. Since two DNA rearrangements (V_H/D and D/J_H) seem to be required to generate a complete V_H gene, the stochastic model predicts a higher frequency of abortive rearrangements in heavy chains as compared to light chains, which require only the single V/J rearrangement.

Since the frequency of λ light chain synthesis is much lower than κ light chain synthesis in mice, it is likely that V_λ/J_λ joining occurs less readily than V_κ/J_κ joining. Indeed, the recognition sites for joining in λ gene segments seem to be more divergent from the consensus sequence than those in κ gene segments (Table 1). The stochastic model for allelic exclusion predicts that κ genes should be abortively rearranged in λ-producing myelomas, but λ genes would not be rearranged in κ-producing myelomas. The limited data available suggest that this prediction may be correct (74-76).

If allelic exclusion is the result of a specific controlling event, then unrearranged immunoglobulin genes in a B cell would not be expected to be transcribed. Interestingly, however, unrearranged C_κ genes are transcribed at a fairly high level in several myeloma tumors (35). The same is true of unrearranged C_μ genes in cells of myeloid and lymphocyte lineages (77,78; D. Kemp, personal communication). Abortively rearranged immunoglobulin genes are also frequently transcribed, although translation products, if any, may be comparatively short-lived (75,76,79). A simple stochastic model of variable region gene rearrangement could account for the data summarized above. It would be useful, though, to determine whether allelic exclusion occurs in a B cell with two functionally rearranged alleles.

STRUCTURE AND EVOLUTION OF THE IMMUNOGLOBULIN GENE FAMILIES

Conserved amino acid residues and three-dimensional structure show that all immunoglobulin domains, V and C, of both light and heavy chains, share a common ancestor (80). The evolution of immunoglobulin genes has apparently proceeded by duplication and divergence of this ancestral gene and this process is reflected in the structures of the immunoglobulin genes as they exist now.

The current picture of gene organization in the three immunoglobulin families is shown in Figure 3. The most notable feature common to all these genes is that one protein domain is encoded by one exon (81,82). The protein domains apparently are functional as well as structural units. Thus, the V domain binds antigen and the C_H2 domain of IgG molecules plays a role in complement activation. Light chains contain only two domains, V and C, which are encoded by separate exons even after V/J fusion (Figure 4). Heavy chain C regions are a tandem array of three or four domains, each encoded by a separate exon (81-87). Some heavy chains contain a short hinge region, which may be the remnant of an ancient domain and is encoded by a separate exon (82). The one domain/one exon structure of immunoglobulins illustrates an evolutionary model suggested by Gilbert (88). This model proposes that one advantage of the split-gene organization of eukaryotes, which require RNA splicing to generate mature mRNAs, is that it readily permits the evolution of compound proteins by allowing relatively inexact fusions to occur in intron DNA between originally independent exons.

Some major gaps still exist in the current maps of the immunoglobulin gene families (Figure 3). In none of the gene families is the distance from V to C genes known in germline DNA, although classical genetics suggests that this distance may be on the order of 10^6 base pairs in the heavy chains (18). Pairs of related V_H gene segments have been linked in mouse and human DNAs: they are 4 to 16 kilobases apart (51,89; S. Crews and H. Huang, personal communication). If V gene segments have evolved by duplication and divergence, related V gene segments can be expected to occur as a cluster, perhaps corresponding to a group of V regions as defined by protein sequencing (4). More extensive analysis of V gene segments will show whether this is the case and may provide a more accurate determination of the numbers of V gene segments. The numbers and location of the germline D gene segments in the heavy chain family are presently unknown, although presumably they will be found between the germline V_H and J_H gene segments which are bridged by D segments in rearranged V_H genes.

Within the C_H cluster, the organization of J_H, C_μ and C_δ has been defined by overlapping clones (90-92). The C_μ and C_δ genes are only 2.5 kilobases apart. The C_γ, C_ϵ and C_α genes have also been linked in a series of genomic clones (49,127,128). The gene order is $C_{\gamma 3}$, $C_{\gamma 1}$, $C_{\gamma 2b}$, $C_{\gamma 2a}$, C_ϵ, C_α. Distances between these genes are 34 kilobases between $C_{\gamma 3}$ and $C_{\gamma 1}$, 21 kilobases between $C_{\gamma 1}$ and $C_{\gamma 2b}$, 15 kilobases between $C_{\gamma 2b}$ and $C_{\gamma 2a}$ and 15 kilobases between $C_{\gamma 2a}$ and C_ϵ.

HEAVY CHAIN CLASS SWITCHING

The first class of immunoglobulin synthesized by a B lymphocyte is IgM, which is later often expressed in conjunction with IgD (Figure 2). Subsequently in development, and probably in response to antigen, the B cell or its progeny may switch to the

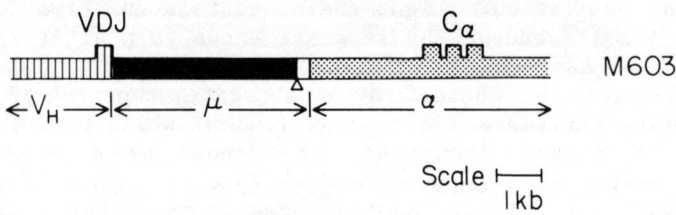

Figure 8. Structure of the rearranged M603 α gene. Sequences derived from the germline J_H-$S\mu$ region and the germline C_α gene are indicated by different patterns of shading. The triangle indicates the point of a deletion in the J_H-$S\mu$ region; the unshaded sequences to the right of the triangle are also derived from the germline $S\mu$ region. It is unclear whether this deletion is related to the mechanism for heavy chain switching. Raised boxes denote exons.

synthesis of a different class of immunoglobulin but retain the same V_H region and the same light chain (7). This process is termed class switching. A possible mechanism for heavy chain class switching by differential RNA processing was eliminated by results which showed that myeloma cells do not contain nuclear transcripts of unexpressed C_H genes (94). Instead, the heavy chain class switch is the result of a second type of immunoglobulin gene rearrangement, apparently unrelated to the rearrangements of V, D and J gene segments.

A deletion model for DNA rearrangements in heavy chain switching was first proposed on the basis of hybridization kinetics experiments to determine copy numbers of individual C_H genes in genomic DNAs from myeloma tumors synthesizing different classes of heavy chains (93). The results suggested that if the order of C_H genes is μ-γ3-γ1-γ2b-γ2a-α, then heavy chain class switching can be explained as the deletion of all C_H genes from C_μ to the C_H gene expressed in any particular myeloma tumor. Thus, an IgA-producing myeloma tumor will have deleted the C_μ and all the C_γ genes. Several investigators later used Southern blots of myeloma DNAs, hybridized with various C_H probes, and got evidence consistent with this deletion model for heavy chain switching (68,69,95-98). These results were often complicated by the presence of DNA rearrangements on multiple chromosomes, perhaps due to the subtetraploid nature of myeloma tumors and to abortive rearrangements (69).

The first clone in which the results of heavy chain class switching at the DNA level were analyzed contained the expressed α gene from the M603 myeloma tumor (99). Comparison of this gene with germline V_H, J_H+C_μ and C_α genes shows that the M603 gene is composed of at least four distinct segments of germline DNA (Figure 8). These are a germline V_H gene segment, a D gene segment, a germline J_H gene segment and sequences originally 3' to it between the germline J_H and C_μ genes, and a germline C_α gene.

The structure of the M603 α gene demonstrates that it must have been formed by at least two separate types of DNA rearrangements (Figure 9). The first type, V-D-J joining, generated a functional V_H gene associated with the C_μ gene. The second type of DNA rearrangement, C_H switching, joined a point in the intron DNA between J_H1 and the C_μ gene to a point in the DNA 5' to the C_α gene. This resulted in the displacement of the C_μ gene by C_α and the formation of the functional M603 α gene. The point at which this second type of DNA joining has occurred in the rearranged gene and the corresponding breakpoints in the germline DNAs are termed the switch (S) sites (100). Other laboratories have obtained similar evidence for switching in $\gamma1$, $\gamma3$ and $\gamma2b$ genes (48,49,100,101).

A one to two kilobase region of DNA (S region) that contains switch sites 5' to C_μ (the S_μ sites) shows at least partial homology to similar regions 5' to the C_μ and C_γ genes, as well as to a region 5' to the human C_μ gene (49). However, the nucleotide sequences of the switch sites themselves are not all identical. Three S_α switch sites have been sequenced and all have been found to occur within a 30 base-pair conserved sequence repeated at least 17 times 5' to the C_α gene (102). This 30 base-pair S_α sequence shares some homology with switch sites 5' to C_γ genes (102) (Figure 10). The S_γ sites also occur within short repeated units 5' to each C_γ gene (49). The repeat units containing $S\gamma1$, $S\gamma3$ and $S\gamma2b$ are all extensively homologous (49). While S_μ sites share some short common sequences with S_γ sites (49), the two S_μ sites sequenced so far are not very homologous to S_α or S_γ sites, although the S_μ sites are homologous with one another.

The current nucleotide sequence data show that heavy chain chain switching is not mediated by the same sequences that are involved in V-D-J joining. Class switching occurs within repeated blocks of nucleotides 5' to each C_H gene, but the sequences of these blocks of nucleotides are markedly different when different C_H genes are compared. In principle, DNA rearrangements in heavy chain switching could be relatively inexact, since they occur in intron sequences between coding regions. In fact, however, switch rearrangements seem to occur within particular blocks of sequences for each C_H gene. The repetition of these blocks of sequences probably increases the frequency of heavy chain switching by increasing the numbers of target sites for switch recombination.

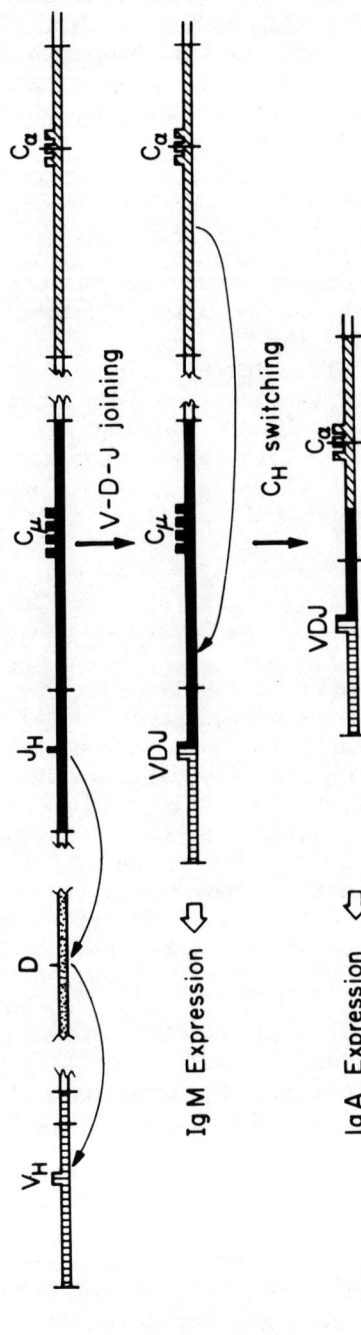

Figure 9. An α gene created by two types of DNA rearrangements: V-D-J joining and heavy chain class switching. V-D-J joining permits a μ chain to be expressed by the differentiating B cell. Heavy chain switching replaced C_μ with C_α, permitting α chain synthesis. (Redrawn from ref. 99).

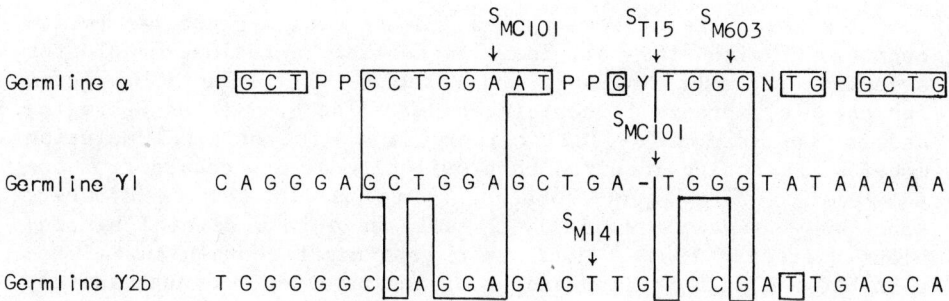

Figure 10. A comparison of the switch sequences from germline 5' flanking sequences for the C_α, $C\gamma_1$ and γ_{2b} genes. Boxes in the germline α sequence denote the conserved bases of the 30-nucleotide recognition sequences. Boxes in the germline $\gamma 1$ and $\gamma 2b$ sequences denote nucleotide identities to the germline C_α sequence. Dash indicates a gap. (From ref. 102.)

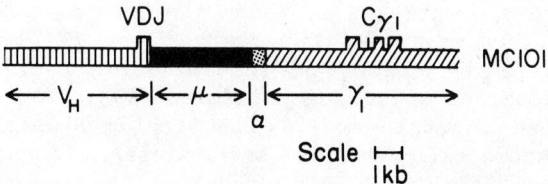

Figure 11. Structure of the rearranged MC101 $\gamma 1$ gene. Sequences derived from the germline V_H segment, the germline J_H-S μ region, the germline S_α region and the germline $C\gamma_1$ gene are indicated by different patterns of shading. Raised boxes denote exons. (Redrawn from ref. 49.)

One possible interpretation of these observations is that the relative conservation of switch sites adjacent to the same C_H gene, compared to those adjacent to another C_H gene, may be an indication of a specific control mechanism for heavy chain switching. This mechanism would allow a specific external stimulus to cause a B cell to switch to a certain class of heavy chain (56,102).

The structure of some cloned immunoglobulin genes may be inconsistent with the original sequential deletion model for switching proposed by Honjo and Kataoka (93). The MC101 gene, for example, appears to contain an S_α region between an S_μ region and an $S_{\gamma 1}$ region (49,102) (Figure 11). The original deletion model of switching required that switching always occurs from one C_H gene to a C_H gene located 3' to it in the C_H cluster, since any C_H gene originally 5' will have been deleted by previous switching (93). Thus, an α gene might contain an $S_{\gamma 1}$ region between an S_μ region and an S_α region if it had successively switched from C_μ to $C_{\gamma 1}$ to C_α, but a $\gamma 1$ gene should never contain an S_α region. A cloned $\gamma 2b$ myeloma cell line has also been observed to switch back to $\gamma 1$ synthesis (103) although the $C_{\gamma 1}$ gene is 5' to the $C_{\gamma 2b}$ gene (49).

Three possible models can be advanced to account for apparently paradoxical sequences of heavy chain switching (49,56). Rather than looping-out DNA for deletion between the V_H gene and a C_H gene (93), recombination might occur between sister chromatids during mitosis (49). If an initial recombination were to occur between, for example, the S_α and S_μ regions, the daughter cell that continues to express IgM would have a chromosome with a silent C_μ gene 3' to S_α sequences. Subsequent switching to this C_μ gene would generate an expressed μ gene with S_α sequences in the J_H-C_μ intron. One more switch to the $C_{\gamma 1}$ gene would generate an expressed $\gamma 1$ gene with both S_μ and S_α sequences in the J_H-$C_{\gamma 1}$ intron, as is observed for MC101 (Figure 11). Various other such schemes of recombination can be devised to account for apparent switching paradoxes (49).

The second possible model of switching simply assumes that homologous chromosomes, rather than sister chromatids, undergo switch recombination. This reduces the number of V_H-C_H switches to two: V_H to C_α on one chromosome, and then V_H to the $C_{\gamma 1}$ gene remaining on the other chromosome. However, existing data show that expressed V_H and C_H regions are derived from the same chromosome homolog (104,105), although perhaps there are exceptions.

A third model for apparently paradoxical switching allows looping-out to occur for DNA between the V_H and C_H genes but suggests that the looped-out DNA may remain in the cell as an episome. If this episome were to reintegrate by switch recombination with the expressed heavy chain gene, the switch order could be reversed from the order in germline genes (56).

SURFACE AND SECRETED IMMUNOGLOBULINS

B lymphocytes can synthesize immunoglobulins either as receptor molecules bound to the cell surface or as secreted antibodies (Figure 2). The predominant surface receptors for antigen on resting lymphocytes are IgM and IgD (105). In response to antigen, B lymphocytes proliferate and enlarge, and begin to secrete large quantities of IgM or one of the other classes of immunoglobulin.

The complete amino acid sequence of a secreted mouse μ chain was determined from a myeloma IgM molecule (106). The secreted C_μ region contains four domains plus a 20 amino acid carboxy-terminal segment which lies past the fourth C_μ domain. Secreted IgM molecules exist in serum as pentamers of subunits such as the one illustrated in Figure 1, held together by disulfide bonds from a cysteine residue in the carboxy-terminal segment and a J chain polypeptide (107). This J chain is unrelated to the J gene segments discussed earlier. Limited quantities of surface IgM have prevented complete protein sequence analysis, although partial characterizations have been made of this molecule. Unlike secreted IgM, surface IgM is a monomer. Surface and secreted IgM share serological markers indicating that the secreted (μ_s) and membrane-bound (μ_m) heavy chains are very similar. Initial efforts to determine whether the μ polypeptide chains in the two forms of IgM are identical gave conflicting results. Some analyses showed that the tyrosine present at the carboxy-terminus of the μ_s chain was absent from the μ_m chain (8), while other investigators reached the opposite conclusion (9). More recently, cells synthesizing surface IgM were found to contain a larger μ polypeptide than those synthesizing only secreted IgM (108,109).

Analyses of μ mRNAs in several cell types demonstrated the existence of at least two and sometimes more species with different sizes (110). The presence of a 2.7 kilobase μ mRNA is correlated with μ_m synthesis by a cell, while a 2.4 kilobase μ mRNA is correlated with μ_s synthesis (10,110,111).

When μ cDNA clones synthesized from a myeloma tumor producing both 2.7 kilobase and 2.4 kilobase μ mRNAs were sequenced, two distinct 3' ends were found for the two μ mRNAs (10). Both μ mRNAs encode all four C_μ domains and are identical up to the 3' end of the coding sequence for the fourth C_μ domain ($C_\mu 4$) (10, 111). The remainder of the 2.4 kilobase μ_s mRNA encodes the 20 carboxy-terminal amino acids of the secreted μ polypeptide chain while the 2.7-kilobase μ_m mRNA contains an alternative 3' end encoding a 41 amino acid hydrophobic carboxy-terminal segment (10). Recent characterizations of the μ_m chain are consistent with the protein sequence deduced from the nucleotide sequence of the μ_m cDNA (11). These results indicate that surface IgM molecules are anchored to the cell membrane by a hydrophobic carboxy-terminal segment not found in secreted IgM and that secreted IgM

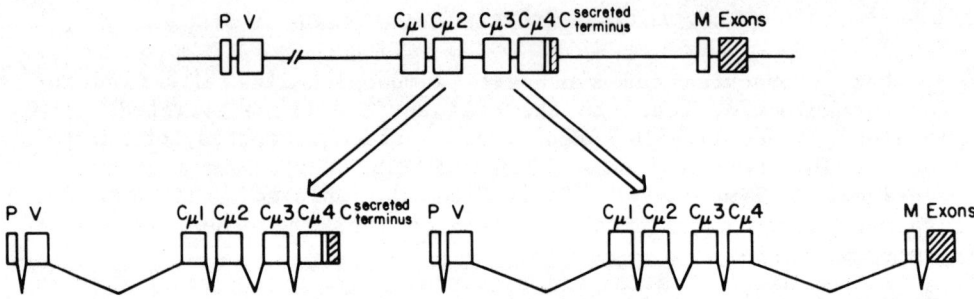

Figure 12. Splicing patterns of μ_m and μ_s mRNAs. Raised boxes indicate exons. Shading indicates 3' untranslated sequences (different for μ_m and μ_s mRNAs). The M exons form the 3' portion of μ_m (the second M exon encodes two amino acids). The $C_{terminus}^{secreted}$ sequences form the 3' portion of μ_s mRNA. P refers to the signal peptide exon and V to the rearranged V_H-D-J_H exon. Bent lines indicate RNA splicing between exons. (From ref. 12.)

instead has a hydrophilic carboxy-terminal segment specialized for pentamer formation.

Southern blots of genomic DNAs (68,69) and analyses of independently isolated genomic $C\mu$ clones demonstrate that the mouse genome contains only one $C\mu$ gene (87). Sequence analysis of the $C\mu$ gene shows that both μ_m and μ_s mRNAs can be derived from the same gene by different pathways of RNA splicing (Figure 12) (12). The 3' sequence specific to μ_s mRNA is immediately adjacent to the sequence encoding the $C_\mu 4$ domain in the $C\mu$ gene. The μ_m 3' sequence, however, is present in two exons about 1.8 kilobases 3' to the μ_s 3' sequence. The junction between the sequences encoding the $C_\mu 4$ domain and the μ_s carboxy-terminus contains an RNA splice signal similar to sequences found at other splicing junctions (112). RNA splicing can occur at this point to generate μ_m mRNA by joining the sequence encoding the $C_\mu 4$ domain to the sequence encoding the hydrophobic μ_m carboxy-terminal segment (Figure 12). Alternatively, no splicing occurs at this point in the synthesis of μ_s mRNA.

While it is possible that alterations in the ratio of synthesis of the two forms of μ mRNA involve regulation of the RNA splicing apparatus, a simpler hypothesis is that the termination/polyadenylation sites of μ transcripts are developmentally regulated (12). If a transcript is polyadenylated at a site 187

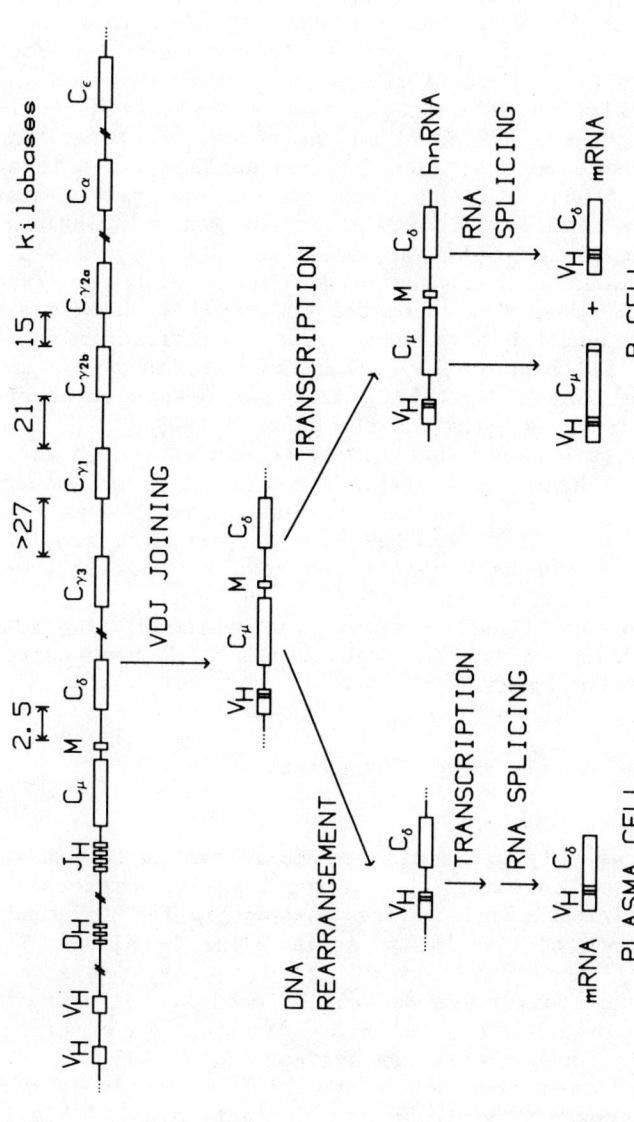

Figure 13. Alternative mechanisms for expression of IgD in B cells and plasma cells. M refers to the membrane exons of the C_μ gene. The three segments (V_H-D-J_H) of the rearranged V_H gene are indicated by vertical lines. (Redrawn from ref. 92.)

nucleotides 3' to the $C_\mu 4$ sequence, it must become μ_s mRNA since it does not contain the specific μ_m sequences. However, if a transcript is polyadenylated at a site about 2.4 kilobases 3' to the $C_\mu 4$ sequence, RNA splicing can occur between the $C_\mu 4$ sequence and the μ_m carboxy-terminal sequence generating μ_m mRNA. Other heavy chains also exist in surface or secreted forms and these C_H genes may be expected to have a similar mechanism for synthesis of alternative forms of mRNA.

The transcription of the C_μ locus is probably more complex than already suggested by the μ_m and μ_s mRNAs. Most early B lymphocytes synthesize both surface IgM and surface IgD with a common V_H region (105). In rats and mice, surface IgD usually disappears along with surface IgM after antigen stimulation.

The C_δ gene is located about 2.5 kilobases 3' to the μ_m carboxy-terminal exons in mouse germline DNA (90,91). Analyses of genomic DNA from mouse B lymphocytes or B-cell lymphomas synthesizing both IgM and IgD have shown that no rearrangement of the C_δ gene occurs in these cells (92). However, in a rat myeloma secreting IgD without IgM, the C_μ gene has been deleted and DNA rearrangement has occurred 5' to the C_δ gene (92).

These results suggest that IgD synthesis can be an exception to the pattern of heavy chain switching seen in other immunoglobulin classes (Figure 13). While DNA rearrangement does occur in the rat myeloma to switch from IgM to IgD synthesis, simultaneous IgM and IgD synthesis is probably the result of RNA processing. Apparently RNA splicing can join the V_H exon to either the first C_μ exon or the first C_δ exon. The synthesis of δ mRNA may involve a more complex type of regulation of RNA processing than needed for μ_m and μ_s mRNA synthesis.

T-CELL RECEPTORS

To date, the only genes of the immune system that have been studied are those encoding the immunoglobulins expressed by B cells. Molecular biologists are just beginning to study the other major branch of the immune system, the T cells. T cells may be the most important part of the immune system since they play central regulatory roles as well as mediate cellular immunity (113-116). Individual T cells are known to have surface receptors specific for one or a few antigens (117-119).

T-cell receptors have not been well characterized, although it has been suggested that they may be antigenically similar to immunoglobulins (119,120). Some investigators have suggested that T cells or T-cell lymphomas contain immunoglobulin RNAs (121), although this seems well documented only for μRNA (77, 78). The function of μRNAs in these T-cell lymphomas is unclear since they apparently are not translated (129). Although DNA rearrangements that are consistent with V_H-D-J_H joining have been observed in Southern blots of DNAs from some (but not all) T-cell lymphomas (122,123), no such rearrangements were observed in two cloned cell lines with known T-cell functions (124). It seems unlikely, therefore, that any of the known C_H or J_H genes encode the T-cell receptors.

There is serological evidence that V_H regions may be present on T cells (120,125,126). T cells may thus be able to use the repertoire of diverse V_H gene segments and perhaps D gene segments, in combination with a hypothetical set of "J_T" gene segments and "C_T" genes, to generate a novel set of surface receptors that nonetheless share some of the antigen-binding sites of immunoglobulins. Work is currently under way in several laboratories to test this possibility by using probes encoding B-cell V_H regions apparently expressed by certain T-cell lines.

Acknowledgments: We thank Mark Davis and Mitch Kronenberg for their comments on the manuscript and Bernita Larsh for her expert typing.

REFERENCES

1. Dreyer, W.J. and Bennett, J.C. (1965) Proc. Nat. Acad. Sci. U.S.A. 54, 864-869.
2. Brach, C., Hirama, M., Lenhard-Schuller, R. and Tonegawa, S. (1978) Cell 15, 1-14.
3. Cancro, M.P., Gerhard, W. and Klinman, N.R. (1978) J. Exp. Med. 147, 776-787.
4. Hood, L., Loh, E., Hubert, J., Barstad, P., Eaton, B., Early, P., Fuhrman, J., Johnson, M., Kronenberg, M. and Schilling, J. (1976) Cold Spring Harbor Symp. Quant. Biol. 41, 817-836.
5. Pernis, B., Chiappino, G., Kelvs, A.S. and Gell, P.G.H. (1965) J. Exp. Med. 122, 853-876.
6. Cebra, J.J., Colberg, J.E. and Dray, S. (1966) J. Exp. Med. 123, 547-558.
7. Gearhart, P.J., Sigal, N. and Klinman, N.R. (1975) Proc. Nat. Acad. Sci. U.S.A. 72, 1707-1711.
8. Williams, P.B., Kubo, R.T. and Grey, H.M. (1978) J. Immunol. 121, 2435-2439.
9. McIlhinney, R.A.J., Richardson, N.E. and Feinstein, A. (1978) Nature 272, 555-557.
10. Rogers, J., Early, P., Carter, C., Calame, K., Bond, M., Hood, L. and Wall, R. (1980) Cell 20, 303-312.
11. Kehry, M., Ewald, S., Douglas, R., Sibley, C., Raschke, W., Fambrough, D. and Hood, L. (1980) Cell 21, 393-406.
12. Early, P., Rogers, J., Davis, M., Calame, K., Bond, M., Wall, R. and Hood, L. (1980) Cell 20, 313-319.

13 Mage, R., Lieberman, R., Potter, M. and Terry, W.D. (1973) in The Antigens (Sela, M., ed.), Vol. 1, p. 299, Academic Press, New York, NY.
14 Williamson, A. (1976) Annu. Rev. Biochem. 45, 467-500.
15 Hengartner, H., Meo, T. and Muller, E. (1978) Proc. Nat. Acad. Sci. U.S.A. 75, 4495-4498.
16 Gibson, D.M., Taylor, B.A. and Cherry, M. (1978) J. Immunol. 121, 1585-1590.
17 Swan, D., D'Eustachio, P., Leinwald, L., Seidman, J., Keithley, D. and Ruddle, F.H. (1979) Proc. Nat. Acad. Sci. U.S.A. 76, 2735-2739.
18 Riblet, R.J. (1977) in ICN-UCLA Symposia on Molecular and Cellular Biology, Vol. 6, pp. 83-89, Academic Press, New York, NY.
19 Stavnezer, J. and Huang, R.-C.C. (1971) Nature New Biol. 230, 172-176.
20 Swan, D., Aviv, H. and Leder, P. (1972) Proc. Nat. Acad. Sci. U.S.A. 69, 1967-1971.
21 Brownlee, G.G., Harrison, T.M., Matthews, M.B. and Milstein, C. (1972) FEBS Lett. 23, 244-248.
22 Potter, M. (1972) Physiol. Rev. 52, 631-719.
23 Joho, R. and Weissman, I.L. (1980) Nature 284, 179-181.
24 Milstein, C., Brownlee, G.G., Cartwright, E.M., Jarvis, J.M. and Proudfoot, N.J. (1974) Nature 252, 354-359.
25 Cowan, N.J., Secher, D.S. and Milstein, C. (1976) Eur. J. Biochem. 61, 355-368.
26 Tonegawa, S., Hozumi, N., Mattyssens, G. and Schuller, R. (1976) Cold Spring Harbor Symp. Quant. Biol. 41, 877-889.
27 Hozumi, N. and Tonegawa, S. (1976) Proc. Nat. Acad. Sci. U.S.A. 73, 3628-3632.
28 Tonegawa, S., Hozumi, N., Brack, C. and Schuller, R. (1977) in ICN-UCLA Symposia on Molecular and Cellular Biology, Vol. 6, pp. 43-55, Academic Press, New York, NY.
29 Tonegawa, S., Brack, C., Hozumi, N. and Schuller, R. (1977) Proc. Nat. Acad. Sci. U.S.A. 74, 3518-3522.
30 Brack, C. and Tonegawa, S. (1977) Proc. Nat. Acad. Sci. U.S.A. 74, 5652-5656.
31 Berget, S.M., Moore, C. and Sharp, P.A. (1977) Proc. Nat. Acad. Sci. U.S.A. 74, 3171-3175.
32 Chow, L.T., Gelinas, R.E., Broker, T.R. and Roberts, R.J. (1977) Cell 12, 1-8.
33 Gilmore-Hebert, M., Hercules, K., Komaromy, M. and Wall, R. (1978) Proc. Nat. Acad. Sci. U.S.A. 75, 6044-6048.
34 Gilmore-Hebert, M. and Wall, R. (1979) J. Mol. Biol. 135, 879-891.
35 Perry, R.P., Kelley, D.E., Coleclough, C., Seidman, J.G., Leder, P., Tonegawa, S., Matthyssens, G. and Weigert, M. (1980) Proc. Nat. Acad. Sci. U.S.A. 77, 1937-1941.
36 Tonegawa, S., Maxam, A.M., Tizard, R., Bernard, O. and Gilbert, W. (1978) Proc. Nat. Acad. Sci. U.S.A. 75, 1485-1489.

37 Milstein, C., Brownlee, G.G., Harrison, T.M. and Mathews, M.B. (1972) Nature New Biol. 239, 117-120.
38 Blobel, G. and Dobberstein, B. (1975) J. Cell Biol. 67, 835-851.
39 Bernard, O., Hozumi, N. and Tonegawa, S. (1978) Cell 15, 1133-1144.
40 Sakano, H., Hüppi, K., Heinrich, G. and Tonegawa, S. (1979) Nature 280, 288-294.
41 Seidman, J.G., Leder, A., Edgell, M.H., Polsky, F., Tilghman, S.M., Tiemeier, D.C. and Leder, P. (1978) Proc. Nat. Acad. Sci. U.S.A. 75, 3881-3885.
42 Seidman, J.G. and Leder, P. (1978) Nature 276, 790-795.
43 Lenhard-Schuller, R., Hohn, B., Brack, C., Hirama, M. and Tonegawa, S. (1978) Proc. Nat. Acad. Sci. U.S.A. 75, 4709-4713.
44 Seidman, J.G., Max, E.E. and Leder, P. (1979) Nature 280, 370-375.
45 Max, E.E., Seidman, J.G. and Leder, P. (1979) Proc. Nat. Acad. Sci. U.S.A. 76, 3450-3454.
46 Early, P., Huang, H., Davis, M., Calame, K. and Hood, L. (1980) Cell 19, 981-992.
47 Bernard, O. and Gough, N.M. (1980) Proc. Nat. Acad. Sci. U.S.A. 77, 3630-3634.
48 Sakano, H., Maki, R., Kurosawa, Y., Roeder, W. and Tonegawa, S. (1980) Nature 286, 676-683.
49 Honjo, T., Kataoka, T., Yaoita, Y., Shimizu, A., Takahashi, N., Yamawaki-Kataoka, Y., Nikaido, T., Nakai, S., Obata, M., Kawakami, T. and Nishida, Y. (1980) Cold Spring Harbor Symp. Quant. Biol. (in press).
50 Schilling, J., Clevinger, B., Davie, J.M. and Hood, L. (1980) Nature 283, 35-40.
51 Rabbitts, T.H., Bentley, D.L., Dunnick, W., Forster, A., Matthyssens, G.E.A.R. and Milstein, C. (1980) Cold Spring Harbor Symp. Quant Biol. (in press).
52 Matthyssens, G. and Rabbitts, T.H. (1980) Proc. Nat. Acad. Sci. U.S.A. (in press).
53 Simon, M., Zieg, J., Silverman, M., Mandel, G. and Doolittle, R. (1980) Science 209, 1370-1374.
54 Weigert, M., Perry, R., Kelley, D., Hunkapiller, T., Schilling, J. and Hood, L. (1980) Nature 283, 497-499.
55 Hood, L. and Early, P. (1980) in Immunoglobulin Genes and B-Cell Differentiation (Battisto, J.R. and Knight, K.L., eds.), Elsevier, New York, NY (in press).
56 Hood, L., Davis, M., Early, P., Calame, K., Kim, S., Crews, S. and Huang, H. (1980) Cold Spring Harbor Symp. Quant. Biol. (in press).
57 Seidman, J.G., Leder, A., Nau, M., Norman, B. and Leder, P. (1978) Science 202, 11-17.
58 Weigert, M., Gatmaitan, L., Loh, E., Schilling, J. and Hood, L. (1978) Nature, 276, 785-790.

59 Kabat, E.A., Wu, T.T. and Bilofsky, H. (1978) Proc. Nat. Acad. Sci. U.S.A. 75, 2429-2433.
60 Joho, R., Weissman, I.L., Early, P., Cole, J. and Hood, L. (1980) Proc. Nat. Acad. Sci. U.S.A. 77, 1106-1110.
61 Tonegawa, S. (1976) Proc. Nat. Acad. Sci. U.S.A. 73, 203-207.
62 Seidman, J.G., Leder, A., Edgell, M.H., Polsky, F., Tilghman, S.M., Tiemeier, D.C. and Leder, P. (1978) Proc. Nat. Acad. Sci. U.S.A. 75, 3881-3885.
63 Wilson, R., Miller, J. and Storb, U. (1979) Biochemistry 18, 5013-5021.
64 Davis, M., Early, P., Calame, K., Livant, D. and Hood, L. (1979) in Eucaryotic Gene Regulation (Axel, R., Maniatis, T. and Fox, C.F., eds.), pp. 393-406, Academic Press, New York, NY.
65 Rabbitts, T.H., Matthyssens, G. and Hamlyn, P.H. (1980) Nature 284, 238-243.
66 Valbuena, O., Marcu, K.B., Weigert, M. and Perry, R.P. (1978) Nature 276, 780-784.
67 Pernis, B. (1967) Cold Spring Harbor Symp. Quant. Biol. 32, 333-341.
68 Coleclough, C., Cooper, D. and Perry, R.P. (1980) Proc. Nat. Acad. Sci. U.S.A. 77, 1422-1426.
69 Cory, S. and Adams, J.M. (1980) Cell 19, 37-51.
70 Schnell, H., Steinmetz, M., Zachau, H.G. and Schecter, I. (1980) Nature 286, 170-173.
71 Choi, E., Kuehl, M. and Wall, R. (1980) Nature 286, 776-779.
72 Seidman, J.G. and Leder, P. (1980) Nature 286, 779-783.
73 Schibler, V., Marcu, K.B. and Perry, R.P. (1978) Cell 15, 1495-1509.
74 Ono, M., Kawakami, M., Kataoka, T. and Honjo, T. (1977) Biochem. Biophys. Res. Commun. 74, 796-802.
75 Alt, F.W., Enea, B., Bothwell, A.L.M. and Baltimore, D. (1979) in Eucaryotic Gene Regulation (Axel, R., Maniatis, T. and Fox, C.F., eds.) pp. 407-420, Academic Press, New York, NY.
76 Alt, F.W., Enea, V., Bothwell, A.L.M. and Baltimore, D. (1980) Cell 21, 1-12.
77 Kemp, D.J., Harris, A.W., Cory, S. and Adams, M.J. (1980) Proc. Nat. Acad. Sci. U.S.A. 77, 2876-2880.
78 Kemp, D.J., Wilson, A., Harris, A.W. and Shortman, K. (1980) Nature 286, 168-170.
79 Kuehl, W.M. and Scharff, M.D. (1974) J. Mol. Biol. 89, 409-421.
80 Hood, L., Campbell, J.H. and Elgin, S.C.R. (1975) Annu. Rev. Genet. 9, 305-353.
81 Early, P.W., Davis, M.M., Kaback, D.B., Davidson, N. and Hood, L. (1979) Proc. Nat. Acad. Sci. U.S.A. 76, 857-861.

82 Sakano, H., Rogers, J.H., Hüppi, K., Brack, C., Traunecker, A., Maki, R., Wall, R. and Tonegawa, S. (1979) Nature 277, 627-633.
83 Kataoka, T., Yamawaki-Kataoka, Y., Yamagishi, H. and Honjo, T. (1979) Proc. Nat. Acad. Sci. U.S.A. 76, 4240-4244.
84 Tucker, P.W., Marcu, K.B., Newell, N., Richards, J. and Blattner, F.R. (1979) Science 206, 1303-1306.
85 Miyata, T., Yasunaga, T., Yamawaki-Kataoka, Y., Obata, M. and Honjo, T. (1980) Proc. Nat. Acad. Sci. U.S.A. 77, 2143-2147.
86 Gough, N.M., Kemp, D.J., Tyler, B.M., Adams, J.M. and Cory, S. (1980) Proc. Nat. Acad. Sci. U.S.A. 77, 554-558.
87 Calame, K., Rogers, J., Early, P., Davis, M., Livant, D., Wall, R. and Hood, L. (1980) Nature 284, 452-455.
88 Gilbert, W. (1978) Nature 271, 501.
89 Kemp, D.J., Cory, S. and Adams, J.M. (1979) Proc. Nat. Acad. Sci. U.S.A. 76, 4627-4631.
90 Liu, C.-P., Tucker, P.W., Mushinski, J.F. and Blattner, F.R. (1980) Science 209, 1348-1353.
91 Tucker, P.W., Liu, C.-P., Mushinski, J.F. and Blattner, F.R. (1980) Science 209, 1353-1360.
92 Moore, K.M., Rogers, J., Hunkapiller, T., Early, P., Nottenburg, C., Weissman, I., Bazin, H., Wall, R. and Hood, L.E. (1980) Proc. Nat. Acad. Sci. U.S.A. (in press).
93 Honjo, T. and Kataoka, T. (1978) Proc. Nat. Acad. Sci. U.S.A. 75, 2140-2144.
94 Marcu, K.B., Schibler, U. and Perry, R.P. (1979) Science 204, 1087-1088.
95 Cory, S., Jackson, J. and Adams, J.M. (1980) Nature 285, 450-456.
96 Rabbitts, T.H., Forster, A., Dunnick, W. and Bentley, D.L. (1980) Nature 283, 351-356.
97 Coleclough, C., Cooper, D. and Perry, R.P. (1980) Proc. Nat. Acad. Sci. U.S.A. 77, 1422-1426.
98 Yaoita, Y. and Honjo, T. (1980) Nature 286, 850-853.
99 Davis, M.M., Calame, K., Early, P.W., Livant, D.L., Joho, R., Weissman, I.L. and Hood, L. (1980) Nature 283, 733-739.
100 Kataoka, T., Kawakami, T., Takahashi, N. and Honjo, T. (1980) Proc. Nat. Acad. Sci. U.S.A. 77, 919-923.
101 Maki, R., Traunecker, A., Sakano, H., Roeder, W. and Tonegawa, S. (1980) Proc. Nat. Acad. Sci. U.S.A. 77, 2138-2142.
102 Davis, M.M., Kim, S.K. and Hood, L. (1980) Science 209, 1360-1365.
103 Radbruch, A., Liesegang, B. and Rajewsky, K. (1980) Proc. Nat. Acad. Sci. U.S.A. 77, 2909-2913.
104 Gally, J.A. and Edelman, G.M. (1972) Annu. Rev. Genet. 6, 1-46.
105 Goding, J.W. (1978) Contemp. Topics Immunobiol. 8, 203-243.
106 Kehry, M., Sibley, C., Fuhrman, J., Schilling, J. and Hood, L.E. (1979) Proc. Nat. Acad. Sci. U.S.A. 76, 2932-2936.

107 Koshland, M.E. (1975) Advan. Immunol. 20, 41-69.
108 Vassalli, P., Tedghi, R., Lisowska-Bernstein, B., Tartakoff, A. and Jaton, J.-C. (1979) Proc. Nat. Acad. Sci. U.S.A. 76, 5515-5519.
109 Sibley, C.H., Ewald, S.J., Kehry, M.R., Douglas, R.H., Raschke, W.C. and Hood, L.E. (1980) J. Immunol. (in press).
110 Perry, R.P. and Kelley, D.E. (1979) Cell 18, 1333-1339.
111 Alt, F.W., Bothwell, A.L.M., Knapp, M., Siden, E., Mather, E., Koshland, M. and Baltimore, D. (1980) Cell 20, 293-301.
112 Lerner, M.R., Boyle, J.A., Mount, S.M., Wolin, S.L. and Steitz, J.A. (1980) Nature 283, 220-224.
113 Mitchell, G.F. and Miller, J.F.A.P. (1968) J. Exp. Med. 128, 821-837.
114 Cantor, H., Bayse, E.A. (1975) J. Exp. Med. 141, 1390-1399.
115 Gershon, R.K. (1974) Contemp. Topics Immunobiol. 3, 1-40.
116 Golstein, P., Wigzell, H., Blomgren, H. and Svedmyr, E.A.J. (1972) J. Exp. Med. 135, 890-906.
117 Kindred, B. and Corley, R.B. (1978) Eur. J. Immunol. 8, 67-71.
118 von Boehmer, H., Hengartner, H., Nabholz, M., Lernhardt, W., Schreier, M. and Haas, W. (1979) Eur. J. Immunol. 9, 592-597.
119 Szenberg, A., Marchalonis, J.J. and Warner, N.L. (1977) Proc. Nat. Acad. Sci. U.S.A. 74, 2113-2117.
120 Rajewsky, K. and Eichmann, K. (1977) Contemp. Topics Immunobiol. 7, 69-112.
121 Storb, U. (1978) Proc. Nat. Acad. Sci. U.S.A. 73, 2905-2908.
122 Forster, A., Hobart, M., Hengartner, H. and Rabbitts, T.H. (1980) Nature 286, 897-899.
123 Cory, S., Adams, J.M. and Kemp, D.J. (1980) Proc. Nat. Acad. Sci. U.S.A. 77, 4943-4947.
124 Kronenberg, M., Davis, M.M., Early, P.W., Hood, L.E. and Watson, J.D. (1980) J. Exp. Med. (in press).
125 Binz, H. and Wigzell, H. (1977) Contemp. Topics Immunobiol. 7, 113-177.
126 Cosenza, H., Julius, M.H. and Augustin, A.A. (1977) Immunol. Rev. 34, 3-33.
127 Shimizu, A., Takahashi, N., Yamawaki-Kataoka, Y., Nishida, Y., Kataota, T. and Honjo, T. (1980) Nature (in press).
128 Roeder, W., Maki, R., Traunecker, A. and Tonegawa, S. (1980) Proc. Nat. Acad. Sci. U.S.A. (in press).
129 Walker, I.D. and Harris, A.W. (1980) Nature 288, 290-292.

THE USE OF CLONED DNA FRAGMENTS TO STUDY HUMAN DISEASE

Stuart H. Orkin

Division of Hematology and Oncology
Children's Medical Center
The Sidney Farber Cancer Institute
Department of Pediatrics
Harvard Medical School
Boston, Massachusetts 02115

Recombinant DNA methods have revolutionized molecular genetics in the past several years. They now afford the opportunity to isolate, characterize and modify individual genes, and finally introduce them into living cells or into in vitro systems. In effect, the "new genetics" permits construction in the laboratory of selected mutants that may facilitate correlation of gene structure and function. An alternate approach towards the same ultimate goals is to exploit naturally-occurring mutations. Such mutations are distributed throughout human populations and often lead to severe phenotypic consequences in inherited diseases. A central aim in many laboratories using the new molecular techniques is to characterize underlying genetic defects in human DNA in the hope that this strategy will provide not only a fuller understanding of the diseases themselves but also insights into normal gene expression and development. In addition, the methods employed to date have already resulted in development of new, specialized means for the prenatal detection of genetic diseases that many families wish to avert.

The use of cloned DNA segments to study human disease is merely in its infancy. As the most significant advances have occurred so far in the analysis of inherited disorders of hemoglobin synthesis, these will be the focus of much of this review. Yet, virtually any class of disease is in principle amenable to molecular attack with current techniques. Therefore, I will broaden the perspective of this review to include general methods for the analysis of disease and linkage states and suggest those areas in which intensive research may soon be profitable.

TECHNICAL CONSIDERATIONS

Molecular study of human diseases is made possible by recent advances in the analysis and cloning of DNA segments. The most useful techniques in this regard are, 1) restriction endonuclease mapping and 2) molecular cloning of DNA sequences complementary to mRNAs (cDNAs), and particularly cloning of chromosomal regions. They will be briefly reviewed as they play a central role in the approaches described below.

Restriction endonuclease mapping, commonly referred to as gene mapping, signifies the fragmentation of cellular DNA with restriction enzymes and the probing of electrophoresed digestion fragments for specific gene sequences. This approach, first described by Southern (1), is exceedingly sensitive and can be used to detect single-copy genes within total human DNA. In practice, approximately 10 to 20 µg of total cell DNA is digested to completion with a restriction enzyme, e.g. EcoRI. The digestion products are then electrophoresed in agarose slabs, transferred to nitrocellulose filter sheets by the blotting method of Southern (1), and finally hybridized with a highly radioactive DNA probe (either cDNA or cloned cDNA). Upon autoradiography, those DNA fragments that anneal with the labeled probe may be identified. Experiments of this kind provide information of two sorts. First, they indicate whether a DNA sequence is, in fact, present in the DNA sample tested. Second, they demonstrate the size of the DNA fragment containing the gene sequence of interest. Clues to the underlying nature of a disorder are provided if differences between normal and patient DNA samples can be identified in this type of experiment. In addition, naturally occurring DNA differences in normal human populations (polymorphisms) may also be detected in this manner. The gene mapping approach has also permitted construction of physical maps for specific normal human DNA regions. These are particularly useful when attention turns to individuals with genetic lesions that are thought to lie within one of these mapped regions. The physical arrangement of the major human hemoglobin genes was established on the basis of gene mapping alone (2-4) prior to molecular cloning.

Gene cloning may involve either isolation of synthetic copies of mRNA species (cDNAs) in plasmids (5) or the isolation of genomic DNA segments (6-8). The former is particularly useful in the preparation of specific DNA probes for gene mapping studies or for genomic cloning. Cloned cDNAs are available for several human gene products, including the major globins (9,10) and selected hormones (insulin, growth hormone, chorionic somatomammotropin) (11-13). Analysis of cDNA clones permits establishment of the primary structure of mRNA and may in certain instances provide insights into a molecular disease.

The cloning of chromosomal DNA segments offers the opportunity to isolate normal and mutant genes from human specimens (7,8). The development of suitable bacteriophage vectors (14) and in vitro packaging methods (15,16) have facilitated progress in this area, as have methods for the screening of large numbers of phage for specific sequences by hybridization (17). Chromosomal segments may be cloned either by introduction of partially purified restriction fragments into vector DNA (6) or by construction of gene banks or libraries that statistically include all DNA sequences (7,8). In theory, any region of the genome can be isolated in pure form by either or both of these approaches. Through the work of Maniatis et al. (7), human DNA libraries have been constructed and have been utilized to clone several gene regions. Most impressive in this regard has been the isolation of the entire α-globin (18) and β-globin (19) linkage groups from normal human DNA. The methods for cloning selected chromosomal regions are sufficiently powerful that single-copy human genes of interest may be isolated from the quantities of DNA that may be conveniently obtained from 50 to 100 cc of peripheral blood cells. Therefore, DNA mutations occurring in human diseases may be examined without the necessity for acquisition of tissue samples.

DIRECT ANALYSIS OF HUMAN DISEASE BY GENE MAPPING AND CLONING

The most straightforward approach to the analysis of genetic diseases is examination of the cellular DNA for the underlying genetic alteration (20). Gene mapping provides a highly sensitive and specific method for such analysis.

Many diseases result from single nucleotide changes in the DNA. Examples in this class are those conditions in which an abnormal protein product with a single amino acid substitution is found. Most of the time, molecular probes for the mutated gene sequence (i.e., its normal counterpart) are not available. Extensive background knowledge of hemoglobin disorders coupled with the use of globin cDNA probes for each major globin chain now permits detection of some globin chain mutations directly. An excellent example of this approach is afforded by the Hb O^{Arab} mutation, resulting from a substitution at amino acid 121 of the human β-globin chain. When inherited with a sickle cell β chain mutation on the opposite chromosome, a potentially severe condition, like that of sickle cell anemia, may develop. The mutation in the Hb O^{Arab} β chain occurs at the DNA sequence recognized by the restriction enzyme EcoRI (GAATC). Normally, gene mapping gels of human DNA probed with β-cDNA yield two hybridizing bands originating from the β-globin gene, one of 5.2 kilobases containing the 5'-portion and another of 3.6 kilobases containing the 3'-portion. In the Hb O^{Arab} gene, the restriction site is deleted, such that a new band of about 8.8 kilobases is formed (3). Several other known hemoglobin mutations occur at cleavage sites for restriction enzymes within globin genes. Many of these mutations are rare and are of little clinical importance. However, the very common and clinically significant mutation that leads to sickle cell anemia (a mutation at the sixth codon of the β-globin chain) also occurs at a restriction enzyme recognition site for the enzyme MnlI (21). For technical reasons though, it has not yet been possible to use this enzyme successfully in gene mapping experiments to demonstrate the presence of the sickle cell mutation directly from total DNA. In theory, any genetic disorder due to a single nucleotide change at a restriction enzyme recognition site could be detected directly by these rather simple gene mapping procedures. Most mutations, nevertheless, will not occur at enzyme cleavage sites or at positions that are of use in gene mapping experiments.

An elegant example of the latter situation is provided by the demonstration of a nonsense or chain termination mutation in one patient with β^0-thalassemia (22). In conditions called thalassemias there is deficient or absent production of a globin chain from an affected locus (23,24). In the case of β^0-thalassemia, by definition, no β-globin chains are produced from the mutant gene. The molecular basis of this entity is heterogeneous (see below). In one Chinese patient, a nonsense mutation was demonstrated in the β-globin gene by analysis of small amounts of β-globin RNA present in erythroid cells. By sequencing cDNA transcripts primed with fragments isolated from a cloned (normal) β-globin cDNA, Chang et al. (22) detected the single nucleotide substitution that resulted in β^0-thalassemia. Although direct sequence analysis of the mutant mRNA or cDNA might be theoretically possible, their use of cloned DNA fragments to prime cDNA synthesis from the mutated mRNA at specific positions facilitated characterization of the lesion in this instance. It may be anticipated that a similar approach will be successful in identifying additional mutations of this kind in human globin genes and other gene products.

Many diseases are probably due to small deletions of DNA within the genome. Again, the thalassemias have provided the first examples of this situation. About five years ago, it was demonstrated by hybridization of sheared cell DNA with cDNA probes that several forms of thalassemia were the result of deletions affecting globin structural genes (25-29). Deletions were found to be associated with most α-thalassemias and with rare β-thalassemias affecting more than one linked β-like gene product ($\delta\beta$-thalassemias and hereditary persistence of fetal hemoglobin syndrome). More recent gene mapping studies have added further definition to these findings. In addition, gene mapping surveys of DNA samples of patients with various thalassemia syndromes have added new variations on the theme. Various deletions that

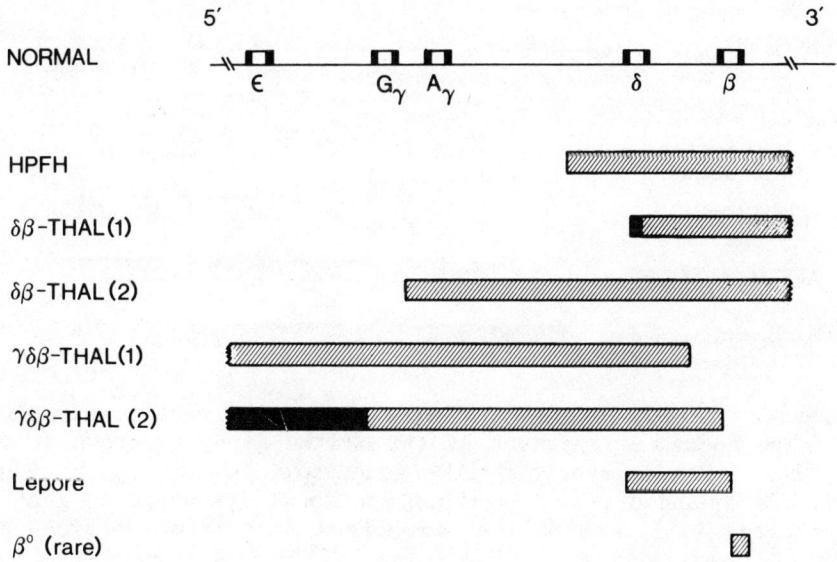

Figure 1. Deletions of DNA within the β-like globin gene cluster. At the top is the arrangement of β-like globin genes in normal human DNA (19). The structural genes are the embryonic ϵ gene, the nonallelic γ-globin genes ($^G\gamma$ and $^A\gamma$), the minor β-like gene δ and the adult β-globin gene. Each has two intervening sequences, the larger of which is depicted as the open area within the structural gene. Gene deletions in various thalassemia syndromes are indicated. Cross-hatched areas are certain to be missing. Stippled areas are likely to be deleted. HPFH = hereditary persistence of fetal hemoglobin syndrome. Cases summarized here are from refs. 3, 30, 31, 34-36, 74 and 75.

have been mapped in the human α-like and β-like gene clusters in this manner are depicted in Figures 1 and 2. Deletions may involve the complete absence of a structural gene, as in most α-thalassemias, and the hereditary persistence of fetal hemoglobin syndrome, or may involve only a segment of a gene, such as in one unusual type of β^0-thalassemia, a rare form of α-thalassemia and the common type of $\delta\beta$-thalassemia. Specific examples of these situations and new insights into human diseases will now be considered for illustrative purposes.

The smallest deletion detected so far by direct analyses is that found in rare patients with β^0-thalassemia (30,31). These deletions, all found in patients of Asian Indian origin, were discovered in gene mapping experiments of Orkin et al. (30) and Flavell et al. (31) by the presence of a β-globin gene containing a DNA fragment that was about 600 base pairs shorter than normal. After further mapping studies, it appeared that this dele-

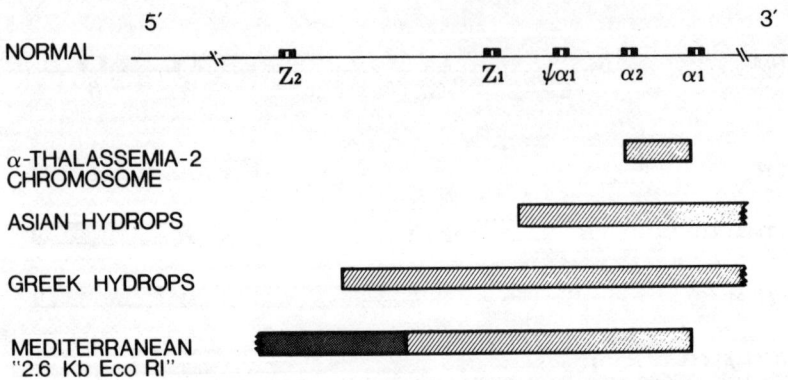

Figure 2. Deletions of DNA within the α-like globin gene cluster. The normal arrangement of the α-like genes is shown at the top (18). The embryonic α-like genes are Z_1 and Z_2; the adult genes are α_1 and α_2. $\psi\alpha 1$ is a nonfunctional (pseudogene) gene in the cluster (18). Cases are summarized from refs. 38,39,44 and 76.

tion involved the 3'-ends of the β-globin gene, but its position could not be determined more precisely due to the lack of available restriction sites near the end of the normal globin gene or within its large intervening sequence located about two-thirds into the gene. In order to define the nature of this deletion more accurately we isolated the β-globin gene region from one such patient using bacteriophage cloning methods (74). With the cloned mutant gene now in phage, it was possible to examine its structure by electron microscope techniques. These experiments involved hybridization of the mutant gene to β-mRNA (R-loop analysis) and to fragments of a normal β-globin gene cloned from a gene library (heteroduplex analysis). The deletion of the terminal coding portion was evident from these R-loop experiments. Heteroduplex analysis in which corresponding mutant and normal gene fragments were annealed revealed a deletion loop (Figure 3) whose position with respect to restriction sites that defined their ends established the location of the deletion within the cloned gene. In this situation, the 600 base-pair deletion removed the terminal third of the large intervening sequence (the last coding block) and about 150 base pairs past the end of the normal β-globin gene. Direct examination of the cloned chromosomal gene, in this manner, led to definitive characterization of this genetic lesion. Deletion of a coding segment as well as the junction of an intervening sequence with a coding region appear

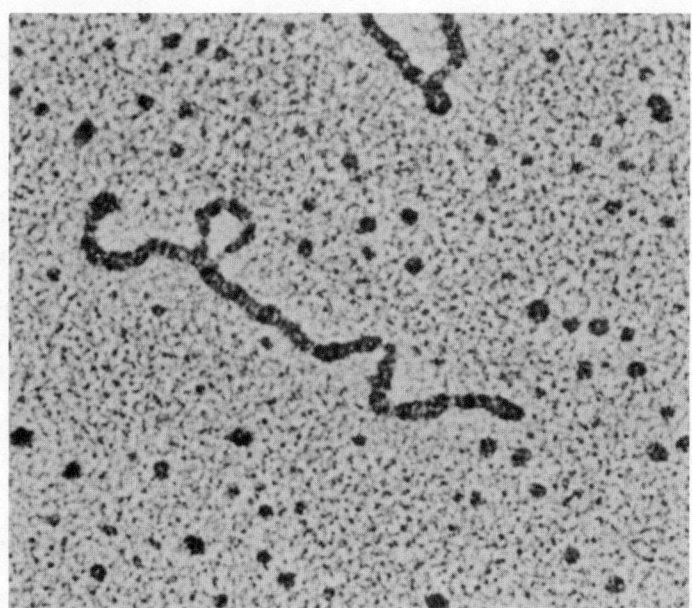

Figure 3. Direct demonstration of DNA deletion in cloned human DNA. A restriction fragment of a cloned β^0-thalassemia globin gene of an Asian Indian patient was annealed with the corresponding fragment cloned from a normal β-globin gene. The deletion loop seen by electron microscopy identified the deleted region, about 600 base pairs long, situated at the 3'-end of the gene (see Orkin et al. (74) for specific details).

to provide adequate explanation for the failure of β-globin production from the affected gene.

Another use of direct analysis for the study of human disease has been the attempts to correlate gene deletions with phenotypic differences in various thalassemia syndromes (see ref. 32). The analyses of the deletions within the β-like globin cluster are particularly illuminating in this regard. In humans, the predominant β-like globin produced during fetal life is that from the nonallelic γ-globin genes. About the time of birth, there is a gradual and then an accelerated dependence on β-globin chain production. If the β-globin gene is defective, usually the γ-globin genes remain relatively shut off. However, in syndromes associated with deletions in the β-like gene cluster, γ-globin gene expression may be considerably higher than normally expected. For example, in the syndrome called hereditary persistence of fetal hemoglobin there is no β- (or δ-) globin production due to deletion of the structural genes. However, γ-globin gene expression almost fully compensates, allowing the affected individuals to be entirely well, clinically. In contrast, in $\delta\beta$-

thalassemia, $\beta-$ (and $\delta-$) globin production is also absent, but γ-globin gene expression, although increased above that seen in common β-thalassemias, does not fully compensate. Affected individuals are generally mildly to moderately affected with anemia. A number of years ago, it was proposed that specific gene sequences located just upstream from the δ-globin gene might be responsible for extinguishing γ-globin gene expression after birth (33). The deletion in the HPFH syndrome, but not $\delta\beta$-thalassemia, might lead to continued high output γ-globin gene expression in the former, but not the latter. With gene mapping, a number of cases of each type of syndrome have been analyzed. It is not yet clear whether the notion of the switch box is at all correct. For example, in one form of $\delta\beta$-thalassemia studied in only one individual ($^G\gamma/\delta\beta$-thalassemia), this putative control region was deleted as well as the $^A\gamma$ gene, yet the disorder was not of the HPFH type (34,35). Therefore, it would seem premature to make definitive statements regarding the applicability of this model. For example, the phenotypes could be explained equally as well by an alternate hypothesis in which chromosomal domains are important in controlling relative outputs from the linked globin genes (36). This latter proposal was advanced by van der Ploeg et al. (36) to explain the findings in one patient with $\gamma\delta\beta$-thalassemia in which β-globin expression from the affected chromosome was absent although the mapped deletion did not extend to the β gene itself. Therefore, it is possible that it is not so much the specific intergenic sequences that are deleted, but how the deletion alters the conformation of the chromatin that leads to the differences in gene expression. What can be said at this time is that we have a fairly good, broad picture of the different extents of DNA deletion in various disorders of hemoglobin synthesis. This is merely descriptive, however, and does not establish as yet cause and effect relationships.

Molecular analysis of those thalassemias associated with defective α-globin production (the α-thalassemias) has provided some surprising insights into the human genome. Normally, the α-globin genes are duplicated (2,18,37) and physically linked within 4 kilobases of DNA. In some individuals with α-thalassemia, a chromosome with only a single copy of the α-globin sequence is present (38,39). This chromosome, called α-thalassemia-2 chromosome, arises due to unequal crossing-over within the α-globin region itself. This was first hypothesized on the basis of gene mapping studies of total cell DNA samples (38). More recently, Embury et al. (39) have shown that the α-thalassemia-2 chromosome can arise from two similar, yet different crossovers. These recombination events appear to be mediated by extensive homology throughout a 4 kilobase stretch within the α-globin gene region, demonstrated by electron microscopic observation of cloned normal α-globin genes isolated by Lauer et al. (18) from a cloned human library. Propagation of recombinant phage containing the duplicated α-globin genes in E. coli leads to deletion events that mimic those that appear to occur in human populations (18). Thus, study of recombinant phage in vitro may provide a new approach to the study of DNA alterations evident in human populations.

The belief that crossover events in the α-globin gene region lead to some deletions seen in α-thalassemic individuals has been strengthened by discovery of individuals who bear three α-globin genes in tandem (40,41). This chromosome arrangement, found in phenotypically normal individuals by gene mapping, is predicted by the crossover model for α gene deletion.

Analysis of recombinant DNA containing α sequences has provided additional insights into specific α-thalassemias associated with the presence of abnormal α-globin chains. Some individuals with α-thalassemia have an abnormal hemoglobin, Hb Constant Spring, in which the α-globin chain of the hemoglobin tetramer has an extension of 31 amino acids at the C-terminal end of the molecule (see. ref. 32). These elongated α-globin chains result from mutation of the normal terminator for translation and subsequent translation of the 3'-untranslated portion of the α-mRNA. Previous sequence analysis of α-mRNA yielded a 3'-untranslated sequence that was compatible with the peptide sequence of the Constant Spring chain (42,43). However, when a partially deleted α-globin gene found in a minority of α-thalassemic individuals (44) as well as the more 3' of the normal α-globin genes was examined by DNA sequencing, a significantly different 3'-untranslated region was identified (45). Thus, by study of recombinant DNA clones of chromosomal material it was documented that the duplicated α-globin genes actually differ in their 3'-untranslated portions. Furthermore, these results showed that the elongated α-globin chain mutants arise specifically from mutations occurring in the more 5' of the normal duplicated α-globin genes.

Direct examination of cloned β-globin genes is currently under way in several laboratories in an effort to investigate the nature of the genetic lesions responsible for typical β-thalassemias. Hybridization of globin cDNA clones with pulse-labeled or electrophoresed nuclear RNA samples has now demonstrated that $β^+$-thalassemia, a disorder in which insufficient numbers of β-globin chains are synthesized, is associated with defective or inefficient processing of β-globin mRNA precursors (46,47). Since these precursors normally contain transcripts of the intervening sequences within the β-globin gene, it is currently thought that inefficient processing results from specific mutations occurring within the intervening sequences (48,49). Attempts to demonstrate this directly involve the isolation of mutant β-globin genes and their characterization by DNA sequence or functional analysis. The problem is formidable, though, as there appears to be some individual variation within intervening sequences that may make identification of the true thalassemic mutations difficult. We can expect nevertheless that comparison of DNA sequences of several thallassemic genes cloned in different laboratories may eventually lead to a resolution of these issues.

Direct analysis of human DNA, both by gene mapping and cloning techniques, has provided new insights into the molecular basis of one class of disorders, the thalassemia syndromes. At present, no other diseases have been studied in similar depth.

INDIRECT STUDY OF HUMAN DISEASE BY DNA ANALYSIS

Although most DNA sequences are conserved among individuals, the differences from one individual to the next indicate that some DNA sequences differ. Because of linkage disequilibrium, we would predict that some specific DNA sequences might become linked to particular genes of medical significance. Kan and Dozy (50) provided the first and most dramatic example of this phenomenon. They reported that a restriction enzyme site 3' to the β-globin gene was usually (but not always) different from normal in chromosomes bearing β-globin genes with the sickle mutation. In their initial studies, this HpaI polymorphism was linked approximately 70 to 80% of the time with the β^S gene. In other studies, the frequency of this linkage seemed somewhat lower (51). Nevertheless, in family studies, Kan and Dozy demonstrated that restriction enzyme cleavage sites can be utilized as markers of a specific gene associated with human disease. This was a milestone in the molecular analysis of disease because in theory any genetic condition could be detected by finding an appropriately linked restriction enzyme polymorphism in the population. In theory, it is not even necessary to have prior knowledge of the specific gene that is mutated in a genetic disorder.

Extension of this indirect method for detection of specific hemoglobin gene mutations has occurred in two related directions. First, Kan and his co-workers (52) demonstrated that another restriction enzyme polymorphism (for BamHI) could be utilized to exclude the presence of a β^0-thalassemia gene in individuals of Sardinian background. This indicated that the polymorphic approach to the identification of mutant genes was suitable for loci other than the β^S gene. This particular polymorphism, however, is of infrequent utility for gene exclusion due to the distribution of the BamHI polymorphism in the normal Sardinian population. Another polymorphism of perhaps greater utility is one for the enzyme HindIII that is present in the intervening sequences of the nonallelic γ-globin genes (53-56). Different cleavage patterns in gene mapping gels for HindIII γ-globin specific fragments are distributed widely throughout human populations. Therefore, within a single family, members may differ with respect to their digestion patterns. From a study of the parents and offspring in a family in which sickle cell anemia or thalassemia is found, it is frequently possible to identify the chromosome bearing a β-globin mutation by examination of the cleavage sites within the linked γ-globin gene region (55,56). By using this approach coupled with Hpa polymorphism reported by Kan and Dozy (50), Phillips et al. (56) reported that nearly 90% of families in which sickle cell anemia is present would be suitable for assignment of the relevant genotypes by DNA analysis. As will be summarized below, the use of these restriction enzyme polymorphisms should greatly extend the use of restriction enzyme mapping for prenatal diagnosis of hemoglobin disorders.

The recognition of these restriction enzyme-site polymorphisms has also permitted examination of the evolution of the sickle cell mutation (57). By studying the association of the β^S gene and the HpaI polymorphism in several populations worldwide, Kan and Dozy (50) have hypothesized that the β^S mutation is multicentric in origin and the HpaI polymorphism they first detected arose (or was selected for) in West Africa. The analysis of DNA polymorphisms then provides a new means to explore population evolution and genetics.

NEW APPROACHES TO GENE LINKAGE ANALYSIS

In theory, use of recombinant DNA probes for other chromosomal DNA segments would be of great utility for mapping the genome and identifying markers for specific genetic diseases. Several recently designed stategies offer great promise in this direction.

First, Gusella et al. (58) have reported a rather simple method for the isolation of cloned DNA segments for specific human chromosomes. Their approach combines somatic cell genetics with recombinant DNA technology in the following way. The DNA of a rodent-human hybrid cell containing a specific human chromosome is cloned in phage. The resulting phages are then screened for human DNA inserts by hybridization with labeled, middle-repetitive human DNA. Since repetitive sequences are interspersed throughout the human genome (59) and interspecific hybridization does not occur under fairly stringent washing conditions, it is possible to identify phage bearing human DNA segments specific for the chromosomes contained in the original hybrid cells. In this manner, Gusella et al. (58) obtained a collection of DNA fragments specific for regions of chromosome 11. If hybrid cells containing only portions of a specific chromosome are utilized, phage clones containing only DNA from these regions can be obtained. This method should be useful for the construction of linkage maps of entire human chromosomes. Systematic dissection of the human genome may be envisioned with this strategy by assembly of phage banks containing selected human chromosomes.

White and his colleagues (60,61) have described a related approach to map the genome and perhaps provide markers for genetic disorders. In their studies, human DNA libraries are screened with repetitive sequence probe and rare phage are identified that contain no repeated DNA sequences. These cloned DNAs are then used directly as hybridization probes for total cell DNA in gene mapping gels. By assembling a collection of such clones, these workers believe that DNA probes for essentially all linkage groups of the human genome may be obtained. It should be feasible to choose informative families with specific disorders with a bank of such cloned probes to search for new markers of specific diseases. Although this may appear at the outset to be a fishing expedition, rough calculations demonstrate that such an approach should be tractable for detection of new markers of mutant genes, if on the order of about 100 probes are developed. Again, it should be stressed that it is not necessary to have prior knowledge regarding the specific gene that is mutated in particular diseases. The indirect approach to mutant gene identification, in theory, obviates these requirements.

A number of human diseases appear to be associated with the inheritance of specific histocompatibility markers (HLA alleles). A very recent development that may provide new insights into this relationship as well as the genetics of the HLA loci is the cloning of a cDNA probe for an HLA sequence by Ploegh et al. (62). Gene mapping with this cloned human sequence should reveal the degree of DNA sequence polymorphism within the HLA cluster on chromosome 6 and may also serve to extend the applicability of DNA analysis for the detection of human diseases.

APPLICATION OF MOLECULAR ANALYSIS TO THE PRENATAL DIAGNOSIS OF DISEASE

The consequences of many genetic disorders are such that families may elect, if given the opportunity, to avoid the birth of affected offspring. Use of amniotic cells for chromosome or enzymatic assays has been widespread for the past 10 years (63). More recently, acquisition of fetal blood cells has permitted analysis of the phenotype of fetal blood cells themselves in utero. This latter approach has been of great utility in the prenatal detection of hemoglobinopathies and thalassemias, and more recently in the diagnosis of hemophilia (64). Because the DNA of any tissue cells is potentially useful for the analysis of nearly all genetic conditions, the examination of amniotic cells shed from the fetus during life in utero offers promise for accurate and safe early detection of disease (20). Sufficient cell DNA can be obtained during the second trimester of pregnancy by simple aspiration of cells present within the amniotic fluid (65). To perform multiple gene mapping assays it is often necessary (and prudent) to culture the amniotic cells for two to three weeks in vitro. Naturally, those conditions in which single nucleotide changes in the DNA lead to restriction pattern alterations or structural gene sequence deletions are most suitable for this direct approach to the prenatal diagnosis of disease. In addition, when restriction enzyme polymorphisms are shown to be linked to a particular gene of interest, the indirect gene mapping method for prenatal diagnosis can be performed. At present, selected families with thalassemia and many with sickle cell anemia have the appropriate genotypes for prenatal diagnosis by DNA analysis.

Due to the potential complexity of polymorphisms within a family, especially when use of HindIII and HpaI polymorphisms is considered for sickle cell anemia diagnosis, it is wise to study family members as early as possible, preferably prior to pregnancy. Once the DNA linkage markers in a family are established, the optimal approach to prenatal diagnosis can be used. Lack of such planning may inevitably lead to experimental errors or failures and undue anxiety. It is clear that further application of DNA polymorphism, often in combination with one another, will extend the clinical utility of this approach in the near future. At this time, no human disease outside the hemoglobin disorders can be examined prenatally in this manner.

STUDY OF OTHER HUMAN GENETIC DISORDERS

This review has by necessity been weighted heavily towards examination of the use of new molecular techniques to probe the disorders of hemoglobin synthesis. Part of the success realized in the study of these conditions has been the product of extensive prior family and genetic investigations of human hemoglobins and the rapid acquisition of recombinant DNA molecules containing cDNA and chromosomal segments bearing globin sequences. The power of recombinant DNA methodologies provides the tools with which to approach other interesting human gene products at the molecular level. Several human hormone genes have already been cloned from normal individuals. These include the insulin and growth hormone structural genes. The primary structures of these loci have been established. Naturally occurring polymorphism with the insulin gene has already been documented (66,67). At least one patient with diabetes is known to have a primary abnormality in insulin, documented by study of biologically inactive insulin in this individual (68). It is possible that additional forms of diabetes may be due to single site mutations within the insulin gene itself. These may eventually be established by study of cloned insulin gene sequences from affected individuals. In general, however, it would appear that only a small minority of individuals with diabetes may have such defects within the insulin gene. Mutations in other genes, perhaps those affecting the function of the pancreatic β cell itself, will be important in this regard.

We should anticipate progress soon in the analysis of immunoglobulin genes in humans as well. Extensive studies in murine systems have now demonstrated elegantly that somatic recombination occurs during immunogenesis to assemble the active immunoglobulin genes (69,70). Numerous inherited and acquired human disorders of immunoglobulin synthesis have been described (71). Although many of these may involve defects in cellular differentiation of the B-lymphocyte, some may reflect mutations that prevent normal recombination during immunogenesis. Once probes for human immunoglobulin genes, particularly the heavy chains, are available we can anticipate extensive investigation of immune disorders at the molecular level.

Large genes encoding proteins with multiple repeating units, such as the collagen genes, may also be associated with inherited disorders (72). Recent research into connective tissue diseases has focused on the structure of the collagen molecule. Collagen is a large protein assembled from polypeptide chains dominated by multiple repeating units. The collagen gene, as isolated from sheep DNA (73), is complex with numerous intervening sequences. If related collagen genes are linked in the DNA as seems likely to be the case, recombination events from unequal crossing-over as found in the globin gene families may occur and produce aberrant collagen chains that might function abnormally in the connective tissue space. Although hypothetical at this time, these possibilities will be tested in the future.

CONCLUDING REMARKS

Recombinant DNA technology has opened new vistas in the analysis of human disorders. As exemplified by study of the diseases of hemoglobin synthesis, the new techniques now permit direct assessment of gene mutations and deletion in DNA samples, the identification of sequence polymorphisms associated with mutant loci and novel approaches to prenatal diagnosis. The capacity to isolate probes for other human DNA sequences and chromosomal regions will lead to rapid progress in the use of linkage analysis to dissect the human genome and to identify the presence of mutant genes. Hormone, immunoglobin and collagen disorders would appear fertile areas for future consideration of the molecular basis of selected conditions. Since the relationship between the human diseases and the respective protein products may be less direct than in the hemoglobinopathies, molecular approaches may have some dead ends and, perhaps, some surprising turns. In any case, we are now on the verge in many areas of defining human diseases by their primary genetic alterations in the DNA rather than by their phenotypes alone. In time, this is certain to have an impact on the diagnosis and, perhaps, the management of human genetic diseases.

Acknowledgments: I thank Dr. David G. Nathan for his continued support and advice and Ms. Catherine Lewis for the preparation of this manuscript. Studies in my laboratory were supported by grants of the National Institutes of Health and the National Foundation-March of Dimes and by a Research Career Development Award.

REFERENCES

1. Southern, E.M. (1975) J. Mol. Biol. 98, 503-577.
2. Orkin, S.H. (1978) Proc. Nat. Acad. Sci. U.S.A. 75, 5950-5954.
3. Flavell, R.A., Kooter, J.M., DeBoer, E., Little, P.F.R. and Williamson, R. (1978) Cell 15, 25-41.
4. Little, P.F.R., Flavell, R.A., Kooter, J.M., Annison, G. and Williamson, R. (1979) Nature (London) 278, 227-231.
5. Maniatis, T., Kee, S.K., Efstratiadis, A. and Kafatos, F.C. (1976) Cell 8, 163-182.
6. Tilghman, S.M., Tiemeier, D.C., Polsky, F., Edgell, M.H., Seidman, J.G., Leder, A., Enquist, L.W., Norman, B. and Leder, P. (1977) Proc. Nat. Acad. Sci. U.S.A. 74, 4406-4410.
7. Maniatis, T., Hardison, R.C., Lacy, E., Lauer, J., O'Connell, C., Quon, D., Sim, G.K. and Efstratiadis, A. (1978) Cell 15, 687-701.
8. Smithies, O., Blechl, A.E., Denniston-Thompson, K., Newell, N., Richards, J.E., Slighton, J.L., Tucker, P.W. and Blattner, F.R. (1978) Science 202, 1284-1289.
9. Wilson, J.T., Wilson, L.B., deRiel, J.K., Villa-Komaroff, L., Efstratiadis, A., Forget, B.G. and Weissman, S.M. (1978) Nucl. Acids Res. 5, 563-581.
10. Little, P.F.R., Curtis, P., Contelle, C., van den Berg, J., Dalgleish, R., Malcolm, S., Courtney, M., Westaway, D. and Williamson, R. (1978) Nature (London) 273, 640-643.
11. Bell, C.I., Swain, W.P., Pictet, R., Cordell, B., Goodman, H.M. and Rutter, W. (1979) Nature (London) 282, 525-529.
12. Martial, J.A., Hallewell, R.A., Baxter, J.D. and Goodman, H.M. (1979) Science 205, 602-607.
13. Fiddes, J.C., Seeburg, P.H., DeNoto, F.M., Hallewell, R.A., Baxter, J.D. and Goodman, H.M. (1979) Proc. Nat. Acad. Sci. U.S.A. 76, 4294-4298.
14. Blattner, F.R., Williams, B.G., Blechl, A.E., Denniston-Thompson, K., Faber, H.E., Furlong, L.A., Grunwald, D.J., Kiefer, D.O., Moore, D.D., Schumm, J.W., Shelton, E.L. and Smithies, O. (1977) Science 196, 161-169.

15 Blattner, F.R., Blechl, A.E., Denniston-Thompson, K., Faber, H.E., Richards, J.E., Slighton, J.L., Tucker, P.W. and Smithies, O. (1978) Science 202, 1279-1284.
16 Sternberg, M., Tiemeier, D. and Enquist, L. (1977) Gene 1, 255-280.
17 Benton, W.D. and Davis, R.W. (1977) Science 196, 180-182.
18 Lauer, J., Shen, C.-K. and Maniatis, T. (1980) Cell 20, 119-130.
19 Fritsch, E.F., Lawn, R.M. and Maniatis, T. (1980) Cell 19, 559-572.
20 Orkin, S.H., Alter, B.P., Altay, C., Mahoney, M.J., Lazarus, H., Hobbins, J.C. and Nathan, D.G. (1978) N. Engl. J. Med. 289, 166-172.
21 Nienhuis, A.W. (1978) N. Engl. J. Med. 299, 195-196.
22 Chang, J.C. (1979) Proc. Nat. Acad. Sci. U.S.A. 76, 2886-2889.
23 Maniatis, T., et al. (1980) Annu. Rev. Genet. (in press).
24 Orkin, S.H. and Nathan, D.G. (1980) Adv. Human Genet. (in press).
25 Ottolenghi, S., Lanyon, W.G., Paul, J., Williamson, R., Weatherall, D.J., Clegg, J.B., Pritchard, J., Pootrakul, S. and Boon, W.H. (1974) Nature (London) 251, 389-392.
26 Taylor, J.M., Dozy, A., Kan, Y.W., Varmus, H.G., Lie-Injo, L.E., Ganesan, J. and Todd, D. (1974) Nature (London) 251, 392-393.
27 Ottolenghi, S., Comi, P., Giglioni, G., Tolstoshev, P., Lanyon, W.G., Mitchell, G.J., Williamson, R., Russo, G., Musumeci, S., Schiliro, G., Tsistrakis, G.A., Charache, W., Wood, W.G., Clegg, J.B. and Weatherall, D.J. (1976) Cell 9, 71-80.
28 Kan, Y.W., Holland, J.P., Dozy, A.M., Charache, S. and Kazazian, H.H. Jr. (1975) Nature (London) 258, 162-163.
29 Forget, B.G., Hillman, D.G., Lazarus, H., Barell, E.F., Benz, E.J. Jr., Caskey, C.T., Huisman, T.H.J., Schroeder, W.A. and Housman, D. (1976) Cell 7, 323-329.
30 Orkin, S.H., Old, J.M., Weatherall, D.J. and Nathan, D.G. (1979) Proc. Nat. Acad. Sci. U.S.A. 76, 2400-2404.
31 Flavell, R.A., Bernards, R., Kooter, J.M., DeBoer, E., Little, P.F.R., Annison, G. and Williamson, R. (1979) Nucl. Acids Res. 6, 2749-2760.
32 Weatherall, D.J. and Clegg, J.B. (1979) Cell 16, 467-479.
33 Huisman, T.H.J., Schroeder, W.A., Etremer, G.D., Dume, H., Miller, A., Brodie, A., Shelton, J.R., Shelton, J.B. and Apell, G. (1974) Ann. N.Y. Acad. Sci. 232, 107-122.
34 Fritsch, E.F., Lawn, R.M. and Maniatis, T. (1979) Nature (London) 279, 598-603.
35 Orkin, S.H., Alter, B.P. and Altay, C. (1979) J. Clin. Invest. 64, 866-869.

36 van der Ploeg, L.H.T., Konigs, A., Oort, M., Roos, D., Bernini, L. and Flavell, R.A. (1980) Nature (London) 283, 637-642.
37 Embury, S.H., Lebo, R.V., Dozy, A.M. and Kan, Y.W. (1979) J. Clin. Invest. 63, 1307-1310.
38 Orkin, S.H., Old, J., Lazarus, H., Altay, C., Gurgey, A., Weatherall, D.J. and Nathan, D.G. (1979) Cell 17, 33-42.
39 Embury, S.H., Miller, J.A., Dozy, A.M., Kan, Y.W., Chan, V. and Todd, D. (1980) J. Clin. Invest. (in press).
40 Goossens, M., Dozy, A.M., Embury, S.H., Zachariades, Z., Hadjiminas, M.G., Stamatoyannopoulos, G. and Kan, Y.W. (1980) Proc. Nat. Acad. Sci. U.S.A. 77, 518-521.
41 Higgs, D.R., Old, J.M., Pressley, L., Clegg, J.B. and Weatherall, D.J. (1980) Nature (London) 284, 632-635.
42 Proudfoot, N.J., Gillam, S., Smith, M. and Longley, J.I. (1977) Cell 11, 807-818.
43 Wilson, J.T., Wilson, L.B., Reddy, V.B., Cavallesco, C., Ghosh, P.K., deRiel, J.K., Forget, B.G. and Weissman, S.M. (1980) J. Biol. Chem. 255, 2807-2815.
44 Orkin, S.H. and Michelson, A. (1980) Nature (London) 286, 538-540.
45 Michelson, A. and Orkin, S.H. (1980) Cell 22, 371-377.
46 Maquat, L.E., Kinniburgh, A.J., Beach, L.R., Honig, G.R., Lazerson, J., Ershler, W.B. and Ross, J. (1980) Proc. Nat. Acad. Sci. U.S.A. 77, 4287-4291.
47 Kantor, J.A., Turner, P.H. and Nienhuis, A.W. (1980) Cell 21, 149-157.
48 Tilghman, S.M., Curtis, P.J., Tiemeier, D.C., Leder, P. and Weissmann, C. (1978) Proc. Nat. Acad. Sci. U.S.A. 75, 1309-1313.
49 Ross, J. (1976) J. Mol. Biol. 106, 402-420.
50 Kan, Y.W. and Dozy, A.M. (1978) Proc. Nat. Acad. Sci. U.S.A. 75, 5631-5635.
51 Feldenzer, J., Mears, G., Burns, A.L., Natta, C. and Bank, A. (1979) J. Clin. Invest. 64, 751-755.
52 Kan, Y.W., Lee, K.Y., Furbetta, M., Angius, A. and Cao, A. (1980) N. Engl. J. Med. 302, 185-188.
53 Tuan, D., Biro, P.A., deRiel, J.K., Lazarus, H. and Forget, B.G. (1979) Nucl. Acids Res. 6, 2519-2544.
54 Jeffreys, A.J. (1979) Cell 18, 1-10.
55 Little, P.F.R., Annison, G., Darling, S., Cleamson, R.W., Camba, L. and Modell, B. (1980) Nature (London) 285, 144-147.
56 Phillips, J.A. III, Panny, S.R., Kazazian, H.H. Jr., Boehm, C., Scott, A.F. and Smith, K.D. (1980) Proc. Nat. Acad. Sci. U.S.A. 77, 2853-2856.
57 Kan, Y.W. and Dozy, A.M. (1980) Science 209, 388-390.

58 Gusella, J., Varsanyi-Breiner, A., Kao, F.-T., Jones, C., Puck, T.T., Keys, C., Orkin, S. and Housman, D. (1979) Proc. Nat. Acad. Sci. U.S.A. 76, 5239-5243.
59 Jelinek, W.R., Toomey, T.P., Leinwand, L., Duncan, C.H., Biro, P.A., Chondary, P.V., Weissman, S.M., Rubin, C.M., Houck, C.M., Deininger, P.L. and Schmid, C.W. (1980) Proc. Nat. Acad. Sci. U.S.A. 77, 1398-1402.
60 Botstein, D., White, R.L., Skolnick, M. and Davis, R.W. (1980) Amer. J. Hum. Genet. 31, 601-619.
61 Wyman, A.R. and White, R. (1980) Proc. Nat. Acad. Sci. U.S.A. 77, 6754-6758.
62 Ploegh, H.L., Orr, H.T. and Strominger, J.L. (1980) Proc. Nat. Acad. Sci. U.S.A. 77, 6081-6085.
63 Nathan, D.G., Alter, B.P. and Orkin, S.H. (1979) Clin. Perinatol, 6, 275-291.
64 Alter, B.P., Orkin, S.H. and Nathan, D.G. (1980) in Laboratory Investigation of Fetal Disease (Barson, H.J., ed.) (in press).
65 Kan, Y.W. and Dozy, A.M. (1978) Lancet ii, 91-93.
66 Ullrich, A., Dull, T.J., Gray, A., Brosius, J. and Sures, I. (1980) Science 209, 612-615.
67 Bell, G.I., Picket, R.L., Rutter, W.J., Cordell, B., Tischer, E. and Goodman, H.M. (1980) Nature (London) 284, 26-32.
68 Tager, H., Given, B., Baldwin, D., Mako, M., Markese, J., Rubenstein, A., Olefsky, J., Kobayashi, M., Kolterman, O. and Poucher, R. (1979) Nature (London) 278, 122-125.
69 Seidman, J.G., Leder, A., Nau, M., Norman, B. and Leder, P. (1978) Science 202, 11-17.
70 Sakano, H., Maki, R., Kurosawa, Y., Roeder, W. and Tonegawa, S. (1980) Nature (London) 286, 676-683.
71 Rosen, F.S. (1980) in Hematology of Infancy and Childhood (Nathan, D.G. and Oski, F.A., eds.), 2nd Edition (in press).
72 Molgaard, H.V. (1980) Nature (London) 286, 657-658.
73 Boyd, C.D., Tolstoshev, P., Schafer, M.P., Trapnell, B.C., Coon, H.C., Kretschmer, P.J., Nienhius, A.W. and Crystal, R.G. (1980) J. Biol. Chem. 255, 3212-3220.
74 Orkin, S.G., Kolodner, R., Michelson, A. and Husson, R. (1980) Proc. Nat. Acad. Sci. U.S.A. 77, 3558-3562.
75 Orkin, S.H., Goff, S. and Nathan, D.G. (1980) J. Clin. Invest. (in press).
76 Pressley, L., Higgs, D.R., Clegg, J.B. and Weatherall, D.J. (1980) Proc. Nat. Acad. Sci. U.S.A. 77, 3586-3589.

PHYSICAL MAPPING OF PLANT CHROMOSOMES BY

IN SITU HYBRIDIZATION

J. Hutchinson, R.B. Flavell and J. Jones

Plant Breeding Institute
Trumpington, Cambridge CB2 2LQ
England

INTRODUCTION

The use of in situ hybridization to identify chromosomal regions complementary to RNA or DNA sequences is now over 10 years old (1). In plants, the technique has been used to identify the chromosomal location of a limited number of highly repeated sequences but with few exceptions the nucleic acid probe molecules were impure, therefore limiting the resolution and the interpretations. The advent of the purification of DNA sequences by the molecular cloning of DNA fragments in bacterial plasmids has, however, opened up the possibility of obtaining large numbers of pure probes, thereby making in situ hybridization potentially much more useful to cytogeneticists and others interested in chromosome structure and chromosome identification.

Chromosome mapping in eukaryotes has been predominantly achieved by establishing genetic linkage groups and by cytological descriptions of metaphase or pachytene chromosomes. The two approaches have been united in some species by using specially characterized stocks containing chromosomal deletions, additions, substitutions or translocations. The determination of genetic linkage groups by scoring the segregation of phenotypic characters after making sexual crosses is time consuming while the characterization of chromosomes by morphological appearance is crude and often equivocal. This means that techniques, such as in situ hybridization, that combine the speed of the cytological approach with resolution similar to the gene mapping approach, are particularly appealing for mapping chromosomes. Staining techniques that produce bands on metaphase chromosomes should also be included in appraisals of chromosome mapping as they have proved to be particularly useful for mapping mammalian chromosomes. However, they have been less rewarding in studies of plant chromosomes because fewer bands are generally obtained.

In this chapter, we describe how the first few genomic clones of repeated sequences isolated from wheat and rye are being used to map plant chromosomes physically. We also discuss some of the applications of in situ hybridization for the identification of whole genomes or even single chromosomes and show how this technique may be used to answer questions about chromosome structure, chromosome behavior and the relationships between genomes.

METHODOLOGY

Most of the techniques used are based on those developed by Gall and Pardue (1) (for review, see refs. 2-6), with only minor modifications. Although this methodology was largely developed for animal chromosomes, it has been used with equal success with plant chromosomes. In outline, the technique consists of first preparing squashes on glass slides of the cells to be investigated. The immobilized nucleic acids in the cells are then denatured before being allowed to hybridize with radioactively-labeled DNA or RNA. After hybridization, the slides are washed to remove any unhybridized or poorly hybridized radioactive RNA or DNA. Finally, the slides are coated with photographic emulsion and the autoradiographs exposed before developing. The sites of labeled DNA or RNA hybridization are marked by blackened silver grains seen under the microscope. Various aspects of the technique, with particular reference to plants, are discussed in more detail below.

Preparation of Cell Squashes

Most in situ studies of plant cells have been made on mitotic root-tip preparations although some have also been made on pollen mother cells at various stages of meiosis (7-10). Polytene suspensor cells of Phaseolus species have also been used (11-14). In most experimental protocols, the tissue is fixed in 1:3 acetic acid:ethanol and then squashed in 45% acetic acid on acid-washed slides. This treatment may remove some of the histones and possibly makes the chromosomal DNA more susceptible to denaturation (15). In order to produce the very flat, well-spread chromosomes necessary for good in situ hybridization, it is best to dissect out the desired cells from the plant tissue using watchmaker's forceps under a dissecting microscope and remove all unwanted tissue debris (16). The usual technique for producing well-spread chromosomes, which includes prolonged hot acid hydrolysis, may not be used for in situ hybridization because the DNA of the chromosomes is too extensively depurinated. After squashing, the slides are frozen in liquid N_2 or on dry-ice so that the cover-slips may be easily removed, leaving the cells firmly adhering to the slide. The cells are then dehydrated by placing the slides in absolute ethanol, then air-dried. They may be stored for a few weeks under desiccating conditions at -20°C although it has been observed that in situ hybridization is less successful if the slides are stored for a long period of time (17).

Prior to the in situ hybridization, endogenous RNA is usually removed by treatment with RNase (e.g., 100 μg/ml pancreatic RNase in 2 x saline-citrate (SSC) at 37°C for 1 hr (SSC = 0.15 M NaCl, 0.015 M Na citrate)) (1). This is especially important when ribosomal DNA sites are being determined.

Denaturation of Chromosomal DNA

Various methods have been used to denature the DNA of plant chromosomes. These include treatment with alkali (e.g., 0.07 N NaOH for 2 min at room temperature) (12), acid (e.g., 0.2 N HCl for 20 to 30 min (7,18)), heat (e.g., 0.1 x SSC at 100°C for 0.5 min) (19,20) or a combination of heat and formamide (e.g., 3xSSC: 50% formamide (v/v) at 67°C for 30 sec) (10). The method employed is usually a compromise between obtaining maximum denaturation and maintaining the structure of the chromosomes (21-24).

Hybridization of Probe Molecules to Chromosomes

In early in situ experiments, hybridization of denatured radioactive RNA or DNA sequences (probes) to the chromosomes was usually carried out in 2 x SSC at a temperature of 10 to 20°C below the Tm of the native duplexes (1). (Tm = temperature at which the native DNA is 50% denatured.) More recently, hybridizations have been frequently carried out in 3 x SSC:50% formamide. By analogy with filter/solution hybridization experiments, the Tm is expected to be reduced by about 0.7°C for every 1% of formamide in the hybridization mixture (25) and therefore the hybridization temperature is appropriately reduced.

The time required for hybridization depends upon the degree of repetition of the sequence in the chromosomes and the amount and specific activity of the probe (23). In practice, hybridizations for highly repeated sequences are carried out overnight for a period of 15 to 16 hrs without substantial loss of chromosome morphology. For low copy sequences, other hybridization conditions may be required.

Post-Hybridization Treatments

With minor modifications, the post-hybridization treatments have followed the schedule outlined originally by Pardue and Gall (1,5). Unbound and poorly-bound radioactive nucleic acids are first removed by a series of stringent washings. The preparations are then dehydrated through 70% and absolute ethanol and air-dried before being dipped in photographic emulsion (e.g., Ilford K2 nuclear track emulsion diluted 1:1 with distilled water). After exposure, the length of which again depends upon the degree of repetition of the sequence and the specific activity of the probe, the autoradiographs are developed and the chromosomes are stained, often with Giemsa, prior to cytological examination.

Labeling of Probes

In the earliest experiments, the common nucleic acid probe was ribosomal 25S + 18S RNA labeled in vivo with tritium. The scope of the in situ hybridizations was therefore often limited because the specific activity of such RNA is usually low (for a review, see ref. 4). More recently, with cloned DNA sequences, the favored method is to use the E. coli enzyme RNA polymerase to produce tritium-labeled RNA complementary to the cloned DNA sequence (3). Alternatively, radioactive DNA may be produced by the method of nick translation (26,27). Since tritium is a fairly weak β emitter, low levels of radioactivity are not always detected very efficiently by the photographic emulsion. As an alternative to using tritium, therefore, the nucleic acid probes may be labeled with ^{125}I (7,8) using the Commerford procedure outlined by Prensky et al. (28). One advantage of using iodinated probes is that with the higher specific activities obtained, the length of time of the autoradiographic exposure is reduced and lower copy sequences can be detected. On the other hand, the more energetic β particles of ^{125}I may result in poorer resolution on the autoradiographs and background labeling may be a problem (17,29).

Types of Nucleic Acid Probes

To date, most of the sequences used to map plant chromosomes by in situ hybridization are highly repetitive. Before molecular cloning of plant DNA sequences was developed, highly repetitive sequences were isolated either by, 1) denaturing sheared DNA, incubating for a very short time (to a low C_0t value; C_0t = moles nucleotide/liter x incubation time in seconds) and recovering the renatured, highly repetitious DNA by hydroxylapatite chromatography (16,30), or 2) taking advantage of the fact that families of highly repeated sequences frequently have unusual base compo-

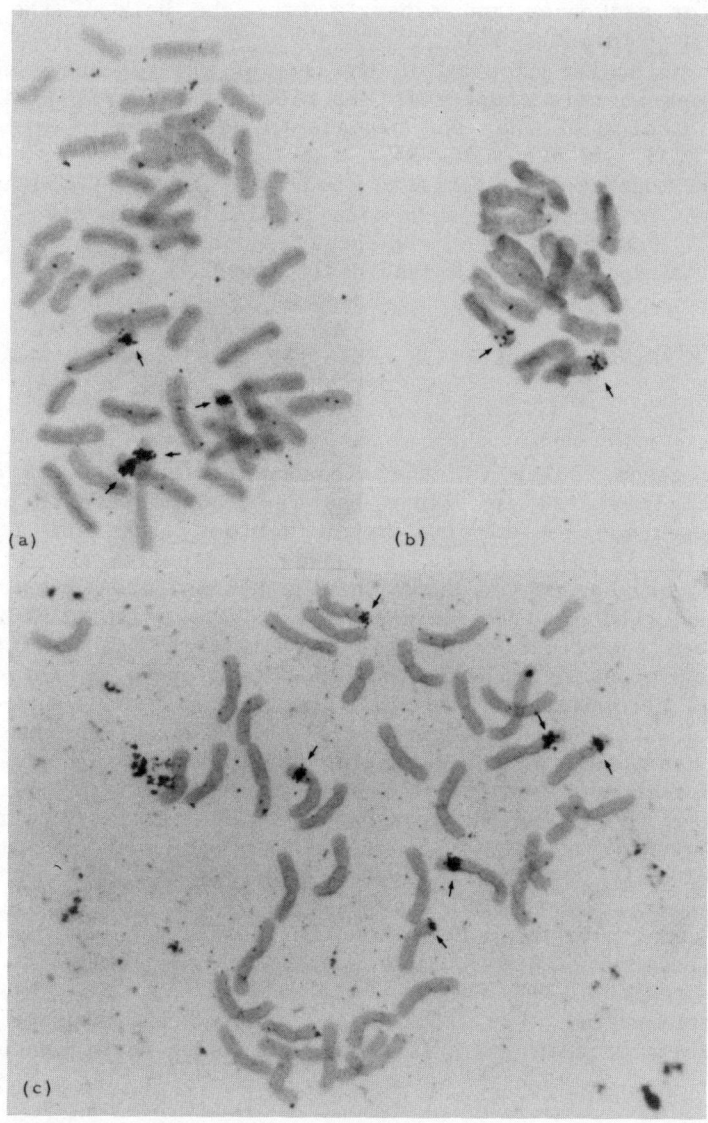

Figure 1. In situ hybridization of the tritium labeled cRNA transcribed by E. coli RNA polymerase from plasmid pTA71. This chimeric plasmid consists of a single wheat ribosomal gene repeating unit (i.e., 18S, 5.8S, 25S rRNA coding sequences and associated spacer DNAs) in the vector plasmid pACYC184 (33). 100,000 cpm of cRNA were hybridized/slide and the slides exposed for one week. The ribosomal rRNA gene sites are designated by arrows. The following species were probed: (a) Triticum aestivum cv. Chinese Spring; (b) Secale cereale cv. King II; (c) T. spelta.

sitions or sequences and separate from the bulk of the DNA in heavy salt gradients (31).

The molecular cloning of the rye highly repeated sequences illustrated in this chapter is described in Bedbrook et al. (32) and the cloning of the rDNA repeats is described in Gerlach and Bedbrook (33; see also ref. 34).

Some moderately repetitive sequences of wheat (35) and also zein genes in maize (36) have been localized by in situ hybridization. However, unique sequences cannot be detected by these methods, unless they are exceptionally long.

THE MAPPING OF REPETITIVE GENES AND DETECTION OF VARIATION IN COPY NUMBER

The genes coding for the ribosomal RNAs (25S, 18S and 5S) are highly repeated in plants, as in other eukaryotes, making their detection by in situ hybridization relatively straightforward (33,37-39). Figure 1a shows the results of hybridizing to wheat chromosomes (cultivar Chinese Spring) cRNA prepared from the cloned ribosomal DNA of wheat (33). Two pairs of hybridization sites are clearly visible, showing that this cultivar has two major clusters of ribosomal DNA (nucleolus organizers) on different chromosomes. Another form of wheat shows three pairs of hybridization sites (Figure 1c). This illustrates the use of in situ hybridization for detecting major variants in clusters of repeated sequences within plant populations. This method is also useful for detecting heterozygosities of clustered arrays of repeats within an individual, as illustrated by the variation in number of silver grains clustered over the two rDNA regions of rye chromosomes in the cultivar King II (Figure 1b) (39). Silver grain counts over many metaphase chromosome spreads have shown mean numbers of 17.0 ± 1.0 and 7.6 ± 0.6, respectively, for the two rDNA regions (39).

Viotti et al. (36) have also reported the mapping of the zein storage protein genes in maize by in situ hybridization.

THE MAPPING OF REPETITIVE SEQUENCES AND THE MOLECULAR ANALYSIS OF HETEROCHROMATIN

Many repeated sequences do not have a single chromosomal location so the precise region of the genome from which they were purified can rarely be inferred (see Figure 2a). However, the most highly repetitive sequences are often clustered in distinct heterochromatic regions and these are easy to discern by in situ hybridization. This is well illustrated (Figure 3b) by the in situ hybridization of members of the four unrelated highly

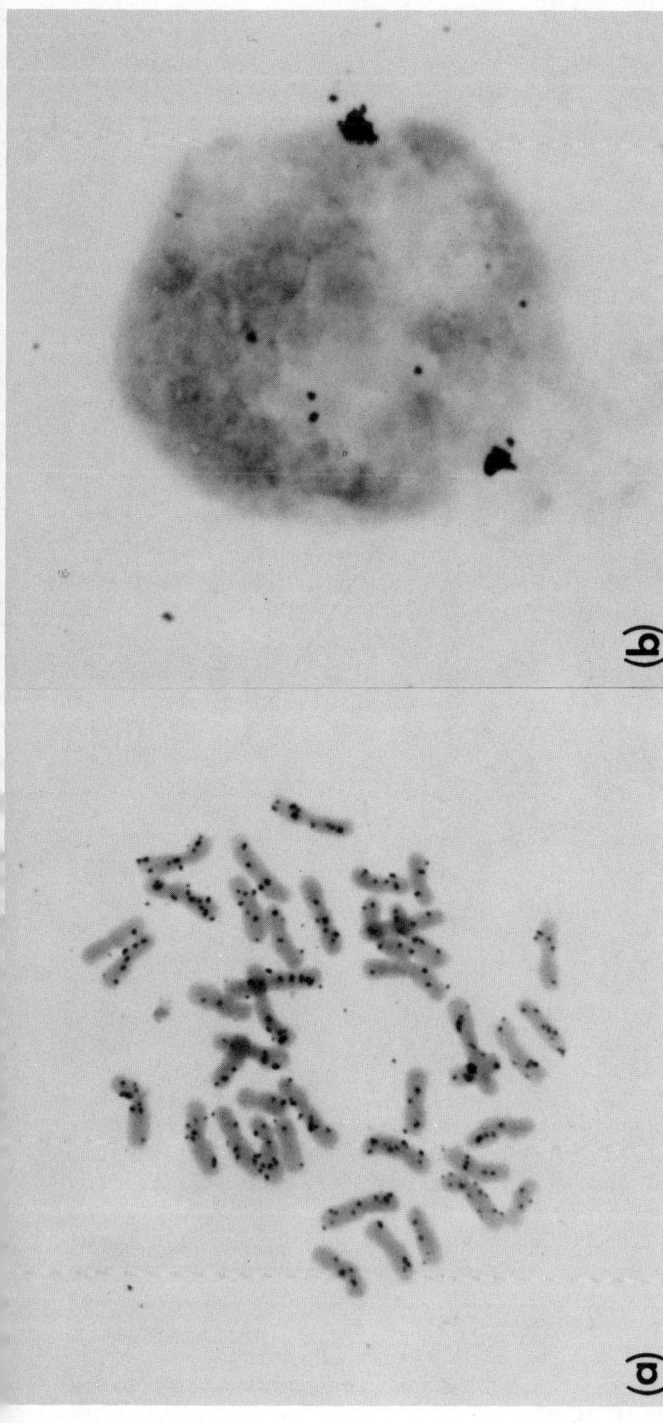

Figure 2. (a) In situ hybridization of a moderately repetitive sequence probe derived from wheat and hybridized to the wheat cultivar Chinese Spring. 100,000 cpm/slide were applied and the slides exposed for 4 weeks. (b) Hybridization of the tritium-labeled cRNA probe transcribed from the plasmid pSC210 to an interphase cell of the 2R addition line of rye cv. Petkus to wheat cv. Chinese Spring; pSC210 consists of a 480 base-pair sequence from rye in the vector plasmid pBR322 (32). The 480 base-pair sequence is highly repeated in rye but not in wheat. The heterochromatic regions of the two 2R chromosomes are clearly labeled by the probe, enabling the presence of rye chromosomes to be inferred even in an interphase wheat-rye addition line cell.

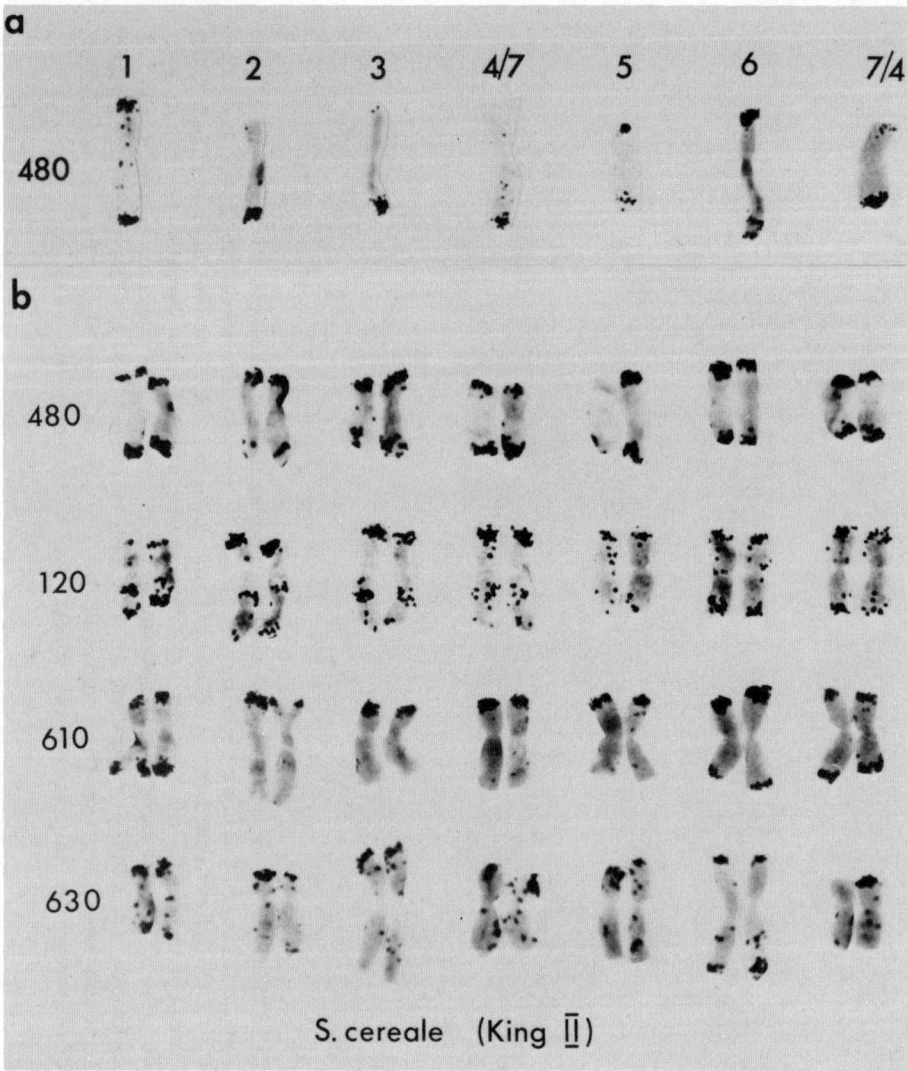

Figure 3. (a) <u>In situ</u> hybridization of the 480 base-pair (bp) sequence to King II rye chromosomes in wheat-rye addition lines. 110,000 cpm were applied/slide, except in the case of the 5R addition line where 65,000 cpm/slide were applied. The autoradiographs were exposed for 14 days. (b) <u>In situ</u> hybridization to mitotic chromosomes of the rye cultivar King II of the four highly repetitive cloned sequences of rye (33). The number of counts applied/slide and the exposure times for each of the four sequences were as follows: 480 bp - 65,000 cpm/14 days; 120 bp - 65,000 cpm/28 days; 610 bp - 60,000 cpm/21 days; 630 bp - 200,000 cpm/35 days. The arrangement of the chromosomes along each row is not ordered.

repeated sequence families cloned from cultivated rye (Secale cereale). Each family is named, for convenience, by the length of its repeating unit, i.e., 480, 120, 610 and 630 base pairs (32). The four cloned sequences hybridize predominantly to the heterochromatin localized on the ends of each rye chromosome (Figure 4), illustrating that the fragments were almost certainly cloned from one of these heterochromatic regions, and that this heterochromatin consists of tandem arrays of a small number of different repeating units (Figure 4). Two of the cloned sequences also hybridize to interstitial bands of heterochromatin, each sequence hybridizing to different bands, illustrating that these regions of heterochromatin differ in their sequence composition. Figure 4 summarizes the molecular dissection by means of in situ hybridization, of the heterochromatin of King II rye revealed by Giemsa C-banding (40,41). The King II rye chromosomes have been numbered 1 through 7 on the basis of results studying wheat-rye addition lines (Figure 3a).

THE USE OF IN SITU HYBRIDIZATION FOR CHROMOSOME IDENTIFICATION, KARYOTYPE ANALYSIS AND EVOLUTIONARY STUDIES, AND ANALYSIS OF MEIOTIC CHROMOSOME PAIRING

The in situ hybridization of sequences which are clustered in one or a few pairs of chromosomal sites provides a means of distinguishing these chromosomes from others in the complement. This is well illustrated by the rDNA examples in Figures 1a, 1b and 1c, and also by hybridization of 5S RNA to cereal chromosomes (37).

The cloned highly repeated sequences of rye are also useful for distinguishing individual chromosomes. The determination of the in situ hybridization patterns of individual rye chromosomes has been simplified by the utilization of plants selected from various wheat-rye hybrids with the full complement of wheat chromosomes and also one homologous pair of rye chromosomes (wheat-rye addition lines, 42). Cells of such plants show only a single pair of labeled chromosomes when hybridized with the 480, 610 or 630 base-pair repeated sequences from rye because these sequences are not highly repeated in wheat. The presence of the added rye chromosomes can be recognized also in interphase cells as illustrated in Figure 2b. It is therefore possible to screen large populations of plants to determine the presence of repeated sequence clusters without necessarily producing good metaphase chromosome spreads. The 480 base-pair sequence hybridization pattern of the chromosomes from King II added to wheat are shown in Figure 3a. Comparison of the pattern with that for the rye cultivar King II (Figure 3b) does, however, show obvious differences. The short arms of chromosomes 2R, 3R, 4R/7R and 7R/4R in the wheat genetic background all lack the major hybridization

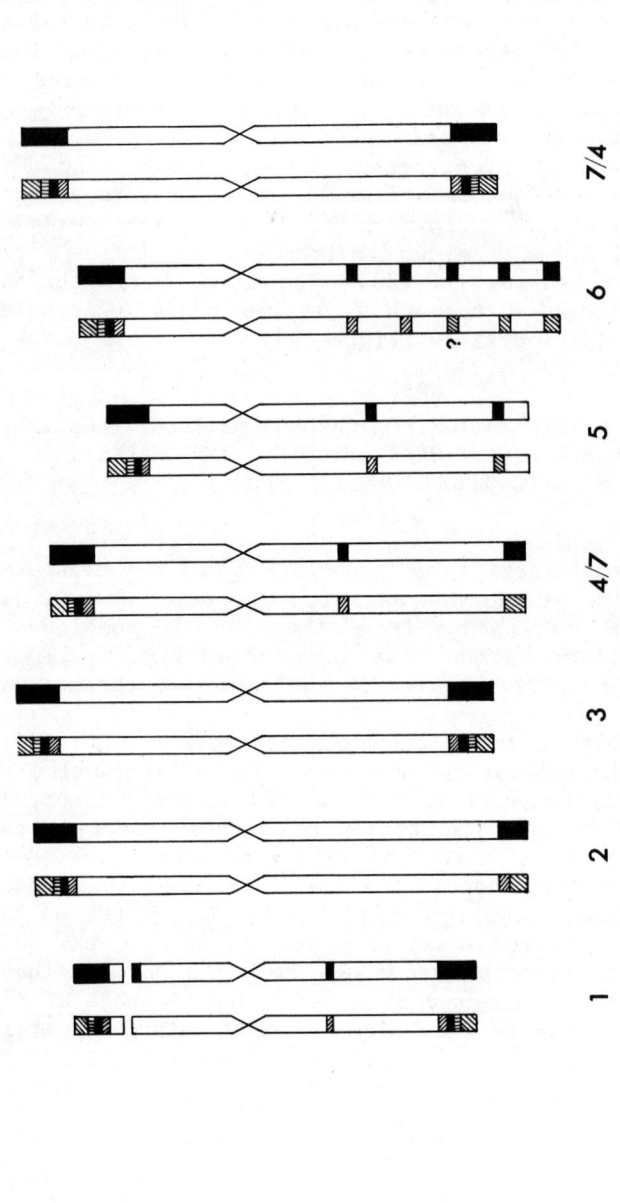

Figure 4. Schematic summary of C bands revealed by Giemsa staining rye chromosomes, cultivar King II (40,41) and the families of repeated sequences present in each C band, as determined by in situ hybridization (Figures 3a and 3b). The right-hand member of each chromosome pair records the C bands while the left-hand member illustrates the repeated sequences. ▨ 120 bp family ▧ 480 bp family ⊞ 610 bp family ■ 630 bp family The arrangement of the arrays of repeats within each telomeric block of heterochromatin may differ from that illustrated. The question mark (?) implies that in other rye cultivars this hybridization site has been observed but not unequivocally in the cultivar King II.

sites found in the parental rye cultivar. These deletions have possibly occurred during the production or maintenance of these chromosome additions to wheat (see ref. 40). This illustrates that in situ hybridization is useful for detecting karyotypic variants. Another variant is obvious in the in situ hybridization pattern of the 480 base-pair sequence to the cultivar King II itself. One chromosome of the fifth pair from the left (Figure 3b) has lost the 480 base-pair repeated sequences.

Although all of the four families of repeated sequences are present on all rye chromosomes, some individual chromosomes can be identified by their pattern of labeling. For example, the 4R/7R, 5R and 6R chromosomes of rye all have unique patterns of interstitial sites with one or two probes (Figure 4). Chromosome 1R has an interstitial site with the 120 base-pair probe like chromosome 4R, but can be distinguished by the nucleolar organizer cluster of rRNA genes. Chromosomes 2R, 3R and 7R are more difficult to distinguish but almost certainly the availability of more probe sequences would bring more resolution to this problem.

A further example of the identification of individual chromosomes by their pattern of labeling has been obtained in studies of wheat and its close relatives (9,31,43-45). With a variety of probes, such as wheat C_0t 10^{-1} DNA (43), an Ag^+/Cs_2SO_4 wheat satellite (31,44) or a satellite sequence from Drosophila melanogaster (density in CsCl 1.705 g.ml^{-1}) (9,45), which all showed similar patterns of hybridization, 9 out of the 21 pairs of chromosomes making up hexaploid wheat could be unequivocally identified. A combination of these wheat probes together with another highly repetitive probe specific for rye chromosomes has permitted the identification of translocated chromosomes resulting from whole arm interchanges between the chromosomes of wheat and rye (9), thereby demonstrating the use of in situ hybridization in plant breeding programs.

The C_0t^{-1} DNA probes and also the wheat and Drosophila satellite probes were found to hybridize predominantly to the regions proximal to the centromeres of the chromosomes (9,31,43-45). A further sequence isolated in this laboratory from the wheat cultivar Chinese Spring and purified by cloning also shows a similar overall pattern of hybridization. Other sequences when hybridized to wheat, however, show a different pattern of labeling with the label mostly in the telomeric regions. Figure 5 shows the 120 base-pair repeat of rye hybridized to the tetraploid wheat T. dicoccum. It will be observed that this sequence hybridizes predominantly to 7 of the 14 pairs of chromosomes making up tetraploid wheat, while 7 pairs of chromosomes show little hybridization. T. dicoccum is an allotetraploid made up of genomes A and B each consisting of 7 pairs of chromosomes. Experiments have shown that with both this 120 base-pair probe and also the other wheat probes mentioned earlier (9,31,43-45), the most highly repetitive sequences of wheat are found in the B genome. This confirms results obtained by other biochemical

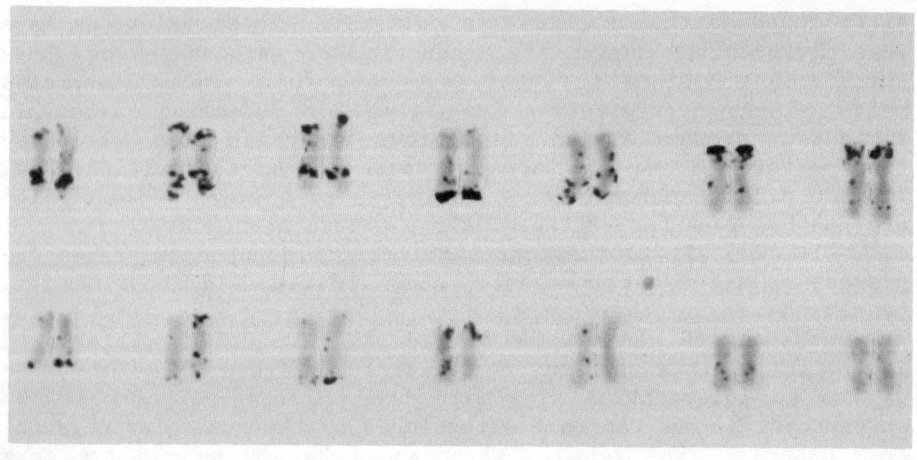

Figure 5. Mitotic metaphase chromosomes from the tetraploid species Triticum dicoccum labeled with the 120 base-pair sequence derived from rye.

methods (30). It is worth pointing out that in the investigation of the distribution of repeated sequence families, less labor is required to carry out in situ hybridizations to chromosomes than to extract DNA and perform filter hybridizations. First indications of the presence of the 120 base-pair family in T. dicoccum were obtained by this means, and the same procedure was subsequently used to show that this family is also present in species of the more distantly related genera Haynaldia, Hysterix, Elymus and Agropyron. It is therefore possible to use in situ hybridization, in conjunction with other methods, to look at the relationship between genomes, and also to examine the possible ancestry of polyploid species such as wheat (31).

The identification of a whole genome is very useful in plant breeding programs that involve the production of interspecific hybrids, because it provides a rapid means of establishing whether or not an interspecific hybrid has been produced. Figure 6a shows mitosis in a tetraploid hybrid between Hordeum chilense and Secale cereale cv. Petkus Somro after hybridization with the 480 base-pair cloned rye repeat. The 14 rye chromosomes are clearly labeled at the telomeres by the probe while the 14 Hordeum chilense chromosomes remain unlabeled because the 480 base-pair family is absent from Hordeum.

In situ hybridization may also be used to examine the pairing behavior at meiosis of the different genomes making up an interspecific hybrid. One such application of this technique was

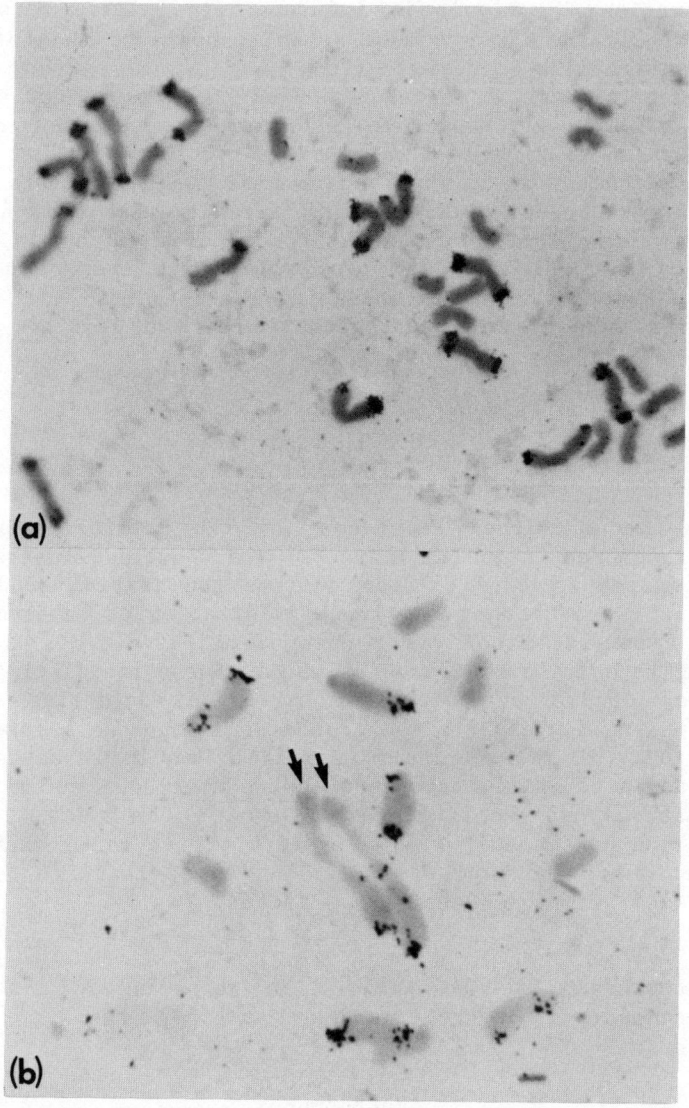

Figure 6. (a) In situ hybridization of the 480 bp sequence derived from rye to the mitotic root-tip chromosomes of a tetraploid 28 chromosome interspecific hybrid between Hordeum chilense and S. cereale cv. Petkus Somro. The 14 rye chromosomes are clearly labeled at the telomeres by the probe. The interspecific hybrid was developed by R.A. Finch, Plant Breeding Institute, Cambridge. (b) In situ hybridization of the 480 bp sequence derived from rye to cells at metaphase I of meiosis of the interspecific hybrid between Ae. caudata and S. cereale cv. Petkus Spring (10). The 2 bivalents, indicated by arrows, each show an association between one rye (labeled) and one Aegilops (unlabeled) chromosome.

an examination of meiosis in interspecific hybrids between the diploid species Aegilops comosa or Ae. caudata or the tetraploid Ae. columnaris with S. cereale (10).

At metaphase I of meiosis in both the diploid and triploid hybrids, a number of chromosome associations were observed. This is illustrated in Figure 6b which shows two such bivalent associations, together with 10 unpaired chromosomes (univalents) in the F_1 hybrid between Ae. caudata and S. cereale. Examination of the bivalents shows that only one chromosome of each bivalent is labeled by the 480 base-pair rye probe while there are also 5 labeled univalents, thus establishing that in this cell, two rye chromosomes were associated together with two Aegilops chromosomes.

CONCLUSIONS

The use of cloned sequences for the physical mapping of plant chromosomes is in its infancy. The results illustrated for cereal genomes in this chapter outline the initial studies but already it is clear that in situ hybridization of labeled cloned DNAs is a powerful tool for mapping repetitive genes and other noncoding repetitive sequences on chromosomes, for distinguishing individual chromosomes within a karyotype, detecting specific chromosomes in new genetic backgrounds and genomes in interspecific hybrids. It is especially useful for rapidly assaying major variations in repetitive sequence copy number between individuals in populations and should therefore prove to be a useful tool in genetics and cytogenetics. The breadth of its value will depend upon the number and type of cloned sequences that are developed and the ease with which they are detected.

Acknowledgment: We would like to thank A.G. Seal for helpful comments about the C banding pattern in rye.

REFERENCES

1 Gall, J.G. and Pardue, M.L. (1971) in Methods in Enzymology (Grossman, L. and Moldave, K., eds.), Vol. 21, pp. 470-480, Academic Press, New York, NY.
2 Hennig, W. (1973) Int. Rev. Cytol. 36, 1-44.
3 Jones, K.W. (1973) in New Techniques in Biophysics and Cell Biology (Pain, R.H. and Smith, B.J., eds.), Vol. 1, pp. 29-66, John Wiley and Sons, New York and London.
4 Wimber, D.E. and Steffensen, D.M. (1974) Ann. Rev. Genet. 7, 205-223.

5   Pardue, M.L. and Gall, J.G. (1975) Meth. Cell Biol. 10, 1-16.
6   Wobus, U. (1976) Biol. Zbl. 95, 1-24.
7   Wimber, D.E., Duffey, P.A., Steffensen, D.M. and Prensky, W. (1974) Chromosoma (Berlin) 47, 353-359.
8   Phillips, R.L., Wang, A.S., Rubenstein, I. and Park, W.D. (1979) Maydica 24, 7-21.
9   May, C.E. and Appels, R. (1980) Theor. Appl. Genet. 56, 17-23.
10 Hutchinson, J., Chapman, V. and Miller, T.E. (1980) Heredity 45, 245-254.
11 Avanzi, S., Buongiorno-Nardelli, M., Cionini, P. and D'Amato, F. (1971) Acad. Naz. Lencei. Rendic. Cl. Sci. Fis. Mat. Nat. Ser. VIII 50, 357-361.
12 Avanzi, S., Durante, M. Cionini, P.G. and D'Amato, F. (1972) Chromosoma (Berlin), 39, 191-203.
13 Brady, T. and Clutter, M.E. (1972) J. Cell Biol. 53, 827-832.
14 Durante, M., Cionini, P.G., Avanzi, S., Cremonini, R. and D'Amato, F. (1977) Chromosoma (Berlin) 60, 269-282.
15 Darzynkiewicz, Z., Traganos, F., Sharpless, T. and Melamed, M.R. (1975) Exp. Cell Res. 90, 411-428.
16 Appels, R., Driscoll, C. and Peacock, W.J. (1978) Chromosoma (Berlin) 70, 67-89.
17 Macgregor, H.C. and Mizuno, S. (1976) Chromosoma (Berlin) 54, 15-25.
18 Peacock, W.J., Lohe, A.R., Gerlach, W.L., Dunsmuir, P., Dennis, E.S. and Appels, R. (1977) Cold Spring Harbor Symp. Quant. Biol. 42, 1121-1135.
19 Timmis, J.N., Deumling, B. and Ingle, J. (1975) Nature 257, 152-155.
20 Ray, J.H. and Venketeswaran, S. (1979) Chromosoma (Berlin) 74, 337-346.
21 Hubbell, H.R., Sahasrabuddhe, C.G. and Hsu, T.C. (1976) Exp. Cell Res. 102, 385-393.
22 Singh, L., Purdom, I.F. and Jones, K.W. (1977) Chromosoma (Berlin) 60, 377-389.
23 Szabo, P., Elder, R., Steffensen, D.M. and Uhlenbeck, O.C. (1977) J. Mol. Biol. 115, 539-563.
24 Shapiro, I.M., Moar, M.H. and Ohno, S. (1978) Exp. Cell Res. 115, 411-414.
25 McConaughy, B.L., Laird, C.D. and McCarthy, B.J. (1969) Biochemistry 8, 3289-3295.
26 Maniatis, T., Jeffrey, A. and Kleid, D.G. (1975) Proc. Nat. Acad. Sci. U.S.A. 72, 1184-1188.
27 Singer, M.F. (1979) in Genetic Engineering (Setlow, J.K. and Hollaender, A., eds.), Vol. 1., p. 8, Plenum Press, New York and London.
28 Prensky, W., Steffensen, D.M. and Hughes, W.L. (1973) Proc. Nat. Acad. Sci. U.S.A. 70, 1860-1864.

29 Hennen, S., Mizuno, S. and Macgregor, H.C. (1975) Chromosoma (Berlin) 50, 349-369.
30 Flavell, R.B., O'Dell, M. and Smith, D. (1979) Heredity 42, 309-322.
31 Gerlach, W.L., Appels, R., Dennis, E.S. and Peacock, W.J. (1978) in Proc. 5th Int. Wheat Genet. Symp (Ramanujam, S., ed.), Vol. 1, pp. 81-91.
32 Bedbrook, J.R., Jones, J., O'Dell, M., Thompson, R.D. and Flavell, R.B. (1980) Cell 19, 545-560.
33 Gerlach, W.L. and Bedbrook, J.R. (1979) Nucl. Acids Res. 7, 1869-1886.
34 Bedbrook, J.R. and Gerlach, W.L. (1980) in Genetic Engineering (Setlow, J.K. and Hollaender, A., eds.) Vol. 2, pp. 1-20, Plenum Press, New York and London.
35 Flavell, R.B., O'Dell, M. and Hutchinson, J. (1980) Cold Spring Harbor Symp. Quant. Biol. 45 (in press).
36 Viotti, A., Pogna, N.E., Balducci, C. and Durante, M. (1980) Mol. Gen. Genet. 178, 35-41.
37 Appels, R., Gerlach, W.L., Dennis, E., Swift, H. and Peacock, W.J. (1980) Chromosoma (Berlin) 78, 293-312.
38 Gerlach, W.L., Miller, T.E. and Flavell, R.B. (1980) Theor. Appl. Genet. 58, 97-100.
39 Miller, T.E., Gerlach, W.L. and Flavell, R.B. (1980) Heredity (in press).
40 Singh, R.J. and Röbbelen, G. (1975) Z. Pflanzenzüchtg. 75, 270-285.
41 Bennett, M.D., Gustafson, J.P. and Smith, J.B. (1977) Chromosoma (Berlin) 61, 149-176.
42 Riley, R. and Chapman, V. (1958) Heredity 12, 301-305.
43 Gerlach, W.L. and Peacock, W.J. (1980) Heredity 44, 269-276.
44 Dennis, E.S., Gerlach, W.L. and Peacock, W.J. (1980) Heredity 44, 349-366.
45 Appels, R. and Peacock, W.J. (1978) Int. Rev. Cytol. Suppl. 8, 69-126.

MUTANTS AND VARIANTS OF THE ALCOHOL DEHYDROGENASE-1 GENE IN MAIZE

Michael Freeling

Department of Genetics,
University of California
Berkeley, California 94720

and

James A. Birchler

Biology Division
Oak Ridge National Laboratory
Oak Ridge, Tennessee 37830

INTRODUCTION

Of the hundreds of specific genes now being investigated in higher organisms, only a few may be approached readily using sophisticated genetics as well as molecular biology. Alcohol dehydrogenase-1 (Adh1; ADH enzyme, E.C. 1.1.1.1.) of maize is one such gene. There are at least four reasons why maize Adh1 relates to the general subject of how to go about "engineering" novel higher organisms. These reasons are summarized below and discussed with citations later in this review.

1) Maize ADHs are among 10 major and 10 minor polypeptides that are synthesized under anaerobic conditions. These anaerobic polypeptides (ANPs) represent a battery of simultaneously-induced genes that probably condition the plant's ability to survive temporary flooding or waterlogging. ADH activity is one way a plant cell can eliminate protons in the absence of O_2 and, in so doing, regenerate NAD^+ oxidizing potential for the glycolytic cycle. The ANPs have been found in widely divergent plant species and are synthesized in most maize organs during anaerobiosis.

2) Anaerobic primary roots yield total or poly(A) RNA preparations that are translated by a reticulocyte lysate cell-free system into ADH, the other ANPs, and not the aerobic polypep-

tides. In one inbred line, 12% of the translatable RNA is ADH1 polypeptide. Recombinant plasmids containing cDNA complementary to RNA sequences in each of the anaerobic messages should be relatively easy to obtain and catalog.

3) Either the presence or absence of ADH activity in a cell or organ can be used successfully for chemical selection. If cells or pollen containing ADH are subjected to allyl alcohol vapor (C=C-C-OH), the oxidized, highly toxic, aldehyde, acrolein (C=C-CHO) is formed and cell death occurs; ADH$^-$ cells survive. This selection system has proven useful for mutant selection in maize pollen. Fortunately, ADH is a dispensable enzyme activity in optimal environments. Conversely, ADH$^+$ or ADH overproducer cells may be selected on the basis of their ability to survive anaerobiosis or their ability to detoxify aldehydes. Selection for ADH$^+$ has not been successful in pollen. Because the NAD-dependent ethanol oxidase reaction of ADH may be monitored by the accumulation of NADH (at 340 nm), or by coupling NADH production to the reduction of nitro-blue tetrazolium to insoluble crystals, ADH activity may be readily assayed in the cuvette, in gels following electrophoresis or histochemically within the plant cells themselves.

Assuming that a functional Adh coding sequence will soon be cloned or constructed, Adh1 might serve as a selectable marker for transforming DNA of general use throughout the plant kingdom.

4) Several species of higher plants efficiently regenerate complete organisms from leaf mesophyll protoplasts. These species are preferred targets for genetic transformation studies. Particularly favorable taxa include the Solanacae (tobaccos, potatoes, petunias and henbanes) and carrots; all share the disadvantage of poorly developed genetics. The most genetically sophisticated plant species are inferior in cell culture. Given the recent advances in recombinant DNA technology, some investigators may be tempted to disregard classical genetics in view of "genetics with isolated genes." Here, a coding sequence for a gene with some or all cis-acting regulatory components could be manipulated in vitro, used to transform protoplasts, integration confirmed molecularly and function could be assessed in the regenerated plant. In this way a specific change in gene function can be related directly to a known change in the gene's nucleotide sequence. We will argue that this important genetic engineering strategy can be considerably enhanced when coupled with classical genetics.

5) The primary contribution of the maize ADH system is in the extensive collection of mutants, variants and chromosomal aberrations affecting the Adhs and a wealth of information at the isozyme/protein level. The genetic development of the Adhs has been accomplished over the last 15 years by Drew Schwartz and those associated with his laboratory. In turn, these studies depended upon over 50 years of classical genetic investigations using maize; no other organism would have sufficed. The assump-

tion underlying this review is that the arsenal of genetic tools affecting the Adhs, when coupled with nucleotide sequence information, will contribute uniquely to our understanding of the gene and of developmental programming in higher organisms.

Our specific purpose in this review is to explain the genetic aspects of the maize Adh gene system. We have assembled a genetic instruction manual for maize Adh. It is our aim to argue by example that classical genetic phenomenology plays an absolutely necessary role in our quest to understand how nucleotide sequences ultimately control physiology and development.

TERMINOLOGY AND GENE-ENZYME RELATIONSHIPS

Alcohol dehydrogenase (ADH) is a dimeric molecule that may be specified by one or both of the Adh genes: Adh1 and Adh2 (1,2).

Adh1 is located 1.5 map units (mu) from lw at position 127 on the long arm of chromosome one (1L) (3). The dominant markers, Kn (knotted) and Mpl (Miniplant; Anther ear), are <0.1 mu and about 3 mu from Adh1, respectively (Freeling, unpublished data). With translocations between 1L and the supernumary B chromosome and pairs of reciprocal translocations, Adh1 has been physically located to a chromosomal interval between 80 and 90% of the distance from the centromere to the 1L telomere (4). When Adh1 is the predominant Adh gene expressed, as is the case in the mature scutellum (diploid, embryonic organ) or pollen grain, its polypeptides homodimerize to the active, ~80,000 dalton protein (2) ADH1·ADH1, which contains at least one Zn^{++} that is released during chemical dissociation (5). The SDS-dissociated ADH1 polypeptide has been estimated to be 38,000 or 40,000 daltons (6,7), which is consistent with recent amino acid composition data (7).

Adh2 has been mapped to the short arm of chromosome 4 at approximately 20 mu distal to su (Dlouhy and Freeling, unpublished data). Unlike Adh1, Adh2 is not known to be expressed without the simultaneous expression of Adh1. For example, anaerobiosis induces ADH activity in seedling roots; this ADH is the reflection of three isozymes: ADH1·ADH1, ADH1·ADH2 and ADH2·ADH2 (2,6,8). Figure 1 diagrams these dimers and their subunit constitutions. Evidence that dimerization is random, and the relationship between activity and protein for each isozyme has been presented (9). SDS-dissociated ADH2 polypeptides are apparently 2,000 to 3,000 daltons larger than ADH1 polypeptides (6,10) as deduced from their slightly slower electrophoretic migration rate in Laemmli gels. Some, but not all, of the antibodies specified against homogeneous ADH1·ADH1 active dimers also cross-react with active ADH2·ADH2 (11). Although the Adh2 gene is not the focus of this review, it is important to keep its existence in mind. A probe for Adh1 coding sequence might well hybridize with the Adh2

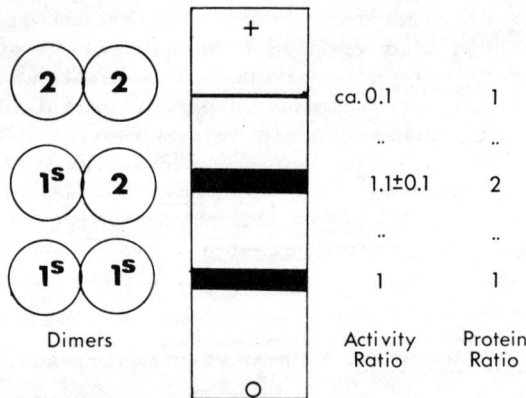

Figure 1. The Adh gene system in maize. There are two anaerobically inducible Adh genes specifying polypeptides that dimerize randomly -- in vivo and in vitro -- yielding three sets of enzymes. The electrophoretogram exemplifies the electrophoretic separation of the three types of isozyme dimers, 1·1, 1·2 and 2·2 in the standard activity ratio of 1:1.1:~0.1 characteristic of the StdS inbred line (Adh1-S/Adh1-S, Adh2-2N/Adh2-2N) is actually a 1:2:1 at the protein level. "O" denotes the origin. 1^S denotes ADH1-S subunits, etc. (1,2,8,9).

coding sequences. In fact, without the Adh1 and Adh2 mutant and variants that are available, it could be difficult ever to identify unequivocally any Adh nucleotide sequence.

Thus far, we have considered the Adh genes as formal loci on chromosomes, but in reality genes exist as particular alleles. The genic nomenclature used with the Adhs is best given by example. The inbred line used routinely for mutant isolation in the Freeling laboratory is called Inbred 1s2p. We know that it is homozygous for the naturally-occurring allele Adh1-S, where the hyphen separates the gene designation from the allelic designation. One of the mutants induced in the Adh1-S gene is Adh1-S3034, where 3034 is an arbitrary mutant number. This allele is mutable, and a number of stabilized derivative alleles have been recovered. These are called Adh1-S3034a, b,...z, aa, etc. The abbreviation for a derivative is S3034b, etc. If lesion(s) 3034b affect, but do not eliminate, the primary structure of the protein product of the gene, then the polypeptide designation would be ADH1-S3034b, abbreviated S3034b. If, however, it were proved that the lesions making up derivative allele S3034b did not involve coding sequences, then the product designation would remain ADH1-S, abbreviated S. No genic nomenclature has been adopted for derivatives of derivatives (i.e., a new mutant lesion in a revertant allele) or for intragenic recombinants between two different alleles. Also no nomenclature has been adopted for genomic fragments that carry some or all of Adh1.

The genic nomenclature described above differs slightly, but not ambiguously, from that used by Drew Schwartz. One example of a mutant lesion in Adh1-S that eliminates all product is 5657; Schwartz has named it Adh1-05657, where the "0" denotes lack of product. In our terminology, this mutant is Adh1-S5657 in order to clearly identify the naturally-occurring progenitor allele, Adh1-S. The allele designations used in the tables of this review are, of necessity, inconsistent in some details.

THE ANAEROBIC POLYPEPTIDES

Before 1978, antibodies specified against particular ADHs provided the best access to polypeptide-level information. An excellent two-dimensional immunoelectrophoretic method was developed that did not require monospecific antibody (12). Today, the need for immunological assays has been largely replaced by fluorographs of two-dimensional polyacrylamide gels (2-D PAGE), reflecting a pulse of radio-labeled amino acid that had been incorporated into proteins under anaerobic conditions. In 1978, it was shown that primary roots, when deprived of oxygen, repress the hundreds of polypeptides normally synthesized and translate a select few new polypeptides, including ADH1 and ADH2 (13). This result seemed similar to the heat-shock response in Drosophila melanogaster and other animals (14). Shortly thereafter, anaerobic protein induction was independently corroborated, and an improved two-dimensional native-SDS polyacrylamide gel system was developed (6). Summary data and conclusions from two laboratories' work on the anaerobic proteins are described below.

1) The inbred line, Berkeley Fast, responds to anaerobiosis by immediately repressing aerobic protein synthesis at the translational level (10); this was shown by in vitro translation of RNAs isolated after various durations of anaerobiosis. During the first 1.5 hrs of anaerobiosis, a class of proteins with 33,000 dalton polypeptides is translated, and serves unknown functions. Then, the synthesis of 20 anaerobic polypeptides (ANPs) begins and continues in a stable ratio for up to 60 hrs. The ANPs constitute approximately 73% of the label incorporated under anaerobiosis, and of this, 12% comprises ADH1 polypeptides. After 24 hrs of anaerobiosis, in vitro translation of total RNA in a message-dependent rabbit reticulocyte cell-free system yields ANPs with Laemmli migration rates indistinguishable from those obtained in vivo, with one exception. It was concluded that in vivo ANP synthesis reflects translatable mRNA-ANP pools, and that, with one exception, primary polypeptide to subunit processing involved a very few or no amino acids (10). Fig-

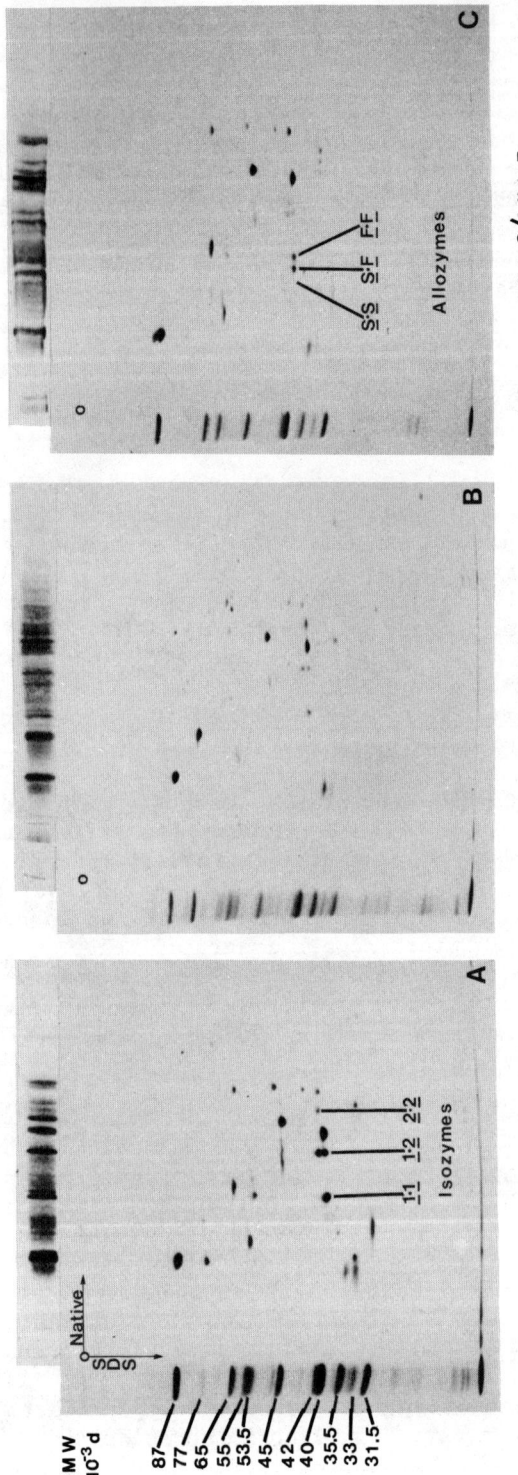

Figure 2. Fluorographs of two-dimensional polyacrylamide gel separations of the anaerobic primary root proteins from three lines of maize following 5 hr pulses of ^{3}H-Leu during 12 to 16 hrs of anaerobiosis. A. Berkeley Fast inbred line (data reproduced courtesy of M.I.T. Press, ref. 10); the ADH·ADH isozymes are labeled. B. Super Gold Pop exotic line (M. Alleman, unpublished data); note the obvious quantitative differences with the ANP profile of Berkeley Fast. C. An Adh1-F/Adh1-S (Adh2-1N/Adh2-1P) heterozygote; the 3 ADH1·ADH1 allozymes, which occur in a 4:4:1 activity ratio, also occur in an approximately 4:4:1 specific radioactivity ratio.

ure 2A, B and C show root ANP profiles for three different genotypes of maize.

2) Focusing specifically on the response of ADH1 and ADH2 translation during root anaerobiosis, it was shown clearly that translatable message levels per cell increase dramatically when anaerobic mRNA is compared with aerobic mRNA (15). IEF-SDS two-dimensional polyacrylamide separations of the products from translation in a reticulocyte lysate were used to identify ADH1 and ADH2 translation products unambiguously. No ADH polypeptide processing was detected (15).

3) Although the seedling, root, mesocotyl, coleoptile, immature endosperm, immature scutellum and mature anther wall all synthesize vastly different patterns of aerobic polypeptides, their anaerobic 2-D profiles are variations on a theme (16). Diagnostic qualitative and quantitative differences do exist; leaves do not incorporate label into any proteins while under anaerobiosis; although wounding aerobic roots did induce some new wound response proteins, the ANPs were not induced (16). Other workers have shown that heat-shock induces some novel proteins, but not the ANPs (17). The synthetic auxin, 2,4-dichlorophenoxyacetate (2,4-D), has been shown to induce ADH in aerobic roots (8), but it seems to induce hundreds of other proteins as well (Freeling and co-workers, unpublished data). We conclude that ANP synthesis is probably a specific adaptation to temporary drowning. This hypothesis had been previously advanced to explain the existence of Adh1 (18).

4) There are over 300 races of maize (19). Although there is general conservation of the 2-D ANP profile among widely divergent genotypes, there is also tremendous quantitative and qualitative variability. This result is demonstrated by comparing the ANP profile for Berkeley Fast (Figure 2A) with that for Super Gold Pop (Figure 2B). Electrophoretic variants are available for several of the ANPs. Further, ANP profiles from barley, sorghum, rice, peas, squash and relatives, turf grasses, tomatoes, teosinte, onion and carrot have all been examined; ANPs seem to be widespread or universal among higher plants (Freeling et al., unpublished data).

From the data summarized above, we can draw at least two conclusions which might be relevant to future research strategies. First, the relative abundance of translatable mRNA for ADH1, ADH2 and at least eight other ANPs should be suitable for direct ANP probe preparation via cDNA cloning in bacterial plasmids. Second, the similarity of ANP response among every plant species tested could mean that a cDNA library from maize might be used to probe the genomes of distant species. Antibodies prepared against ADH1·ADH1 of maize will titrate the ADH activity present in tomato seeds (Freeling and co-workers, unpublished data) and the seeds of many other species as distant as pinon pine (M. Beremand, personal communication). It should be kept in mind that flowering plants, much like mammals, experienced mas-

Table 1

Some Useful Naturally-Occurring Alleles of the Adh1 Gene

Allele	Subunit[a] Charge	Source	Description[b] and Citations
Adh1-S	S=-2	Cornbelt, Std S, 1s2p (1,36)	Standard slow allele, Adh1-1S. Underexpressed in seedlings and recombinogenic as compared to Adh1-F (3).
Adh1-F	F=-3	Cornbelt, Std F, Berkeley Fast (1,10)	Standard fast allele, Adh1-1F. Relatively overexpressed in seedlings (3). No recombination with Adh1-S or Cm (23).
Adh1-Cm	C=-4	Columbian race (1)	~5% of ADH1-F in specific activity (22). No recombination with Adh1-S, F or FkF (23).
Adh1-FCm	F,C=-3,-4	As above	Linked duplication of two structural genes resembling Adh1-F and Adh1-Cm (1).
Adh1-Ct	C=-4	Crossed into corn from a teosinte (1)	Similar in all properties to Adh1-F.
Adh1-FkF	F=-3	Funk G4343 hy-hybrid	Differs from Adh1-F at level of intragenic recombination. Recombines with both Adh1-S & F (23).
Adh1-54S	S=-2	Y. Efron, low line (25)	Extreme reciprocal effect: high in seedling (24;Woodman and Freeling, submitted for publication).
Adh1-33F	F=-3	Super Gold Pop	Extreme reciprocal effect; low in seedling (24;Woodman and Freeling submitted for publication).

[a]That ADH1-S=-2 in charge is arbitrary, given our standard electrophoretic conditions at pH 8.4 (20).
[b]All descriptions confirmed (Freeling et al., unpublished data).

sive adaptive radiations within the last 100 million years. Perhaps the lack of divergence in coding sequence among plants, which are morphologically very different, might present advantages for studies on the molecular level for evolutionarily meaningful changes. The probable function of the ANPs in flood tolerance add ecological and agronomic meaning to the ANPs as a coadapted gene complex.

NATURALLY-OCCURRING ELECTROPHORETIC VARIANTS OF Adh1 AND THEIR PROTEIN PRODUCTS

The Adh gene system in maize first began to be understood in 1966 when Schwartz (1) and Schwartz and Endo (20) discovered that various maize inbred or exotic lines contained electrophoretically distinguishable ADH1·ADH1 dimers. Scandalios (21) also reported variability for Adh1; his terminology was different from that used here. Table 1 lists and describes some of the naturally-occurring variants of Adh1 that are used routinely. At pH 8.4 of the borate-NaOH starch gel electrophoresis protocol (20) used to separate ADH allozymes and isozymes, all of the variants obeyed a strict unit charge rule: if an ADH1-S (S=slow) subunit is arbitrarily assigned a net surface charge of -2 under standard conditions, then an ADH1-F (F=fast) subunit is charged -3 and an ADH1-Ct or ADH1-Cm subunit is charged -4. All hybrid allozymes migrate according to the charges predicted on the basis of the mean charge of their subunits; a few induced mutants violate this rule.

Among the hundreds of maize lines that carry a slow or fast Adh1 allele, one allele of each was selected to be the Adh1-S and Adh1-F and are now carried in the StdS and StdF inbred lines, respectively. Adh1-S and Adh1-F have been compared extensively at several levels. These results are summarized below.

ADH1-S·ADH1-S and ADH1-F·ADH1-F differ in surface charge (1), in thermolability (26) and kinetics of Zn^{++} binding as a monomer (5), but not in specific activity measured in the NAD^+-dependent oxidase direction (27) (see Figure 2C). Following modification of unpublished (Schwartz and co-workers) and published (28) procedures, both of these allozymes have now been purified to homogeneity, their amino acid compositions have been ascertained and excellent tryptic fingerprints on TLC plates have been obtained (7). Twenty-nine of the maximum 30 possible tryptic peptides were resolved; ADH1-S and ADH1-F differed in the position of only one tryptic peptide consistent with a single amino acid substitution (7). The N-terminus is blocked (Freeling et al., unpublished data) as is the case with ADHs from other organisms.

Adh1-S and Adh1-F differ in their quantitative, organ-specific expressions (3). In the dry seed, Adh1-S and Adh1-F are almost equal in expression, yielding about a 1:2:1 allozyme protein ratio (S·S:S·F:F·F) in an Adh1-S/Adh1-F hybrid. However, Adh1-F is expressed to about twice the extent as Adh1-S in the primary root and mesocotyl. The allozyme ratio in these

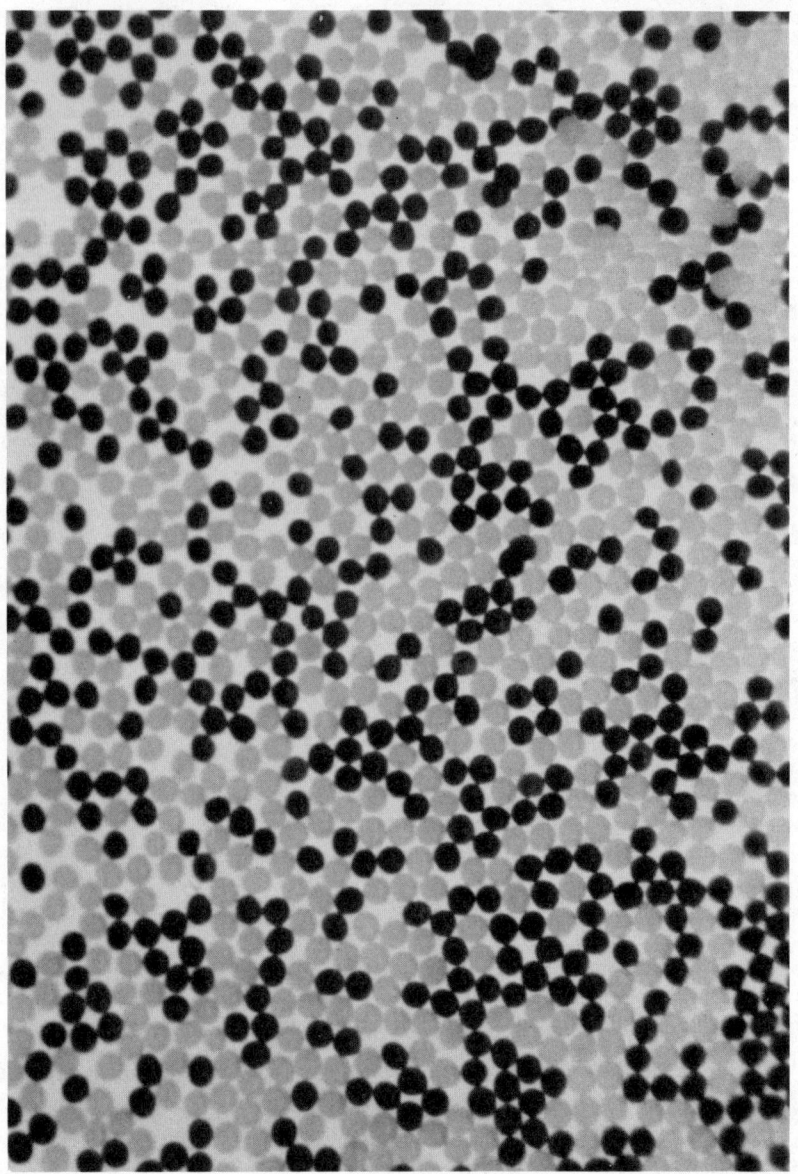

Figure 3. Pollen shed from an Adh1$^+$/Adh1$^-$ plant then histochemically stained for the presence or absence of ADH activity.

seedling organs is 1:4:4. It is known that these allozyme ratios reflect the rates of ADH1 synthesis (8,13) (Figure 2C).

Several pieces of data suggest that the balance between ADH1-S and ADH1-F expression is controlled separately from the total amount of ADH1 expressed per cell. Based on these data, Schwartz (3) advanced the gene competition hypothesis. In this explanation, different Adh1 alleles compete for a factor that limits ADH1 expression; the total amount of ADH1 per cell is directly related to the concentration of limiting factor. However, the balance of allelic expression can vary depending on each alleles' ability to compete. There is also work suggesting that Adh1 and Adh2 may be compensatorily regulated when simultaneously expressed (3,29).

Adh1-S and Adh1-F also differ at the level of intragenic recombination (23,30). If newly shed pollen is frozen and thawed, then incubated in isotonic buffer to allow small molecular substrates for dehydrogenases to dialyze from the grain, the presence or absence of ADH can be visualized by specific staining (30). Figure 3 is a stained pollen sample from a plant heterozygous for an Adh1-dysfunctional mutant: ADH^+ grains stain blue and opaque while ADH^- grains remain yellow and translucent. The 1:1 Mendelian ratio confirms that pollen ADH is a gametophytic (haploid) function, as was predicted previously, since hybrid ADH allozymes did not occur in pollen (3). In situ staining of pollen for ADH has permitted measurements of phenotypic revertant and intragenic recombinant frequencies with a resolution of about 10^{-6}. Mutants in Adh1-S recombine intragenically about 10 times more readily than those in Adh1-F (23). Eight EMS-induced mutants in Adh1-S, when combined in all pairwise combinations, generated a median intragenic recombination frequency of 11.7×10^{-5} ADH^+/pollen grain. (The mean frequency $\pm S_{\bar{x}} \times 10^5 = 16.7 \pm 3.1$ with a range from $1.9-66.3 \times 10^{-5}$.) Homoallelic controls for revertant frequency gave a 0.4×10^{-5} median and were always below 10^{-5} ADH^+/pollen grain. Eight EMS-induced mutants in Adh1-F, when combined in all pairwise combinations, generated a median intragenic recombinant frequency of 1.49×10^{-5} ADH^+/pollen grain (mean $\pm S_{\bar{x}} \times 10^5 = 1.9 \pm 0.3$ with a range from $0.0-8.4 \times 10^{-5}$). Homoallelic controls were always below 0.6×10^{-5} ADH^+/pollen grain. Most of the interesting regulatory-type mutants at Adh1 are in the Adh1-S allele. If they were in Adh1-F, determination of genetic fine structure would prove difficult or impossible.

The extraordinary result was that Adh1-S/Adh1-F heterozygotes do not display intragenic recombination at all (23). When $Adh1^-$ mutants induced in the Adh1-S progenitor allele were crossed to $Adh1^-$ mutants induced in the Adh1-F progenitor allele, the frequency of blue-staining (ADH^+) pollen does not increase over homoallelic control (phenotypic reversion) levels. However, either Adh1-S or Adh1-F will recombine with another naturally-occurring allele, Adh1-FkF, thus excluding large chromosomal aberrations as the mechanism underlying the recombination restriction.

Some of the differences between Adh1-S and Adh1-F expression are certainly reflections of their different amino acid sequences: net surface charge, thermostability and Zn^{++} binding. On the other hand, differences at the level of quantitative, organ-specific expression or intragenic recombination need not involve coding sequences (3,7,23,24,27), although the involvement of coding sequences has not been excluded.

NATURALLY-OCCURRING QUANTITATIVE ALLELES OF Adh1: THE RECIPROCAL EFFECT

The vast majority of Adh1 alleles among the races of maize encode ADH1 subunits that confer either a slow or fast electrophoretic mobility to active dimers in the pH range of 8.4 to 10. Woodman (24; unpublished data) examined 29 maize lines representing at least six races of maize. Plants with alleles that conferred a slow electrophoretic migration rate were crossed to plants carrying the standard Adh1-F (fast) allele. Plants with fast Adh1 alleles were crossed to a line carrying Adh1-S. The F_1 generated three ADH1·ADH1 allozymes: S·S, S·F and F·F. In the simplest case, the ADH activity distribution among three allozymes would be binomially distributed, where [S·S]:[S·F]:[F·F] would be represented by the ratio $p^2:2pq:q^2$ when p=fraction ADH1-"S" subunits and q=fraction of ADH1-"F" subunits and p+q=1. A binomial distribution of allozymes almost always occurred. However, different allelic combinations generated very different ADH activity ratios among the three allozymes. Figure 4 displays the allozyme ratio in three Adh1"S"/Adh1-"F" genotypes (I, II, III) measured independently in two organs, the scutellum and the anaerobic primary root. Enough F_1 electrophoretograms for each of the 29 lines were run and inspected densitometrically to obtain accurate confidence limits on each ratio in each organ. Woodman found considerable quantitative variation in allozyme ratios, and all were accounted for by variation in the Adh1 gene (i.e., cis-acting); there is no evidence for any trans-acting modifier genes affecting allozyme ratio.

For ease in reporting data, an allozyme ratio is quantified as %S subunit (%F is simply 100-%S). The graph of Figure 4 plots %S (allozyme ratio) data for 20 Adh1-"S"/Adh1-"F" heterozygotes of diverse Adh1 genotypes, where the scutellum is related to the root. The result was a clear negative correlation: the pair of Adh1 alleles with the higher %S in the scutellum had the lower %S in the anaerobic root, and vice versa. Every Adh1 allele examined fit this negative correlation, including the standard Adh1-S and Adh1-F. This organ reciprocity of quantitative expression is called the reciprocal effect. The inbred or exotic line source for several alleles in Figure 4 is given in the legend along with an arbitrary Adh1 allelic designation that is recommended for consistency between different laboratories.

Studies seeking a molecular basis for the reciprocal effect have suggested that these allozyme ratios reflect ratios of

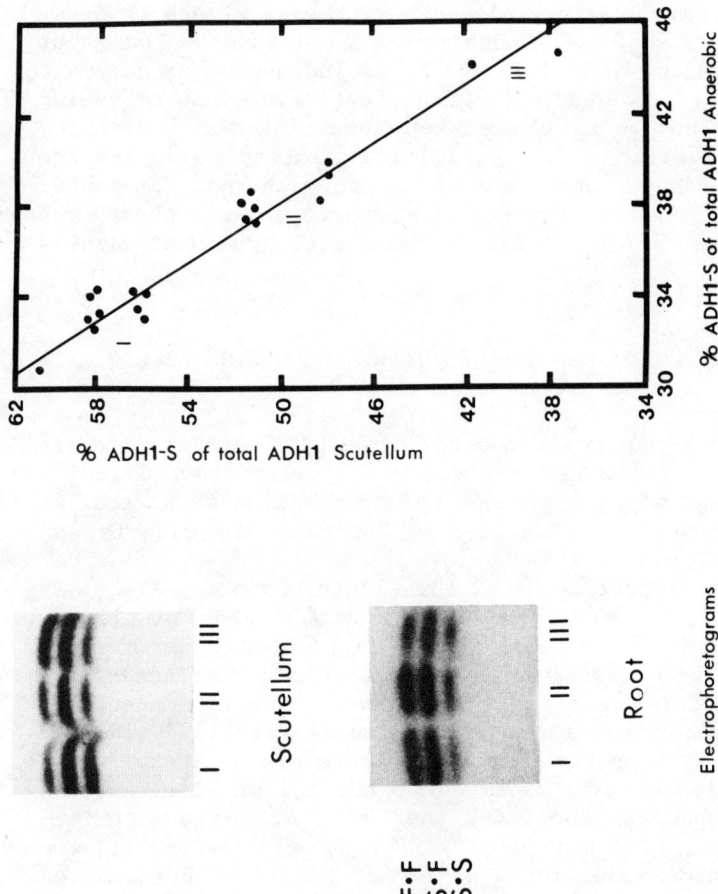

Figure 4. The organ-specific reciprocal effect for the quantitative expression of naturally-occurring Adh1 alleles. The graph plots the %ADH1 of total ADH1 subunits -- the allozyme ratio -- for 20 Adh1-"F"/Adh1-"S" heterozygotes of diverse origin. Note the 3 basic groups: I, II, III. The electrophoretograms are examples of allozyme activity ratios from each group (reproduced by permission of Academic Press, ref. 24). Examples of allelic pairs for each group are as follows. Group I: StdF/Hull-less Pop (1F/34S); StdF/StdS (1F/1S). Group II: StdS/Super Gold Pop (1S/33F); Red Pop/Efron S (6F/54S). Group III: Papago Flour/Efron S (12F/54S); Efron S/Super Gold Pop (54S/33F). (Woodman and Freeling, submitted for publication).

ADH1-S and ADH1-F translational rates per cell of the anaerobic primary root (Woodman and Freeling, submitted for publication). It is difficult to account for the reciprocal effect without evoking one or more quantitative, organ-specific regulatory components at Adh1.

The use of naturally occurring variants -- rather than mutants -- to provide the function for a nucleotide sequence is insufficient in itself. There is no reason to suppose that any two naturally-occurring alleles of a gene differ at only one or a few sites. By definition, variant alleles are the survivors of cryptic evolutionary steps. Although we do have information about the rate of nucleotide substitutions in coding sequences per unit time per taxon, we know nearly nothing of the functions or evolutionary rates of noncoding DNA. When comparing the nucleotide sequences of two variant alleles, there is no way to define the limits of the cis-acting unit (the gene) and much room for confusion if there are several sequence differences between them. The use of induced mutants to reduce function to structure avoids this confusion.

RECOVERING INDUCED Adh1 MUTANTS WITH AN ALLOZYME SCREEN

Several dozen mutants induced by ethyl methanesulfonate (EMS) and at least one mutant recovered following gamma irradiation were recovered after a process called an allozyme screen, as diagramed in Figure 5. Schwartz's methods for mutagenesis and mutant recovery (31,32) are as follows.

Kernels homozygous for one of the electrophoretic alleles in routine use (Table 1) were placed in a 0.08 M EMS solution in water at 25°C for 10 hr and then washed at room temperature according to the procedures of Briggs et al. (33). The seeds were then planted directly or dried at 4°C for subsequent planting. The treated seeds were reared to flowering, detasselled and pollinated by a tester line that had been planted in alternate rows. The tester was also homozygous at Adh1, but for an allele with a different electrophoretic mobility than that of the mutagenized line. For example, if Adh1-S were the target of mutagenesis, the tester would be homozygous for Adh1-F. Therefore, the harvested ears carried seeds that were heterozygous for two electrophoretic alleles such that three allozymes were specified in the scutellum (Figure 5). About 12 seeds scattered about each ear were sacrificed for scutellar allozyme analysis using a standard starch gel electrophoresis protocol (20). (It would have been possible to obtain enough extract for over a dozen electrophoretic analyses per seed without impairing germination by extracting small slivers of scutellum.) A presumptive mutant sector on the ear was identified by the existence of one or more seeds displaying aberrant allozyme profiles: a skewed allozyme ratio, a missing band,

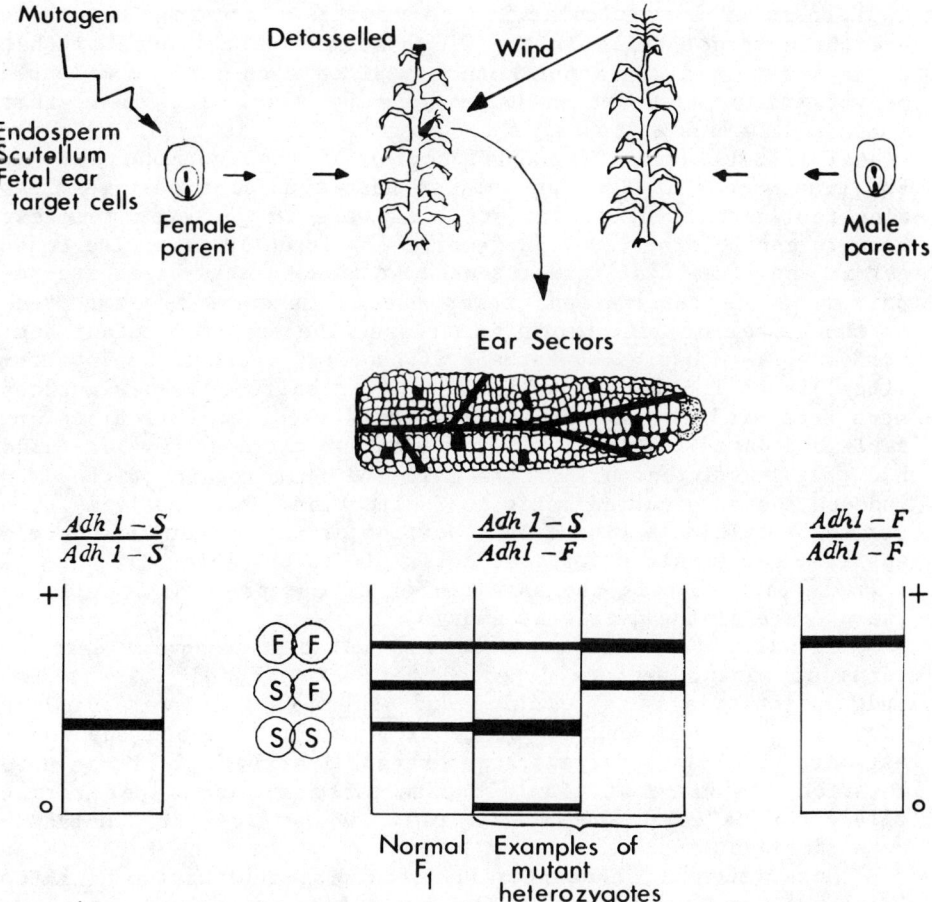

Figure 5. The kernel ADH allozyme screen for detection of Adh1 mutants. This diagram utilizes Adh1-S as the target (♀ ♀) and Adh1-F as the tester (♂ ♂) (31,32). The center electrophoretogram is as expected for an Adh1-S mutant that leads to a reduction in the net negative charge of the ADH1-S subunit. The electrophoretogram on the right indicates a mutant ADH1-S subunit that is both inactive and that dimerizes poorly.

bands with new electrophoretic mobilities, nonbinominally distributed allozymes and the like.

Of the 12 kernels tested, the number expected to show any given mutant turns out to be a complicated calculation. First, we need to know how many ear-primordial cells exist in the quiescent embryo of the treated inbred line. Coe and Neuffer (34) have determined that there are two to four ear-primordial cells in kernels of a particular line of maize. Assuming that there are three target cells in the G_1 phase of the cell cycle, that gives six target genes per kernel. Since each gene is a double helix, and because EMS probably acts by alkylating bases that subsequently misreplicate, then there may be at least 12 mutational targets (single-stranded copies of Adh1) at the time of EMS treatment. If EMS can persist past the first post-germination replication, or if target cells were in G_2, even more ear sectors can be expected. If fewer cells actually give rise to an ear in the lines that Schwartz used or if alkylated bases are repaired before replication, fewer than 12 sectors are expected. As the number of Adh1 targets increase, the expected mutant sector size, as measured by percent of the ear, decreases. In actuality, in an EMS mutagenesis study, over half of 30 new mutants were recovered in one-half of the ear's seed set (Birchler, unpublished data). This result implies that the Adh1-FCm line used had only one G_1 ear-primordial cell and that repair of the EMS-induced lesion occurred prior to replication. Alternatively, the ear might originate from only a portion of the sector that develops from the single primordial cell. It is possible that not all inbred lines contain the same number of ear-primordial cells in the plumule of the quiescent embryo.

Schwartz (32) has reported his mutation recovery data in terms of mutant sectors detected per ear: 3% and 1.5% in two independent trials. The Adh1-S and Adh1-F variants were used as targets. Given an approximately 2% mutation rate and the rough estimate of 12 Adh1 targets, the extraordinarily high EMS-induced mutation frequency of 1.6×10^{-3} detectable mutations per target allele can be calculated. With only two targets, this mutation rate increases to 1%.

More recently, the naturally-occurring duplication Adh1-FCm (Table 1) was used as the target for a large-scale EMS mutagenesis project (35). Of approximately 2500 ears subjected to the allozyme screen, 30 confirmed mutants were recovered, or about 1.2% detectable mutant sectors per ear. Even when this value is halved, to account for the duplicated Adh1 coding sequences, it is still extraordinarily high.

SELECTION OF Adh1 MUTANTS VIA ALLYL ALCOHOL-RESISTANT POLLEN GRAINS

A possible reason why the allozyme screen is so effective at identifying Adh1 mutants is that it is a reflection of an inherent mutability of Adh1. However, this is not the case. Of 4000 unmutagenized kernels of the genotype Adh1-S/Adh1-FkF, none showed an aberrant S·S:S·F:F·F allozyme ratio (36). Further, the forward mutant frequency at Adh1-S or Adh1-F -- measured as ADH⁻ germinating gametophytes/germinating pollen grains -- has been estimated to be below 2×10^{-7} (37). Therefore, it is probable that most or all of the mutants that were identified following EMS treatment were actually induced by EMS. The high frequency at which EMS induces mutations in this system was fortunate because the allozyme screen is not a high-resolution method for mutant recovery.

Allyl alcohol vapor has been used to kill preferentially the ADH+ pollen grains segregating from an Adh1+/Adh1- heterozygote (36-38). Similar schemes were used previously in yeast (39) and Drosophila melanogaster (40). In theory, pollen ADH activity catalyzes the oxidation of allyl alcohol (C=C-C-OH) + NAD+ to the toxic aldehyde, acrolein (C=C-CHO) + NADH. Since acrolein reacts immediately upon production, we do not expect a rare, surviving ADH- pollen grain to be adversely affected by being surrounded by poisoned pollen. The use of the male gametophyte as the point for chemical selection has several advantages. A healthy maize plant sheds 10^6 pollen grains at one time and several million grains in all. Each pollen grain is a tricellular haploid gametophyte composed of one metabolically active tube nucleus and two presumably inert sperm nuclei. Pollen ADH is synthesized after the reduction division of meiosis (3). Therefore, an Adh1-deficient mutation must occur in both complementary strands of DNA in the microspore or before in order to be recovered. Mutagenizing mature pollen grains would be useless.

The step-by-step description that follows is used by David S.-K. Cheng in the Freeling laboratory (36; unpublished data) (Figure 6) and is detailed here for two reasons. First, pollen selection is generally a once-per-year endeavor requiring an acre of detasselled maize plants and much physical labor; improper methods have proved costly. Second, the success of these methods depends on continuing tests for pollen viability (with "David's Bread Loaf" method, ref. 38), moisture content and appropriate readjustments of allyl alcohol concentrations, storage conditions and the like. Because of variable field conditions, particularly changes in temperature and humidity, there is no fool-proof recipe that will yield Adh1-dysfunctional mutants.

Step 1. Large quantities of a healthy line (Funk G4343), homozygous for a fast variant of Adh1, were planted in cycles throughout the season. These testers were detasselled.

Step 2. A highly inbred line, ls2p, was the object of mutagenesis and pollen selection. This line is homozygous for Adh1-S and for a mutant of the Adh2-1N naturally-occurring allele, Adh2-1P, that confers a unique electrophoretic mobility to ADH dimers containing ADH2 subunits. The choice of the Adh1-S gene as the target is important because it is the one allele that is recombinogenic enough to permit genetic fine structure studies (23) The ls2p plants were reared in 10 inch plastic pots to facilitate moving them to and from radiation sources.

Step 3. Two of the most interesting Adh1 mutants have been recovered following irradiation with heavy ions accelerated at

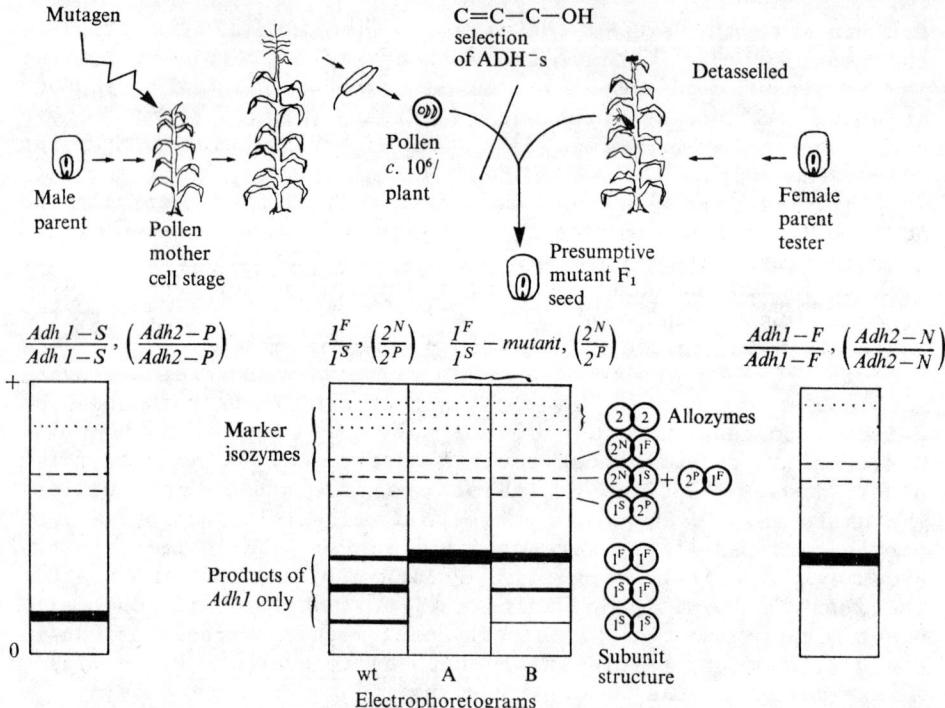

Figure 6. Chemical selection, recovery and classification of Adh1-mutants. The paternal or target Adh1 allele Adh1-S is abbreviated 1^S, etc. The electrophoretograms depict semi-quantitative banding patterns and intensities after native starch gel electrophoresis. o denotes origin; + denotes the anode. The ADH activities reflected in these diagrams are from the wild-type F$_1$ (wt), Adh1-negative heterozygote (A) and Adh1-underproducer heterozygote (B). Extracts were obtained from F$_1$ scutellar slivers. The subunit composition(s) of each band is indicated where subunit 1^S is specified by Adh1-S, etc. The dashed and dotted bands in the marker isozyme region of the gel denote low-activity and usually-not-present, respectively. (Reproduced by permission of Cambridge University Press, ref. 36.)

the Bevelac, Lawrence Berkeley Laboratory. Immature plants at 5 to 6 weeks (10 weeks to shedding) were slit longitudinally at the stem to expose and measure the immature, premeiotic tassel. Once located, the plant was taped closed and moved to the Bevelac, and the entire tassel region was irradiated with approximately 350 rads of plateau LET Ne^{10+} (400 MeV/amu at approximately 50 rads/min) or C^{6+} (500 MeV/amu at approximately 100 rads/min). These doses and dose rates elevate pollen abortion frequencies from

0.02 to about 0.1, but generally do not cause serious stunting of the tassel. Premeiotic damage that generates microspores with chromosomal deficiencies will abort and not confound subsequent analyses. Severe duplications, such as disomic haploid pollen, do not compete well with euploid pollen and might be selected against here.

Step 4 (36). After mutagenic treatment, potted plants were grown outdoors during the summer months. Shortly before anthesis, plants were transported to a greenhouse isolated three miles away from any Adh1-deficient pollen; contamination by mutants already in our collection was eliminated. On days 2 and 3 of shedding, the heaviest days, Lawson 400 pollination bags were closed at 9:00 am and opened again between 10:00 am and 12:00 noon for pollen collection; 5×10^5 grains is a reasonable sample since these plants have relatively small tassels. With Millipore pollen counting techniques, 1 mg of Adh1-S pollen was found to reflect about 2×10^3 grains. Collected pollen was separated from anthers with a folded glassine bag and stored at 4°C for 0.5 to 1 hr. Control experiments show that this cold treatment increases germination in vitro by about 10%. After cold treatment, pollen was dispensed in about 500 mg lots (10^6 grains) onto glassine paper for allyl alcohol vapor treatment. A 500 ml Mason jar with sealed lid containing 20 ml granular $CaSO_4$ desiccant was used; unless specified, we used between 1 and 6 μl allyl alcohol/500 ml chamber for a total treatment time of 20 to 30 min; 1 μl allyl alcohol vapor/500 ml equals 75 μM. After treatment, the seal was broken and the jar cooled in an ice bucket then transported to the field of detasselled plants (Adh1-FkF, Adh2-"N") whose silks had been cut back the day before. Tester ears were pollinated with about 5.8×10^4 pollen grains/ear and a camel hair brush; there are about 400 silks, each leading to an ovum. Each ear was then covered until harvest about 2 months later.

Step 5. On a typical day, if 3 μl of allyl alcohol/500 ml jar for 15 min is the treatment and 1s2p pollen is the target, about 90% of all ears will be barren. This treatment was chosen as a weak selection condition to ensure the survival of ADH-low as well as ADH-negative pollen. The remaining 10% of the ears bore a few seeds. All of these seeds were catalogued and a small sliver of scutellum was removed in order to prepare an extract for allozyme analysis per standard procedures of starch gel electrophoresis and staining (21). The products of all four Adh alleles of the Adh1-FkF/Adh1-S, Adh2-"N"/Adh2-1P F_1 are distinguishable in overstained gels as diagramed in Figure 6. The presence of ADH2-1P product ensures that the 1s2p line is the true male parent.

Step 6. Some F_1 seeds showed aberrant or unexpected allozyme profiles, and also carried the Adh2-1P marker. Only a few percent of screened F_1 kernels appear aberrant. These were planted for further analysis. Of the escapers, many are morphologically atypical suggesting that unhealthy pollen may accumulate relatively low levels of ADH and, for that reason, survive allyl alcohol treatment.

Even after successfully accomplishing all six of the above steps, the great majority of F_1 seeds showing aberrant allozyme profiles were dead, sterile or suffering from severe morphological abnormalities; the Adh1 behavior was unstable or often Adh1 appeared normal in every respect. Four mutants were recovered and characterized (see Table 3, below); about a dozen await confirmation (Freeling et al., unpublished data). Osterman has reported recovering one mutant with a protocol similar to that detailed here (41). We now believe that the low efficiency of mutant recovery is explained by the extremely low frequency with which ionizing radiation or insertional-type mutator systems induce viable mutants at specific genes in the pollen lineage (for discussion, see ref. 36). An unstable mutant allele, Adh1-S3034, was recovered with allyl alcohol selection. This mutant synthesizes only 40% of normal ADH polypeptide in pollen grains. When S3034 pollen is reselected with allyl alcohol vapor as described here, the average nonbarren ear carried a new mutant derivative of S3034. Given this success and the results of carefully executed control experiments on pollen selection (36), we believe that mutagens other than ionizing radiation or insertional-type mutators might generate stable Adh1 mutants with a greater efficiency.

Allyl alcohol selection has proved very useful in constructing chromosomes carrying particular Adh1 mutants flanked by particular genetic markers. For example, we needed to place the dominant marker Kn (Knotted, 0.1 mu from Adh1) next to Adh1⁻ mutant S5657. Pollen was collected from the $Adh^+Kn/Adh1-S5657\ +$ heterozygote, selected with allyl alcohol and crossed to an unmarked tester. The rare but readily identified expression of the dominant Kn phenotype identified an Adh1-S5657 Kn recombinant chromosome.

TEST FOR ALLELISM AND THE PRELIMINARY CHARACTERIZATION OF NEW MUTANTS AFFECTING ADH1 EXPRESSION

Independently of the particular method of Adh1 mutant or variant recovery employed, new Adh1 alleles were almost always recovered as heterozygotes with a tester allele of a different electrophoretic mobility. For Adh1-S mutants recovered, the F_1 was Adh1-F (Standard)/Adh1-S-mutant. A genotyped F_1 seed showing an aberrant allozyme ratio was then reared to flowering, its pollen examined for abortion and stained for the presence of ADH activity. The F_1 plant was self-pollinated to obtain mutant homozygous seed, and pollen from this F_1 plant was test-crossed to a line of maize homozygous for Adh1-Ct, a variant allele that encodes an ADH1 subunit that migrates very fast under standard electrophoretic conditions (Table 1). This test-cross is diagrammed:

$$\frac{Adh1-Ct}{Adh1-Ct} \times \frac{Adh1-F}{Adh1-S \text{ mutant}} \quad .$$

If the Adh1-S mutant specifies zero product due to a point mutation in the Adh1-S gene, then the F_1 pollen should stain 1:1 $ADH^+:ADH^-$ without abortion (e.g., Figure 3), one-fourth of the F_2 kernels should be Adh1-S mutant homozygotes and the progeny of the test-cross should segregate 1:1 for seeds expressing both Ct and F subunits: Ct subunits only. Any mutants with the polypeptide product clearly altered may be shown to be allelic by this simple segregation test.

The examination of F_1 pollen is a powerful tool for mutant analysis. Consider a pollen sample with four pollen phenotypes in an approximately 1:1:1:1 ratio: ADH^+, plump: ADH^-, plump: ADH^+, aborted: ADH^-, aborted. The reason for this particular pollen phenotype has proven to be that an $Adh1^+/Adh^-$ heterozygote was recovered in a background where a reciprocal translocation involving chromosomes other than number one was also present; the adjacent segregations generated aborted pollen. An alternative explanation for the same data might be that one of the breakpoints of a new reciprocal translocation disrupted the Adh1 gene, rendering it Adh1$^-$. In cases where the target allele specifies some active product, an electrophoretic analysis of the allozymes in F_1 pollen can be most informative; the presence of a hybrid allozyme indicates that, for some reason, both maternal and paternal Adh1 alleles are segregating into the same pollen grains. Thus, examination of F_1 pollen gives the first indication of chromosomal abnormalities.

For mutants or variants in the amount or timing of gene expression, the test for allelism needs to be considerably more rigorous. The procedure used here (Freeling and Cheng, submitted for publication) is best given by example. Mutant Adh1-S3034 was recovered in an allyl alcohol screen as an Adh1-FkF/Adh1-S3034 kernel whose S·S:S·F:F·F allozyme balance was shifted toward F product. It appeared that S3034 was only expressed about one-third as much as the progenitor S allele. The F_1 pollen stained 1:1 for dark blue (ADH^+):light blue (ADH-low). The results of the test-cross -- Adh1-Ct/Adh1-Ct x Adh1-FkF/Adh1-S3034 -- also showed that the low-expression behavior always segregated with the Adh1-S allele, as would be expected if the S3034 lesion were part of the Adh1-S complementation group (i.e., cis acting). There is, however, an alternative explanation, which was tested. S3034 could have been a lesion closely linked to Adh1-S acting in trans to increase the expression of the homologous allele. The phenomena of paramutation and transvection provide precedent for this sort of trans effect. One must remember that in all data on S3034, the ratio of allozyme expression was an endpoint rather than any absolute measure of ADH1 levels per cell. This alternative was easily ruled out. The F_2 progeny segregated kernels of the genotypes FkF/FkF, FkF/S3034 and S3034/S3034. When scutella were analyzed for ADH activity per µg protein, a dosage effect was obtained: the mutant homozygote specified only about one-third the specific ADH activity as the FkF homozygote, with the heterozygote showing an intermediate value. This dosage effect result is the needed control that allowed us to assign S3034 to the Adh1-S complementation group.

Once a mutant is confirmed and placed in or out of the Adh1 gene, several genetic and biochemical tests may prove valuable in seeking a mechanism for its abnormal behavior.

During the past few years, several investigators have specifically sought mutant or variant alleles of specific genes in which the polypeptide product appeared to be normal, but the behavior or regulation of the gene appeared to be altered. Alleles have been sought with behaviors not likely to be mediated by the amino acid sequence of the polypeptide product (e.g., 24,42,43). Examples of such alleles include those altered in timing during development, specificity for developmental stage, organ-specificity, rate of expression, stability, fidelity circuitry with co-regulated genes, etc. One way to argue that an allele displaying one or more regulatory-type behaviors resides in regulatory DNA is to collect data suggesting that the protein product is unaltered. Unfortunately, even a complete amino acid sequence comparison between the products of mutant and progenitor alleles cannot absolutely rule out differences in the third base of the codon or differences in the primary polypeptide that might have been processed away. Among variant alleles we expect to find amino acid sequence differences as a function of time diverged, but there is no protein-level way to determine if these differences cause the regulatory behavior under consideration. The genetic strategy of mapping a variant regulatory site outside of a structural gene is also ambiguous, since we have no assurance that variant alleles are sufficiently similar to pair at all or in perfect alignment; perhaps the widespread occurrence of intervening sequences plays a confounding role here. Finally, there are at least two cases -- in a 5S RNA gene of Xenopus borealis (44), and in a tRNA gene of Drosophila melanogaster (45) -- where coding sequences also serve a transcriptional control function. These results break down the absolute distinction between coding and regulatory components of the gene. We suggest that if an allelic difference behaves like a regulatory difference, then it can be defined as such. The next level of mechanical explanation for regulatory behavior lies in our ability to couple mutant behavior to nucleotide sequence information.

SOME Adh1 MUTANTS THAT WERE INDUCED BY EMS OR GAMMA RADIATION AND RECOVERED VIA AN ALLOZYME SCREEN

Table 2 lists several confirmed Adh1 mutant alleles along with data on their origins, protein-level characteristics,

Table 2.

Some Adh1 Mutants Recovered Via Allozyme Screens Following EMS or Gamma Treatment of the Seed

Progenitor Allele	Mutant	Subunit Charge	Protein Level Description and Citations[a]	Coding sequences involved?	Frequency ADH$^+$/pollen[b]
Adh1-S	5657	--	Cross-reacting material negative (CRM$^-$)(48). Maps as a point (23).	perhaps	--
	664	--	CRM$^-$. Point.	perhaps	--
	296	S = -2	Inactive, faulty dimerizer. CRM$^+$. Point.	yes	--
	508	F = -3	Charge change; low specific activity. Point.	yes	--
	W	W = 0	Charge change (31). Low specific activity. Point.	yes	--
	667	S = -2	Inactive, CRM$^+$ subunit. Point.	yes	--
	1015	--	Probably no product. Point.	perhaps	--
	719	variable	Unique complementations; very low specific activity. Point.	yes	--
	1108	S = -2	Temperature sensitive product (49).	yes	unreliable (23)
	586	W = 0	Active subunit, charge change (50)	yes	active
Adh1-F	460	--	CRM$^-$	perhaps	--
	207	--	CRM$^-$	perhaps	--
	748	C = -4	Charge change; inactive product.	yes	--
	217	--	Probably no product.	perhaps	--

Table 2 (contd.)

Some Adh1 Mutants Recovered Via Allozyme Screens Following EMS or Gamma Treatment of the Seed

Progenitor Allele	Mutant	Subunit Charge	Protein Level Description and Citations[a]	Coding sequences involved?	Frequency ADH⁺/pollen[b]
	119	--	Probably no product.	perhaps	--
	63	--	Probably no product.	perhaps	--
	6	C = -4	Charge change; inactive; faulty dimerizer (30).	yes	--
	334	C = -4	Charge change; inactive	yes	active
	7086	C = -4	Fully active subunit (50).	yes	active
	725	U = -1	Active Laemmli variant; 2,000 d smaller (6).	yes	active
	278	F = -3	Inactivated during electrophoresis. Not as reported (47).	yes	active
Adh1-FkF	γ25	--	No protein product, CRM⁻, gamma radiation to seed (23).	perhaps	--
Adh1-?	1097a	variable	Inactive, but dimerizes; unique complementations.(Freeling, unpublished data)	yes	--
Adh1-FCm	64a	UCm = -1,-4	Charge change in F; active	yes	active
	118	UCm = -1,-4	Charge change in F; reduced activity.	yes	active
	253	SCm = -2,-4	Charge change in F; active.	yes	active
	355	FBm = -3,-5	Charge change in Cm; Bm retains Cm's 5% activity.	yes	active

86	Cm = -4	No F product. Called "B86" (23).	perhaps	$>10^{-5}$
91	Cm = -4	No F product. Called "B91" (23).	perhaps	—
97	FCm = -3,-4	Low F activity or expression. Called "B97" (23).	perhaps	—
64b	FCm = -3,-4	Low F activity or expression.	perhaps	—
8	Cm = -4	No F product, or inactive C.	perhaps	$>10^{-5}$
84	Cm = -4	No F product, or inactive C.	perhaps	$>10^{-5}$
≃36	FCm = -3,-4	Low F activity or expression.	perhaps	unreliable
20	FCm = -3,-4	Low F activity; complemented by Cm subunits	perhaps	—

aThe Adh1-S and Adh1-F mutants were induced with EMS, recovered and characterized by Drew Schwartz. 1097a was induced by M.G. Neuffer by EMS treatment of pollen, recovered by accident in F2 and characterized by Freeling (unpublished data). The Adh1-FCm mutants were induced by EMS, recovered and characterized by Birchler (35; unpublished data). All protein-level characterization, except for that of Adh1-F278, were confirmed and sometimes expanded in the Freeling laboratory. Poin= means behaves as a point in intragenic recombination tests (23).

b-=stable Adh1-deficient mutant with ADH$^+$/total pollen from homoallelic plant $<10^{-5}$. Values $>10^{-5}$ imply unequal crossing-over or conversion involving the duplication (Freeling, unpublished data). The instability of Adh1-FCm86 changes a published interpretation (23).

behavior in pollen and other relevant data. All but gamma-induced Adh1-FkF 25 (23) and Adh1-1097a (found by chance in an EMS-induced pale green mutant line from M.G. Neuffer) were induced and recovered by Schwartz et al. This list does not include every Adh1 mutant, but does include representatives of each category and especially those which might prove useful in further studies. All of these induced mutants are most simply explained as alterations in Adh1 coding sequences. Some mutants, like S5657 or F207, do not specify any ADH1 polypeptide or cross-reacting material (CRM). Others definitely specify a product (CRM$^+$), but it is either inactive, low-activity, faulty in its ability to dimerize, thermolabile, dialysis-sensitive, altered in electrophoretic mobility or some combination of the above. For the latter mutants, coding sequences are definitely involved. For the mutants expressing no protein, the lesions could be anywhere in the complementation group. However, the 9 no-product mutants listed in Table 2 recombine intragenically as if they were points (23). Of 30 new mutants induced in the Adh1-FCm naturally-occurring duplication, none simultaneously affected both structural genes (35) as would have been expected of a deletion. The F and Cm components of the duplication are less than 0.01 mu apart; their degree of physical separation is unknown. All in all, the data suggest that EMS induces base substitutions or small lesions preferentially. All of the mutants in Table 2 transmit mutant as well as nonmutant pollen through the male gametophyte. Multigenic deletions are expected to transmit poorly.

The study by Birchler and Schwartz (35), where the Adh1-FCm duplication was the target for EMS, generated data on the percent contribution of any category of mutants to the total. About 33% (11 of 30) specify zero product from either the F or Cm component; another third (8 of 30) specified fully or very active subunits of an altered electrophoretic mobility, and the remaining third (11 of 30) specified subunits that were either low in activity or faulty dimerizers and have not been characterized in depth. Thus, about one-third of all mutants induced by EMS and recognized in an allozyme screen specify zero product or inactive product that does not dimerize.

Over half of the Adh1 mutants induced by EMS are low enough in ADH activity per pollen grain to permit intragenic recombination and reversion tests. The Adh1-S allele is particularly recombinogenic (23). Each of the eight Adh1-S deficient mutants listed in Table 2 recombines with any other in all possible pairwise combinations. This set of point lesions should be useful for deletion mapping.

Some of the mutants listed in Table 1 complement. That is to say, mutant No. 1/mutant No. 2 heterozygotes have more ADH specific activity than either mutant homozygote. The explanation is a subunit-subunit interaction in the heterodimer. Variant Adh1-Cm (Table 1) is very low in activity but is very stable (46); mutants Adh1-S908, Adh1-F748 and Adh1-F334 are inactive but

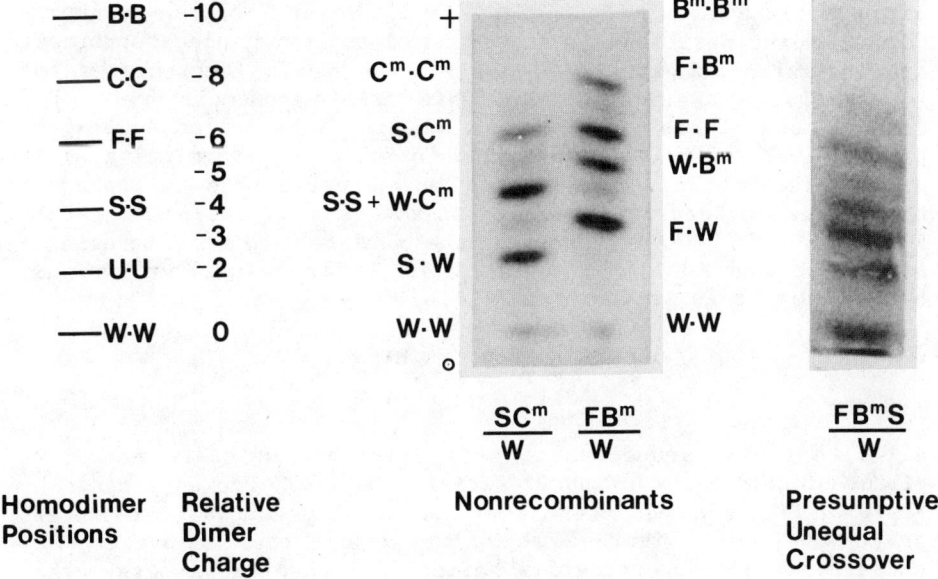

Figure 7. An example of the use of allozymes to mark each of the four coding sequences in mutagenized Adh1-FCm/Adh1-FCm. The electrophoretograms depict the allozymes present in scutellar slivers of the progeny from the cross "SCm"(Adh1-FCm253)/"FBm"(Adh1-FCm355) x "W"/"W"(Adh1-S586 homozygote). Each of the 5 Adh1 structural genes used in this cross can be distinguished because each specifies an active allozyme of a unique electrophoretic mobility. Note that crossover chromatids that arise by normal pairing (SBm/W, FCm/W) are discernible, as well as chromatids of unequal crossovers (e.g., S/W, FBmCm/W, F/W, SFBm/W) or conversions (e.g., CmBm/W, SF/W. The electrophoretogram on the right is the allozyme profile of a presumptive SFBm/W product: a triplication.

participate in dimerization (Table 2). Adh1-W and Adh1-S1108 are mutants that specify low activity as homozygotes because they specify unstable homodimers (Table 2); thus, the pollen staining reaction is ADH⁻. ADH⁺ frequencies in pollen from complementing mutant heterozygotes, like Adh1-Cm/Adh1-W, might be used to quantify diploid or disomic haploid pollen. If protoplasts could be made from lines carrying these alleles, selection for ADH⁺ would augment for cell fusions.

Active allozymes that run at different mobilities in native gels can be extremely useful for a variety of genetic manipulations. If we define an S subunit as having a net surface charge of -2 (i.e., S·S dimer has a -4 charge) then active ADH1 subunits are available as follows: W=0; U=-1; S=-2; F=-3; C=-4; B=-5. Corresponding positions in a starch gel following electrophoresis are indicated as part of Figure 7. (Schwartz incorporates the letter charge designation into his allele nomemclature. For example, one of his "W" mutants from Adh1-S is called Adh1-W586; the progenitor allele is not noted in Schwartz's terminology.)

One example of the use of allozyme markers is in a project that seeks to identify unequal crossovers among mispaired Adh1-FCm loci. Two EMS mutants of the duplication are abbreviated Adh1-"SCm" and Adh1-"FBm" (see Table 2 for actual mutant numbers). The cross is shown as

$$\frac{Adh1-FBm}{Adh1-SCm} \times \frac{Adh1-W}{Adh1-W} .$$

All five coding sequences in this cross are uniquely marked by virtue of the charge of their active subunits. Progeny should be FBm/W or SCm/W unless there are some equal or unequal crossovers or conversions. Figure 7 shows the actual banding patterns of scutellar electrophoretograms reflecting these two progeny classes and an allozyme profile indicating an unequal crossover that generated a presumptive SFBm triplication (Freeling, unpublished data); all subunit designations and positions are marked.

Eight of the 30 EMS-induced mutants of the Adh1-FCm duplication either produce no ADH1-F subunit at all or specify an inactive subunit at the C position (35; Birchler, unpublished data). The pollen shed from plants homoallelic for these mutants generally stain very light blue, rather than nonblue, because of the 5% activity of the Cm component of the duplication. Some of these mutants revert at frequencies below 10^{-5} ADH$^+$ per total pollen, consistent with previous results on homoallelic Adh$^-$ mutants. Others revert at recombinational frequencies, 10^{-5}-10^{-4} ADH$^+$ per total pollen, suggesting a misaligned pairing followed by a Cm allele correcting a lesion in F, or the like (Freeling, unpublished data). Table 2 identifies some of the Adh1-FCm mutants in each category.

The EMS-induced Adh1 mutants that alter the charge of an active subunit fall into categories that are extremely difficult to understand. Schwartz (32) found that all charge changes in the net negative direction (faster electrophoretic mobility under standard pH 8.4 condition) were of one unit. However, of 18 mutants that decreased the net negative charge, 16 were of two charges while only two involved one unit charge. Schwartz (32) pointed out that only two codons, GAA to AAA or GAG to AAG, permit the substitution of a positively charged amino acid (Lys) for a negatively charged one (Glu). The sum of Glu and Gln residues of ADH1-S was found to be 26/40,000 dalton polypeptide (7). Schwartz reported that the double charge change polypeptides do not represent a single primary sequence, but rather each has a different sequence as deduced from data measuring differences in their affinity for binding phytic acid (32). Further, isoelectric focusing in 9 M urea preserved the two unit charge differences (32). On the basis of these results, Schwartz has argued that one or a few EMS hot-spots in Adh1 is an unlikely explanation, as is the hypothesis that there is a flip-flop between two alternative conformations of active ADH that differ by two net surface charges (32).

A particularly useful mutant induced in the Adh1-F allele is U725 (6). (Our preferred terminology would list this as Adh1-F725.) This mutant encodes an active ADH1 subunit with a surface charge of -1 ("U") and an apparent molecular weight (in Laemmli gels) of 2,000 dalton less than normal. This Laemmli variant is a useful marker in SDS-polyacrylamide gel separations of in vitro translation products because it moves ADH1 subunits away from another polypeptide, ANP40C (10), of the same molecular weight. Alternatively, the Laemmli variants of the Anp40C gene could be used (Freeling et al., unpublished data).

The most important conclusion to be drawn from the totality of Adh1 point mutants induced by EMS or gamma radiation is that none of these mutants shows a change in organ-specificity or quantitative expression common among the naturally-occurring alleles (see Table 1 and Figure 4). Since the allozyme screen, unlike allyl alcohol selection, does not require dysfunction or low expression, the mutants recovered are truly unselected. For example, an EMS-induced mutant allele that encoded over-production of activity or one that conferred organ-specificity would have been recognized; they did not occur. One EMS mutant, Adh1-F278, was originally reported to affect the regulation of Adh1 (47), but has subsequently been shown to involve product activity (Schwartz, personal communication). It seems that EMS does not routinely induce alleles that mimic the naturally-occurring variants in quantitative or organ-specific behavior. (For a general review of variant and mutant alleles in higher organisms that do display regulatory properties, and their possible origin by chromosomal breaks, see Freeling and Woodman, 24.)

MUTANTS IN QUANTITATIVE, ORGAN-SPECIFIC AND/OR STABILITY FUNCTIONS OF Adh1: POSSIBLE REGULATORY ALLELES

Perhaps the most valuable mutants of Adh1 are listed and described in Table 3. All but one, the Ds insertion mutant studied by Osterman and Schwartz, were induced, recovered and characterized by Freeling and Cheng (37; submitted for publication). The recovery of all of these mutants followed allyl alcohol selection of pollen, except FkF3037, which was discovered as a spontaneous ear sector.

The Ds mutant of Adh1, Adh1-Fm355, has been described (41). The progenitor allele is an Adh1-"F" allele that occurred in a

Table 3.

Mutants in Quantitative, Organ-Specific or Stability Components of Adh1

Progenitor	Mutant	Mutagen used with Allyl Alcohol Selection	Characteristics[b]
Adh1-S	1951a[c]	400MeV Ne^{10+}, instability[a]	Underexpressed in scutellum but overexpressed in anaerobic root. A reciprocal effect mutant (24).
Adh1-FkF	3037[c]	Spontaneous	Overexpressed in scutellum only.
Adh1-"F"	m335	Ds at bz2+Ac	Low ADH1-F expression due to Ds. Responds to Ac. Protein is altered (41). Progenitor alleles were whatever present in bz2-m, Ac line.
	3020	500 MeV C^{6+}	Usually stable at 2×10^{-6} ADH$^+$/pollen grain, no product, but some plants revert at very high frequencies (Freeling, unpublished data).
	3034[c]	Robertson's Mu (54)	40% normal expression, unstable in both directions.
Adh1-S	3034a		Derivative of S3034; no product.
	3034b		Derivative of S3034; 14% normal expression (Alleman, unpublished data).
	3034 unstable		Very unstable, generating every level of expression from 0-40% normal expression among 50 progeny. Several derivative alleles are very unstable (Alleman, unpublished data).

[a]A stable derivative of an original unstable S1951 allele recognized due to low scutellar expression (36). The unstable allele is clearly associated with chromosomal abnormalities but S1951a is not.

[b]All characterizations (except for Fm335) performed in Freeling lab.

[c]Freeling and Cheng (submitted for publication).

line carrying Ds-suppressed Bz2 (bz2-m) with one or two doses of the Ac controlling element. The phenotype of these kernels was colored aleurone (3n) sectors on a bronze background, where each sector represents a bz2-m to Bz2 reversion. This mutability was dependent on the presence of Ac, as is the case with all examples of the Ac-Ds controlling element system (51,52). The bz2-m allele was chosen because it is about 20 mu from Adh1 and there is evidence that Ds transposes more often over short chromosomal distances (53). Following allyl alcohol selection, with the anthocyanin color genes as contamination control markers, Fm355 was picked in an allozyme screen as a single F_1 kernel which apparently under-expressed ADH1-F product. This allele was shown to be mutable, and high mutability in scutellum and pollen was dependent on the presence of Ac. Osterman has shown that the ADH1-Fm355 subunit is more thermolabile than its ADH1-F progenitor subunit and also than ADH1-F subunits resulting from Ac-mediated reversions of Adh1-Fm355. Therefore, coding sequences are involved in the original Ds insertion, but this fact does not imply that the low-expression behavior of Fm355 is also explained by a protein level mechanism.

A second mutable Adh1 allele, S3034, was recovered following allyl alcohol selection of pollen shed from plants carrying the Robertson's mutator system, Mu (54,55). The data from which the following description derives will be published elsewhere (Freeling and Cheng, submitted for publication). Adh1-S3034 specifies ADH1-S polypeptides at about 40% of the rate specified by Adh1-S (the rate of ^{3}H-Leu incorporation into ADH) and all organs tested showed low ADH1-S production; we have ascertained that the under-production behavior acted in cis to Adh1 coding sequences. Since about 35% of all mutants induced by Mu have displayed instability (54), it was not surprising that S3034 was also unstable. Mutability in the low to normal direction was demonstrated in the allozyme profiles of S3034/Ct scutella and from S3034 pollen for ADH activity. None of these revertants has been recovered. We were able to recover S3034 alleles of lower, null or extreme instability of ADH expression by subjecting S3034 pollen to yet another allyl alcohol treatment. That is, S3034 is unstable in both directions. Table 3 lists two stable derivatives, a null and an ultralow, that behave in a Mendelian manner and display normal morphology as homozygotes.

Another presumably unstable mutant is Adh1-S3020 (Freeling, unpublished data). Homoallelic S3020 plants appeared to specify zero ADH1 product and showed a phenotypic revertant frequency of 2×10^{-6} ADH$^+$/pollen, which is consistent with the great majority of EMS-induced Adh1$^-$ mutants (Table 2). Since the mutant was recovered following accelerated heavy ion irradiation, a deletion was suspected, and the pattern of intragenic recombination was obtained when S3034 was crossed to a series of EMS-induced Adh1-S$^-$ alleles indicating the presence of a chromosomal aberration that partially disrupted pairing. However, about one homoallelic S3020 plant out of 10 generates extraordinarily high ADH$^+$ frequencies: 1/1000 to 1/50; such high frequencies are unique to the S3020 allele. Perhaps a reversible chromosomal event has switched S3020 off.

The remaining two mutants described in Table 3 are more interesting from a developmental viewpoint. FkF3037 specifies an ADH1 polypeptide that accumulates to higher levels than normal in the scutellum only. This organ-restricted behavior is part of the Adh1 gene (i.e., cis-acting). S1951a is a mutant Adh1-S allele which behaves exactly like the reciprocal effect variants described earlier in this chapter. That is, S1951a is under-expressed in the seed but simultaneously over-expressed in the anaerobic root. The mechanism that underlies the reciprocal effect is not known; S1951a should provide a useful tool toward understanding this phenomenon.

Unfortunately, there are two differences, not just one, which might account for the fact that the mutants in Table 2 all behave as small coding sequence alterations while those in Table 3 appear to be regulatory in nature. First, the point-mutations were induced by EMS or, in one case, gamma radiation, while the regulatory-type mutants followed densely-ionizing irradiation or association with insertional mutators. Second, the point mutants were induced in the quiescent embryo while the regulatory-type mutants were in the pollen lineage; it is known that these two types of cells differ in their lesion repair properties (for example, ref. 36). Therefore, the cause for the differences between the mutants of Table 2 and 3 has yet to be determined.

SOME USEFUL Adh2 VARIANTS AND MUTANTS

If antigenic similarity is any indication (11), Adh1 and Adh2 should have considerable sequence homology. Any hybridization probe for Adh1 sequences in DNA or RNA might well form stable double helices with Adh2 DNA or RNA. The mutants described in Table 4 might prove necessary to identify an Adh1 genomic sequence unequivocally.

Adh2 is simultaneously expressed in the form of increased translatable mRNA as a part of the anaerobic response (10,15). Induction of ADH by the synthetic auxin 2,4-D does not induce Adh2 expression (8). However, Adh2 is naturally present as the predominant ADH polypeptide, along with Adh1 product, in the peduncles of the immature ear and in mature node cells of stem and tassel. Adh1 and Adh2 may be regulated in a compensatory fashion (3,29). Naturally-occurring Adh2 alleles of diverse origin have been shown to differ in quantitative, organ-specific expression. In particular, Adh2 alleles that are relatively over-expressed in the scutellum are under-expressed in the anaerobic root (56), a phenomenon at least analogous to the Adh1

Table 4.

Some Adh2 Alleles That May Prove Useful

Allele[a]	Source	Charge and Behavior[b]
Adh2-1N	StdF Inbred	N=-7.5; differs from 2N and 3N in organ-specific, quantitative expression and reaction to ethylene (56).
Adh2-2N	StdS Inbred	As above.
Adh2-3N	--	As above.
Adh2-1P	EMS in 1N (2)	P=-6.5; behaves like 1N and permitted observation of regulatory differences (56).
Adh2-88L	Guanajuato; Tlaltzapan Morelos	L=-8.5 (57); enzymologically active.
Adh2-37P[c]	Sonora, Tlaltzapan Morelos	P=-6.5 (57); enzymologically active.
Adh2-12R[c]	Costa Rica; Tlaltzapan Morelos	R=-5.5 (57); enzymologically active.
Adh2-33	Knobless Wilber Flint	No product (57).
Adh2-90	Ecuador; Anguil Limon Cocha	No product (57).
Adh2-143	Puebla; Tlaltzapan Morelos	No product (57).

[a]In general, Schwartz et al. use an allelic designation like -L88 rather than the -88L used here.

[b]Apparent charge is relative to ADH1-S=-2, deduced from electrophoretic mobility in standard starch gels (20). Alleles 1N, 2N, 1P and 33 have been reconfirmed.

[c]These variants are not necessarily allelic to Adh2 (57).

reciprocal effect. Thus, the coinduction of Adh1 and Adh2 by anaerobiosis is the most obvious regulatory phenomenon with which to correlate altered nucleotide sequences or sequence arrangements.

The function of ADH2 subunits is not known. ADH2⁻ plants appear as healthy as ADH2⁺ plants, even under anaerobiosis, but ADH1⁻, ADH2⁻ (homozygous Adh1-S5657 and Adh2-33) seedlings seem to drown more rapidly than ADH1⁻, ADH2⁺ seedlings, and germinate poorly (Freeling et al., unpublished data). Specific activities of ADH2 subunits relative to ADH1 and proof that dimerization is random for all ADHs have been presented (9).

USE OF RECIPROCAL TRANSLOCATIONS FOR THE STUDY OF Adh1 EXPRESSION

Maize has a distinguished history as a cytogenetic organism (58). The combination of well-developed linkage groups, ease of identifying chromosomal aberrations with pollen abortion, over 1000 translocation stocks (59) and direct examination of paired meoitic chromosomes in pachytene pollen mother cells permit chromosome manipulations at a level of sophistication comparable to that in Drosophila melanogaster. Electrophoretically-unique allozymes specified by variant or mutant Adh1 alleles can be used as markers for the Adh1 region of various segments of 1L. Allozyme markers have proved to be useful tools.

One way to perform a gene dosage effect/compensation test is to construct the 1, 2, 3 chromosomal arm dosage series using the appropriate translocation between the normal A set and a supernumerary B chromosome. These interchanges are called the B-A translocations (60). The only B-A translocations that move the Adh1 region of 1L onto the B centromere (abbreviated B^1) are TB-1La and TB-1Lc, both of which have the 1L breakpoint near 0.20 (Figure 8). The propensity for B centromeres to nondisjoin at the second microspore (post-meiotic) pollen division leads to duplication-deficient sperm. A $1/B^11B$ plant gives rise to two viable (balanced) types of meiotic products: 1 and B^11B. The B^11B chromosomes replicate and disjoin into a germ and a nondividing tube nucleus. Each has an entire chromosome 1 as 1^BB^1. The germ nucleus replicates ($1^B1^BB^1B^1$) and then undergoes B centromere nondisjunction. The result is the production of nonidentical sperm 1^B and $1^BB^1B^1$ within the same pollen grain. During fertilization, the hyperploid sperm preferentially fertilizes the haploid egg at a frequency greater than 50%, leaving the hypoploid sperm for the endosperm nucleus (2n):

$1/1 \times 1/1^BB^1$ ♂ → $1/2$ $1/1$ > $1/3$ $1/1^BB^1B^1$ and < $1/6$ $1/1^B$.

Note that a 1, 2, 3 dosage series of 1L (B^1) occurs among siblings. If the 1/1 ♀ is homozygous recessive for a 1L aleurone (3n) marker, such as bz2 (bronze), then segmentally trisomic embryos (2n) reside in bronze seeds; these seeds also happen to be distinctly smaller owing to endosperm dysfunction. Distinguishing 1/1 from $1/1^B$ is not straightforward without the use of allozyme markers or scutellar anthocyanin-conditioning genes. TB-1La normally carries an Adh1-"F" allele. The S, W(S586), FCm, C(F7086), Cm and Ct alleles have been placed on the B^1 chromosome (50; Birchler, unpublished data). An appropriate cross, with

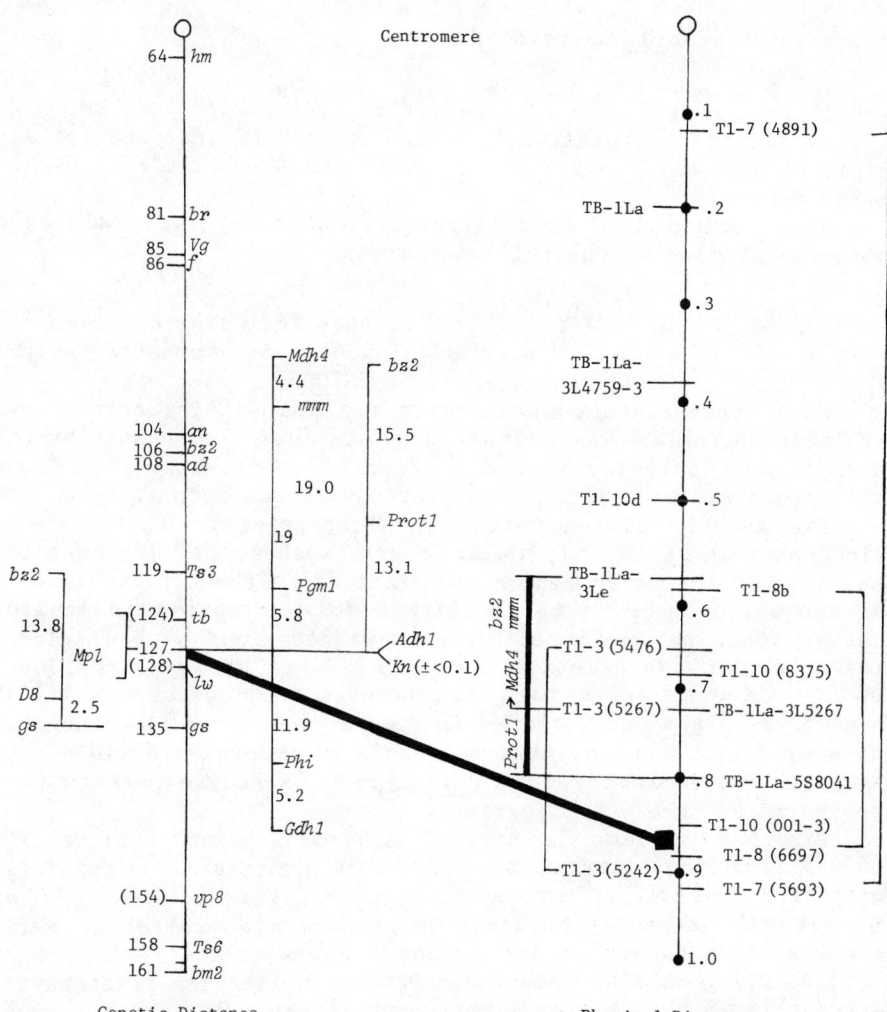

Figure 8. Cytogenetic localizations in the long arm of chromosome 1. Birchler (4) showed the placement of Adh1 on the physical map relative to the breakpoints in 13 reciprocal translocations; pairs of translocations that successfully duplicated Adh1 are connected. Longley (59) gave the cytological locations of breakpoints. The thick vertical bar gives physical locations for bz2 (61) ▒▒▒▒, modifier of mitochondrial MDH (62; Newton, personal communication) and the gene encoding protein 1 (Birchler, unpublished data). Mdh4 is proximal to TB-1La-3L5267 and TB-1La-5S8041 (Newton and Goodman, personal communication; 62). The basic genetic map (63) has been augmented as follows: the bz2-Adh1 linkage data (64); Mdh4-Gdh1 (6 allozyme-marked genes) linkage (Goodman et al., submitted for publication); bz2-gs (65); tight linkage of dominant markers Kn and Mpl (Freeling, unpublished data).

electrophoretic Adh1 markers is

$$1^F/1^F \times 1^C/1^B_B 1^C_B \,\male \to 1/2 \; 1^F/1^C, \; \leq 1/6 \; 1^F/1^B, \; \geq 1/3 \; 1^F/B^C 1^C_B \; .$$

A small sliver of scutellum from each kernel is analyzed for allozyme balance. F/C embryos are distinguishable from F/CC by allozyme ratio.

A 1L segmental disomic:tetrasomic comparison was made from among the progeny of the following cross:

$$1^F/1^B_B 1^S_B 1^S \times 1^C/1^B_B 1^C_B 1^C \to 9 \text{ classes including disomics}$$
(F/C, S/C, FS/-) and the tetrasome FS/CC.

When appropriate controls were run and sibling comparisons were made, Birchler found almost complete dosage compensation for specific ADH1 activity in the 1, 2, 3, 4 1L dosage series (50). This result was consistent with previous TB-1La data (3).

The genetic distance from the breakpoint of TB-1La to 1L telomere is about 100 mu, which comprises about 12% of the total map units of the maize genome and about 7% of the total cytological length. In order to facilitate dosage tests with smaller Adh1 regions, pairs of reciprocal translocations with displaced breakpoints on the same two chromosomes were used. First, Adh1 was localized relative to 13 1L translocation breakpoints (4). These breakpoints are included in Figure 9, where a line connects pairs of translocations that successfully generated haploid eggs segmentally disomic for the Adh1 region. Allozyme markers were used to detect these duplications.

Figure 9 diagrams the pair of reciprocal translocations, T1-3(5267)/T1-3(5242) marked as F/S, that generated the smallest duplicated segment. This region stretches from 0.72 to 0.90 on 1L. The 1L segmental duplication gametes are viable; no deficiencies are produced in the genome. A duplicated region of 3L (0.65 to 0.73) is also produced. Double duplication gametophytes are usually viable, but duplication-deficient gametophytes almost always abort. Figure 9 also diagrams the procedure used to establish a stable segmentally tetrasomic line (Birchler, Alleman and Freeling, submitted for publication). This line is stable through sibling and self-pollinations and is particularly useful because it is duplicated for exactly the same chromosomal regions represented in the original T1-3(5267)/T1-3(5242) heterozygote. With this displaced breakpoint method, it was possible to obtain comparable data from which to build a 2, 3, 4 Adh1 dosage series for the 0.72 to 0.9 region of 1L. The result was a clear 2, 3, 4 dosage effect (Birchler, Alleman and Freeling, unpublished data), in striking contrast to the 1L dosage data generated by TB-1La (50).

In the process of finding combinations of translocation lines that generated duplications for Adh1, much information on the ordering of breakpoints along 1L was obtained (see Figure

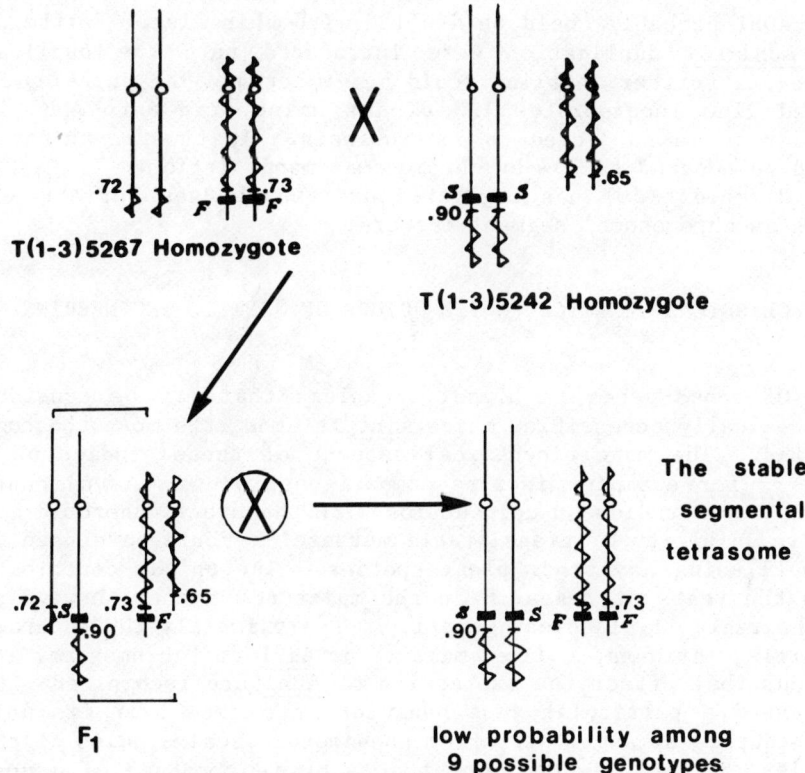

Figure 9. The construction of a stable, segmental tetrasome including Adh1 by means of a pair of reciprocal translocations with breakpoints in the same chromosomal arms. The nine genotypes expected in the F_2 exclude recombinants.

8). By constructing compound translocations, with various A-A translocations, TB-1La and marker loci along 1L, much valuable information relating physical and genetic distance has emerged. Adh1 lies between the 1L breakpoints of T1-5(8041) and T1-3(5242); these are at 0.80 and 0.90, respectively, and are also included in Figure 8. Measured in genetic distance, position 127 on chromosome 1 is about 67% of the map units down from the centromere on 1L; there is considerable noncorrespondence between physical and genetic location of Adh1. This noncorrespondence is graphically represented in Figure 8 by the bold line connecting Adh1 on the genetic map to Adh1 on the physical map.

Since the Adh1 tetrasomic produces a gene dosage effect, the level of ADH1 translation and, presumably, the level of mRNA-ADH1 probably could be doubled with this line. Further, if the Adh1-FCm duplication were introduced onto the duplicated region, a further doubling would be expected. The Berkeley Fast inbred line incorporates 12% of its amino acid into ADH1 when seedlings are subjected to anaerobiosis. In theory, this value could be boosted to 35% by chromosomal manipulation.

Birchler (50) has suggested additional uses for ADH allozymes as chromosomal segment markers.

CLASSICAL GENETICS IN THE FUTURE OF GENETIC ENGINEERING

Of those genes in higher organisms that may be considered biochemically accessible, maize Adh1 is among the most thoroughly studied. The more biochemical aspects of these studies of the Adhs -- for example, isozyme comparisons, protein level characterization, studies on coinduction with the other anaerobic genes and inducibility of translatable message -- could have been done as well using any other plant species. The unique contribution from the years of research on the maize Adh system, however, is in the realm of classical genetics. Approximately 200 naturally-occurring variants, induced mutants of Adh1 and chromosomal aberrations that effect the expression of Adh1 are recognized. Each expressed a particular gene behavior as a result of its unique DNA sequence arrangement. From phenomenological studies of these alleles, it may be concluded that Adh1 is composed of a coding component(s) and also at least three other components controlling regulatory functions of the gene. Variation in these components affect quantitative (Adh1-S3034), organ-specific (Adh1-FkF3037) and allele balance/reciprocal effect (Adh1-S1951a and naturally-occurring variants) behavior.

The allelic variation defining this genic complexity was not genetically engineered. Rather, the variability at Adh1 represents both spontaneous and induced mutations and aberrations of unknown molecular composition. In order to meet the ultimate goal of reducing behavior to molecular structure, genomic clones of these alleles could be sequenced. The genetic engineering approach might also be used to reduce an Adh allele's function to structure. An ADH^+-containing recombinant DNA molecule would be used to transform an ADH^- protoplast. Assuming that enough information was present to express Adh functions in the cell line, as well as in the regenerated plant, every nucleotide in the cloned Adh segment is available for in vitro mutagenesis. Very specific means, such as restriction endonuclease cleavage, would be used to alter the target gene. The transformed protoplasts and the plants into which they eventually develop might then be assayed for Adh expression. The progenitor and the mutated alleles could be compared regarding Adh behavior including quantity, organ-specificity, etc., as well as transmission, position and stability of the inserted sequence.

So, both classical genetics and genetic engineering can generate allelic comparisons where it is known that a particular gene behavior is altered by a specific DNA sequence change. However, there are some differences between these approaches.

1) Since we now have incomplete information as to the size of a typical complementation group, the genetic engineering approach might use only a portion of a gene as the target for mutagenesis. If this genic segment contained the coding sequence and the transcription initiation information, Adh expression might result, thus underestimating the capabilities of an intact gene. The classical genetic approach, because altered function is the starting point, would not suffer from this drawback. The error of confusing a part of a gene for the whole is fundamental.

2) Unlike classical genetics, in vitro mutagenesis can rigorously test chosen representative base substitutions, deletions and rearrangements for alterations in gene behavior in transformants.

3) Directed mutation in vitro cannot properly test insertion mutations because living plant cells generate these aberrations via highly evolved systems (e.g., Ac-Ds, Spm). While the position of an insertion may be important, it is possible that an insert might alter a gene's behavior because of information in its sequence rather than its position in a gene. There would be no way for the engineer to know what sequence to insert without the help of classical genetics. Likewise, position effects occur not due to a change in a gene's structure but due to a change in its position in the genome.

4) If information in a gene's nucleotide sequence is at all co-adapted with that of trans-acting modifier genes, certain aspects of a gene's programming would be organism-specific and would not be expressed in transformed cultures or plants. Genetic engineering is predicated on the desirability of transcending species barriers and treats the gene as an autonomous unit rather than as an integrated, programmed part of a living machine. On the other hand, classical genetics is highly organismal in its approach.

Fortunately, these two approaches are not mutually exclusive, especially not in the plant kingdom. It seems reasonable that genes, including their regulatory components, which have already been studied extensively with classical genetics and molecular biology, should then be used as the target for in vitro mutagenesis followed by the transformation assay. A more reasonable estimate of the size of the gene would be obtained. Actual examples of insertional sequences could be placed at various locations in the gene. The maize Adh1 system has the advantages of both classical genetics and molecular accessibility; it might also serve as a useful, selectable marker for transformations. ADH$^-$ protoplasts might be derived from plants selected as allyl alcohol resistant and ADH^+ transformants would be resistant to anaerobiosis and formaldehyde.

In fact, we know almost nothing about the sort of DNA sequence information that participates, by virtue of natural selection, in biological form (morphology) or physiology. What we fundamentally do not understand are the rules by which genes are regulated as a part of the entire organism's ontogeny and phylogeny; these programs are not merely manifestations of individual genes or isolated DNA sequences. We need to reduce regulatory/programming functions to DNA sequence information of both individual genes and complexes of coordinately regulated gene batteries. The Adhs and the other anaerobic genes comprise one such system developed to an extent that such a reduction could prove successful. Without classical genetics, the era of true genetic engineering will probably remain out of reach.

Acknowledgments: We thank Deverie Bongard and David S.-K. Cheng for help with the graphics and Mary Alleman, Elizabeth K. Porter, Martin M. Sachs and Kathleen Newton for criticism and information. This review and analysis derives indirectly from research grants (to M.F.) from the National Institutes of Health and also U.S. Department of Energy through the Lawrence Berkeley Laboratory.

REFERENCES

1. Schwartz, D. (1966) Proc. Nat. Acad. Sci. U.S.A. 56, 1431-1436.
2. Freeling, M. and Schwartz, D. (1973) Biochem. Genet. 8, 27-36.
3. Schwartz, D. (1971) Genetics 67, 411-425.
4. Birchler, J.A. (1980) Genetics 94, 687-700.
5. Fischer, M. and Schwartz, D. (1973) Mol. Gen. Genet. 127, 33-38.
6. Ferl, R.J., Dlouhy, S.R. and Schwartz, D. (1979) Mol. Gen. Genet. 169, 7-12.
7. Kelley, J. and Freeling, M. (1980) Biochim. Biophys. Acta 624, 102-110.
8. Freeling, M. (1973) Mol. Gen. Genet. 127, 215-227.
9. Freeling, M. (1974) Biochem. Genet. 12, 407-417.
10. Sachs, M.M., Freeling, M. and Okimoto, R. (1980) Cell 20, 761-767.
11. Freeling, M. (1973) Maize Genet. Coop. News Lett. 47, 57-58.
12. Schwartz, D. (1972) J. Chromatogr. 67, 385-388.
13. Sachs, M.M. and Freeling, M. (1978) Mol. Gen. Genet. 161, 111-115.
14. Ashburner, M. and Bonner, J.J. (1979) Cell 17, 241-254.

15 Ferl, R.J., Brennan, M.D. and Schwartz, D. (1980) Biochem. Genet. (in press).
16 Okimoto, R., Sachs, M.M., Porter, E.K. and Freeling, M. (1980) Planta (in press).
17 Laszlo, A. and Sung, Z.R. (1980) Fed. Proc. (Abstract) 39, 1745.
18 Schwartz, D. (1972) Amer. Natur. 103, 477-481.
19 Brown, W.L. and Goodman, M. (1977) in Corn and Corn Improvement (Sprague, G.F., ed.), pp. 49-88, American Society for Agronomy, Inc., Madison, WI.
20 Schwartz, D. and Endo, T. (1966) Genetics 53, 709-715.
21 Scandalios, J.G. (1966) Genetics (abstract) 54, 359.
22 Laughner, W.J. (1970) Ph.D. Dissertation, Indiana University.
23 Freeling, M. (1978) Genetics 89, 211-224.
24 Freeling, M. and Woodman, J.C. (1979) in The Plant Seed (Rubenstein, I., Phillips, R.L. and Green, C.E., eds.), pp. 85-111, Academic Press, New York, NY.
25 Efron, Y. (1970) Science 170, 751-753.
26 Felder, M.J. and Scandalios, J.G. (1971) Mol. Gen. Genet. 111, 317-326.
27 Schwartz, D. (1973) Genetics 74, 615-617.
28 Felder, M.J., Scandalios, J.G. and Liu, E. (1973) Biochim. Biophys. Acta 317, 149-159.
29 Freeling, M. (1975) Genetics 81, 641-654.
30 Freeling, M. (1976) Genetics 83, 701-717.
31 Schwartz, D. (1969) Science 164, 585-586.
32 Schwartz, D. (1980) Environ. Health Perspt. (in press)
33 Briggs, R.W., Amano, E. and Smith, H.H. (1965) Nature 207, 890-891.
34 Coe, E.H. Jr. and Neuffer, M.G. (1978) in The Clonal Basis of Development (Subtelny, S. and Sussex, I.M., eds.), pp. 113-219, Academic Press, New York, NY.
35 Birchler, J.A. and Schwartz, D. (1979) Biochem. Genet. 17, 1173-1180.
36 Freeling, M. and Cheng, D.S.-K. (1978) Genet. Res. 31, 107-129.
37 Freeling, M. (1977) Nature 267, 154-156.
38 Schwartz, D. and Osterman, J. (1976) Genetics 83, 63-65.
39 Megnet, R. (1967) Arch. Biochem. Biophys. 121, 194-201.
40 Sofer, W. and Hatkoff, M.A. (1972) Genetics 72, 545-549.
41 Osterman, J. (1979) Ph.D. Dissertation, Indiana University.
42 Dickinson, W.J. and Caroon, H.L. (1979) Proc. Nat. Acad. Sci. U.S.A. 76, 4559-4562.
43 Berger, F.G. and Paigen, K. (1979) Nature 282, 314-316.
44 Bogenhagen, D.F., Sakonju, S. and Brown, D.D. (1980) Cell 19, 27-35.
45 De Franco, D., Schmidt, O., and Söll, D. (1980) Proc. Nat. Acad. Sci. U.S.A. 77, 3365-3368.
46 Schwartz, D. and Laughner, W.J. (1969) Science 166, 626-627.

47 Schwartz, D. (1976) Proc. Nat. Acad. Sci. U.S.A. 73, 582-584.
48 Schwartz, D. (1971) Proc. Nat. Acad. Sci. U.S.A. 68, 145-164.
49 Schwartz, D. (1971) Genetics 67, 515-519.
50 Birchler, J.A. (1979) Genetics 92, 1211-1229.
51 McClintock, B. (1951) Cold Spring Harbor Symp. Quant. Biol. 16, 13-47.
52 Fincham, J.R.S. and Sastry, G.R.K. (1974) Annu. Rev. Genet. 8, 15-50.
53 Orton, E.R. (1966) Genetics 53, 17-25.
54 Robertson, D.S. (1978) Mut. Res. 51, 21-28.
55 Robertson, D.S. (1980) Genetics 94, 969-978.
56 Schwartz, D. (1978) Genetics 90, 323-330.
57 Dlouhy, S.R. (1980) Ph.D. Dissertation, Indiana University.
58 Carlson, W. (1977) in Corn and Corn Improvement (Sprague, G.F., ed.), pp. 225-304, American Society of Agronomy, Madison, WI.
59 Longley, A.E. (1961) U.S. Dep. Agr. Agr. Res. Serv. 34, 16.
60 Beckett, J.B. (1978) Heredity 69, 27-36.
61 Newton, K.J. and Birchler, J.A. (1980) Maize Genet. Coop. News Lett. 54, 14.
62 Newton, K.J. and Schwartz, D. (1980) Genetics (in press).
63 Coe, E.H. Jr. and Neuffer, M.G. (1977) in Corn and Corn Improvement (Sprague, G.F., ed.), pp. 111-224, American Society of Agronomy, Madison, WI.
64 Schwartz, D. (1979) Mol. Gen. Genet. 174, 233-241.
65 Coe, E.H. Jr. (1980) Maize Genet. Coop. News Lett. 54, 27.

DEVELOPMENTALLY REGULATED MULTIGENE FAMILIES IN DICTYOSTELIUM DISCOIDEUM

R.A. Firtel, M. McKeown, S. Poole, A.R. Kimmel, J. Brandis

Department of Biology, B-022
University of California, San Diego
La Jolla, California 92093

and

W. Rowekamp

Deutsches Krebsforschungs/Zentrum
Institute für Zellforschung, Im Neuenheimer Feld 280
D-6900 Heidelberg 1, West Germany

INTRODUCTION

Dictyostelium as a Developmental System

Cellular slime molds have intrigued biologists since they were discovered at the end of the 19th century. In 1935, Kenneth Raper first reported the isolation of Dictyostelium discoideum and described its developmental cycle (1).

This review will discuss three developmentally regulated multigene families in Dictyostelium examined in our laboratory. In order to give the reader a framework for this analysis, we will first briefly review the biology of Dictyostelium and the structure and organization of the Dictyostelium genome. (For a more detailed analysis, see refs. 2-6).

Dictyostelium represents a good model system for studying various aspects of eukaryotic development. Dictyostelium amoebae grow vegetatively in nature using nonmucoid bacteria as a food source. They can be grown in large numbers in the laboratory using defined and complex axenic media as well as Klebsiella aerogenes. When the food source is depleted and the cells are starved for several required amino acids, the amoebae initiate the developmental cycle. During this period, many biochemical

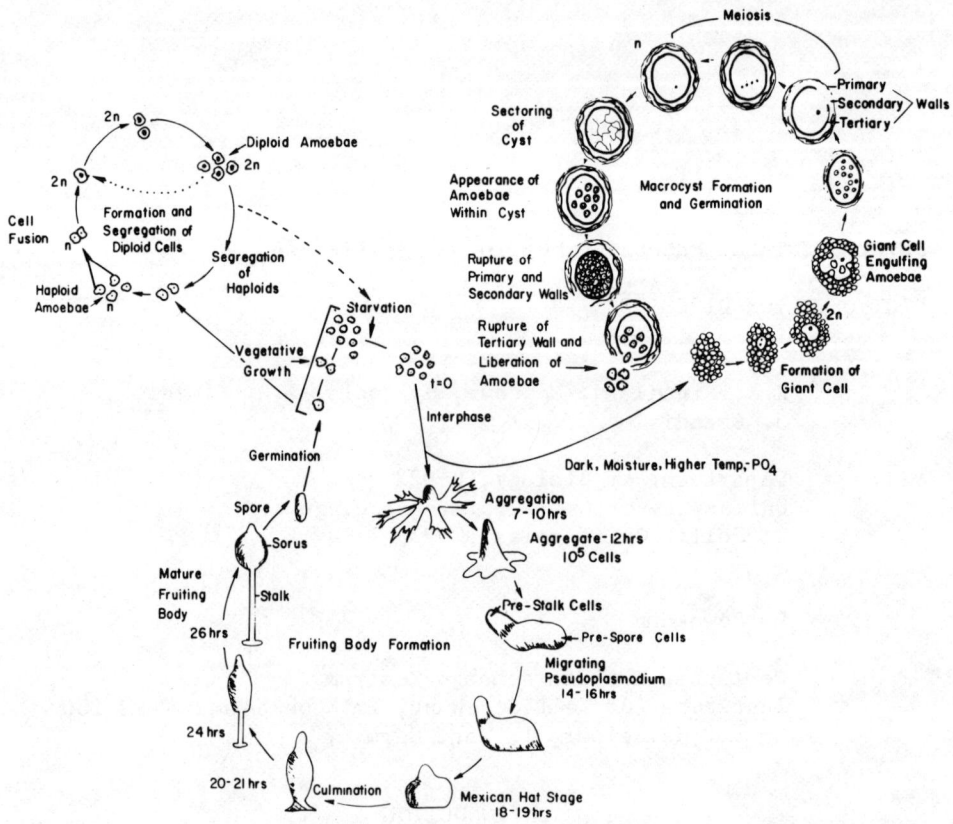

Figure 1. Stages of the slime mold life cycle. Center, the commonly studied aspects of the slime mold life cycle: vegetative growth of haploid amoebae and fruiting body formation. Fruiting body formation occurs when starved amoebae are placed on an air-water interface in the presence of visible light. Fruiting body formation culminates with the differentiation of approximately 80% of the cells into spores. Under appropriate conditions, these germinate and liberate amoebae, which can then enter vegetative growth. Upper right, the formation and germination of macrocysts. Entry into this cycle is favored by starved amoebae in the dark, in the presence of excess moisture or in the absence of phosphate. Two amoebae fuse to form a diploid giant cell. The eventual meiotic divisions that yield four haploid nuclei are not observable cytologically, but are included in the figure because their existence is implied from genetic experiments. Amoebae liberated in macrocyst germination can commence vegetative growth or aggregation. Upper left, the formation and segregation of diploid cells, the cycle that is exploited for parasexual genetics. Diploid cells can grow, as such, vegetatively and also can enter fruiting body formation. In the latter case, diploid spores are found (see refs. 2-5 for details).

and molecular changes have been monitored, including the establishment of the physiological pathways necessary for aggregation. After approximately 6 hrs, the cells begin to migrate towards aggregation centers using cAMP as a chemotactic agent. By 10 hrs after starvation, they form tight aggregates containing approximately 10^5 cells. The aggregates continue to develop and, under certain physiological conditions, form migrating slugs or pseudoplasmodia which are covered by a slime sheath. These slugs are capable of migrating for long periods of time and can be induced to culminate in response to light and several other physiological changes. Approximately 18 hrs after starvation the aggregates start to culminate and form fruiting bodies containing two cell types: the stalks (20 to 25% of the cells) and the spores (70 to 80% of the cells). The developmental cycle is shown in Figure 1. In the laboratory, large quantities of vegetative cells can be grown and synchronized for development.

The work of the molecular biologist, the physiologist and the biochemist have now converged so that a large number of different techniques are being used to study specific developmental phenomena. The aggregation process has been studied in considerable detail. Proteins and other molecules involved in the cell's response to cAMP and the species-specific cell-cell adhesion have been identified. Factors regulating various aspects of Dictyostelium morphogenesis and cell motility are also known. Molecular biologists have been concerned with the structure and organization of the Dictyostelium genome and the developmental changes in RNA and proteins.

Developmental Changes in Protein Synthesis

Experiments initiated by Sussman and co-workers (7,8) have shown that the specific activity (enzymatic activity per mg of total cell protein) of various enzymes changes appreciably during the Dictyostelium developmental cycle. With the drugs cycloheximide and actinomycin D, workers inferred that concomitant protein synthesis and prior RNA synthesis were required for the appearance and the disappearance of these enzyme activities through Dictyostelium development (9,10). By 1970, approximately 15 different enzymes had been studied for maximal periods of activity during the Dictyostelium developmental cycle. Some of these activities increase at the onset of starvation while others do not increase until much later in development, reaching peak activities during culmination. A summary of some of these data is shown in Figure 2. The use of mutants that affect development at specific stages indicates that timing of the expression of the genes encoding these enzymes is tightly coupled to morphogenesis (see refs. 2-6). It was inferred that the genes for these modulated enzymes are induced at set periods during the Dictyostelium developmental cycle and that a significant portion of the control

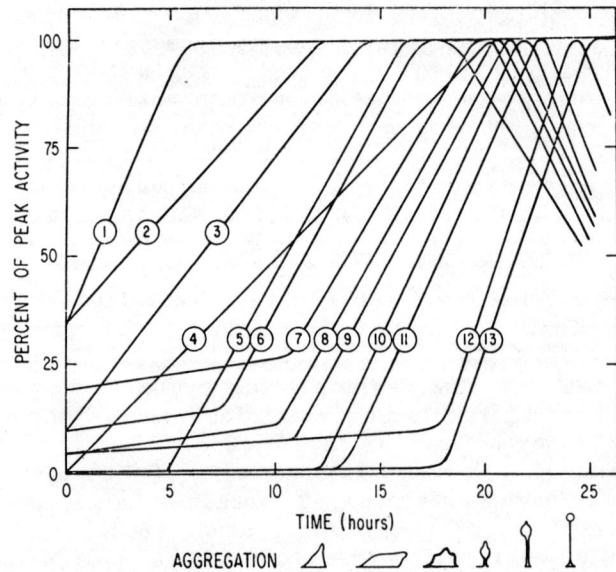

Figure 2. Stage-specific enzymes of Dictyostelium discoideum. Vegetative Dictyostelium cells are plated on an air buffer interface for the initiation of development. Samples are then taken at the appropriate times, and specific activities of the enzymes are determined. 1) alanine transaminase; 2) leucine aminopeptidase; 3) N-acetylglucosaminidase; 4 α-mannosidase; 5) trehalose phosphate synthetase; 6) threonine deaminase-2; 7) tyrosine transaminase; 8) UDPG-pyrophosphorylase; 9) UDPgal polysaccharide transferase; 10) UDPG-epimerase; 11) glycogen phosphorylase; 12) alkaline phosphatase; 13) β-glucosidase-2. (See refs. 3-6 for details). Reproduced from Loomis (3) with permission of Academic Press.

of gene expression may be at the level of transcription. These studies first focused attention on the fact that Dictyostelium discoideum is useful for examining the regulation of gene expression during development.

Characteristics of the Dictyostelium Genome

The genome of Dictyostelium discoideum has been characterized with such techniques as buoyant density centrifugation, thermal denaturation and renaturation kinetics (11,12). The haploid genome size of Dictyostelium discoideum is about 50×10^3 kilobases (kb) or approximately 11 to 12 times the genome size of E. coli as determined by renaturation kinetics. The nuclear DNA

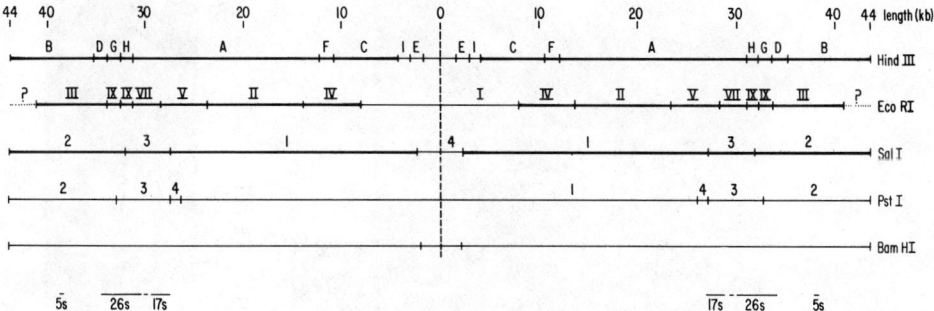

Figure 3. Restriction nuclease map of Dictyostelium rDNA. The location of the regions encoding the 17S, 5.8S, 26S and 5S rRNA is shown. The polarity of the 5S gene is not known. There are several HindIII sites near the center of the dimer and the resulting fragments have not been mapped. The numbers on the map refer to specific restriction fragments visible on digests of Dictyostelium DNA (See Figure 4 below).

is resolved into three peaks with CsCl buoyant density centrifugation. The main band (1.676 g/cc) represents 80 to 85% of the nuclear DNA and has a calculated G+C content of 22 to 23 mole%. This value agrees well with the G+C content determined by thermal denaturation. This average G+C content is one of the lowest of any genome. The base composition of in vivo and in vitro synthesized poly(A)$^+$ cellular RNA is about 30% G+C (13). More recent sequencing of various Dictyostelium genes indicates that the protein coding regions are relatively G+C rich (30 to 40%) but that the 3' and 5' untranslated regions of genes, intervening sequences and the regions 5' and 3' to the genes have a much higher A+T content (5 to 20% G+C) (14-16; Brandis and Firtel, unpublished data). The second CsCl peak (1.687 g/cc) corresponds to the rRNA genes; a third small peak, observed at a density of 1.682 g/cc, contains mitochondrial DNA contamination of the nuclear DNA preparation.

Analysis of the rDNA with restriction endonuclease mapping and recombinant DNA technology has shown that the haploid genome contains about 90 copies of an 88 kb palendromic extrachromosomal dimer carrying two copies of the ribosomal RNA cistrons (17-21). These genes encode a 36S rRNA precursor which is processed into the 5.8S, 17S and 26S rRNA. This palendromic rDNA organization is similar to that of some other protists (e.g., Tetrahymena and Physarum) (22-24). The average G+C content of the mature rRNAs is 42% (13), while the vast majority of the rDNA consists of spacer with a G+C content varying from 22 to 28%. Unlike the majority of protists examined (see ref. 25), the 5S genes in Dic-

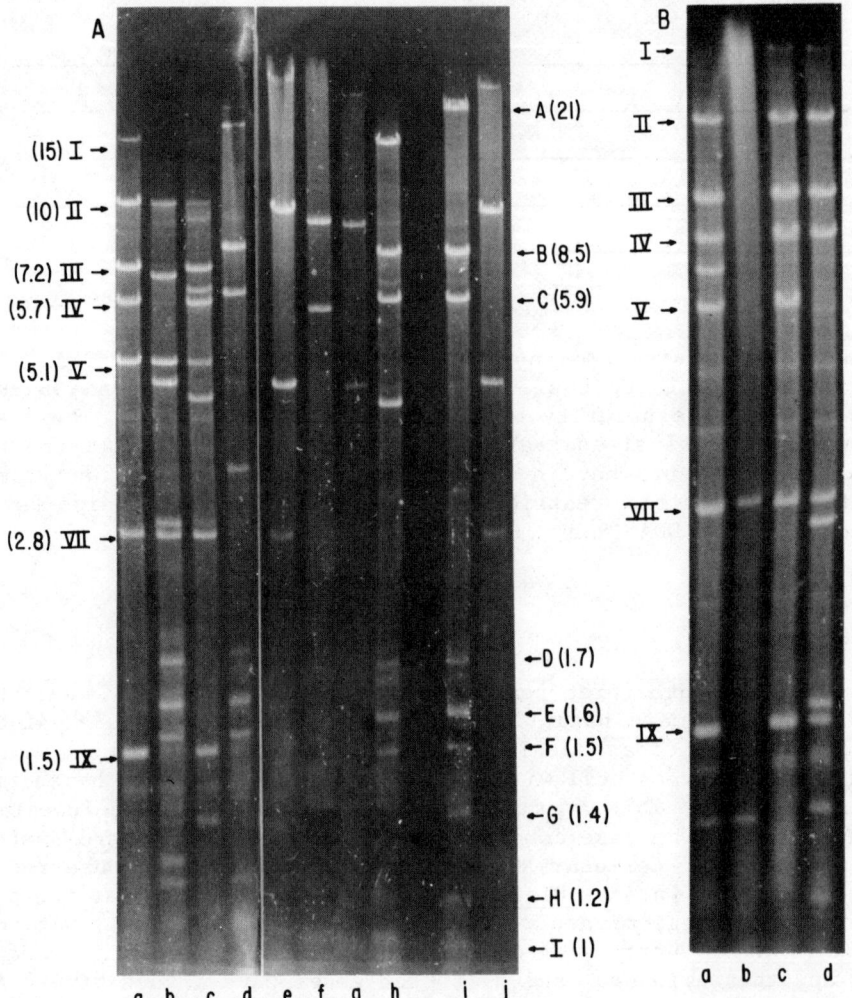

Figure 4. Restriction endonuclease digestion of Dictyostelium DNA. Dictyostelium nuclear DNA was restricted and electrophoresed on 0.8% agarose gels. Molecular weights in kb are given on the side. A. (a) EcoRI; (b) EcoRI+HindIII; (c) EcoRI+SalI; (d) HindIII+SalI; (e) SalI; (f) PstI; (g) PstI+SalI; (h) HindIII+PstI; (i) HindIII; (j) SalI+BamHI. B. (a) EcoRI+BamHI; (b) BamHI; (c) EcoRI; (d) EcoRI+PstI (see refs. 17-19,21).

tyostelium are linked to the 36S rRNA genes with two copies per dimer several kilobases 3' to the 36S rRNA coding region (19). As shown in Tetrahymena (26), the ends of the dimers are heterogeneous in length (Weiner, unpublished data; Firtel, unpublished data). A map of the rDNA is shown in Figure 3.

When Dictyostelium nuclear DNA is digested with various restriction enzymes, the majority of the DNA is cleaved into a heterogeneous pattern of fragment sizes. In addition, there is a series of prominent bands representing 18% of the DNA that can be mapped to the rDNA dimer (see Figure 4 and refs. 17-21). An EcoRI band of approximately 3.8 kb repeated about 50 times per haploid genome is seen. It does not map as part of the rDNA. Other less prominent bands estimated to repeat 5 to 10 times per genome are also observed. When the drug netropsin, which preferentially binds to A+T rich sequences, is used in equilibrium CsCl gradients, the main band DNA shows a much broader band with appreciable skewing towards the higher G+C rich region. These minor fragments are observed in the more G+C rich regions of netropsin-CsCl equilibrium gradients. Such an analysis is shown in Figure 5. By analogy with the genome organization of other species, these fragments may represent tandemly reiterated sequences, but their function is unknown at the present time.

Purified mitochondrial DNA has a density of 1.682 g/cc (28% G+C) and has been shown to be circular by electron microscopy (EM) (12,19). Mitochondrial DNA renatures with homogeneous second order kinetics corresponding to a complexity of about 50 kb. EM analysis indicates that the contour length of the mitochondrial circles is in agreement with this complexity measurement.

When denatured sheared nuclear DNA (300 to 400 nucleotides) is allowed to renature, approximately 70 to 75% of the DNA behaves as single-copy with a complexity of approximately 35×10^3 kb. The remainder of the sequences are reiterated on the average of 50- to 100-fold (12). The rDNA represents a large subset of the reiterated sequences.

The hybridization kinetics of labeled, tracer quantities of nuclear DNA sheared to defined size classes (0.4 to 3.5 kb) to 0.4 kilobase nuclear driver DNA indicated that 40 to 50% of the single-copy DNA is linked to reiterated DNA with an average single-copy sequence length of 1.2 kb (27,28). Approximately 20% of the single-copy DNA is tightly linked to repeat with a repeat-single-copy spacing of 4 to 5 kb. The remaining 30 to 40% of single-copy DNA is not linked to repeat sequences at fragment lengths of 5 kb. Other experiments indicate that approximately 70% of the repeat DNA is at least 1.5 kb while the remainder has a size distribution of 0.2 to 1 kb in length. The majority of the longer length reiterated sequences is rDNA. This pattern of short interspersed repeat-single-copy sequence organization is very similar to that observed in many metazoans (27,28).

Transcription of the Dictyostelium Genome

The general pattern of labeling and sedimentation of pulse-labeled Dictyostelium mRNA has been characterized. When cells are pulse-labeled for 20 min in the presence of a low concentra-

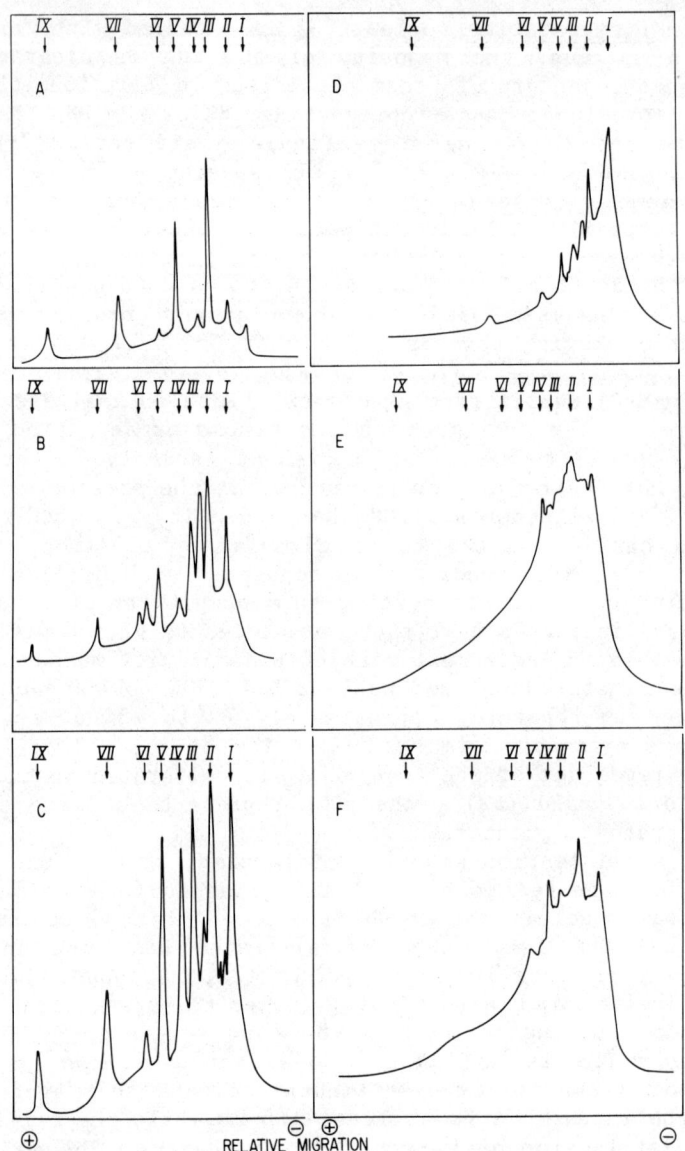

Figure 5. EcoRI mapping of DNA from netropsin/CsCl gradient. A preparative CsCl gradient was fractionated into 6 regions (A to F) with A being from the most G+C rich region and F from the most A+T rich region. The DNA was dialyzed versus restriction buffer, digested with EcoRI and analyzed on agarose gels. The gels were dried and the DNA bands localized by autoradiography. Densitometer tracings were done on X-ray autoradiograms. (a) to (f) Tracings of regions A to F (see ref. 21 for details). Roman numerials refer to EcoRI bands (see Figure 4).

tion of actinomycin D to inhibit rRNA synthesis and the RNA is extracted and analyzed on sucrose gradients, a heterogeneous pattern of newly synthesized RNA is observed with a number average size of 1.2 kb. The majority (>90%) of this label binds to poly(U)-Sepharose indicating most molecules contain a poly(A) sequence. The kinetics of label incorporation indicate that it takes approximately 7 min from the onset of mRNA synthesis for new transcripts to be transported into the cytoplasm. Nuclear RNA is, on the average, 25% larger than mRNA (10,28-30). It is expected that the larger size Dictyostelium nuclear RNA is due to the presence of intervening sequences (introns) found in nuclear RNA. Two separate genes (M4, a gene associated with a short interspersed repeat sequence, and M3, a developmentally regulated duplicated gene) have been shown to have higher molecular weight nuclear precursors and several intervening sequences (15,31,32; Brandis and Firtel, unpublished data). In contrast to the general observation in mammalian cells, these introns are relatively short (90 to 150 nucleotides) and have an extremely high A+T content (about 90%). The consensus sequence of these splice sites is similar to that found in other eukaryotes.

The half-life of total Dictyostelium messenger RNA (and specifically actin mRNA) has been measured by various methods. The majority of the mRNA, including actin mRNA, decays with a half-life of approximately 3.5 to 4 hrs in both vegetative and developing cells. A fraction of the mRNA has a shorter half-life of approximately 1 hr (33,34; Firtel, unpublished data; Jacobson, unpublished data).

Two size classes of poly(A) tracts are seen in Dictyostelium mRNA and hnRNA. One of approximately 25 nucleotides (poly(A)$_{25}$) is encoded in the genome; a second, longer sequence of 50 to 125 nucleotides (125 nucleotides in newly synthesized hnRNA) is added post-transcriptionally (28,30,35). These are shown in Figure 6. In polysomal RNA the two size classes are present in approximately equal molar amounts and thus on the average each message contains one sequence of each class. Studies have suggested that the two classes of poly(A) are linked within 1 to 10 nucleotides. In contrast to polysomal mRNA, hnRNA has approximately 4 to 5 moles of small poly(A) for each mole of poly(A)$_{125}$. This is interpreted to mean that a large fraction of poly(A)$^+$ hnRNA contains the short, transcribed poly(A) sequence while approximately 20 to 25% of the molecules, as a steady state level, have the longer poly(A)$_{125}$. Approximately 70 to 75% of the nuclear RNA is conserved and is transported to produce polysomal RNA.

Analysis of oligo(dT) stretches by depurination shows that there are 15,000 poly(dT)$_{25}$ sequences in the Dictyostelium genome (28,35). With sequencing, we have found several Dictyostelium genes with a poly(dA)$_{25}$ sequence at the 3' end. From the examination of nuclear and cytoplasmic RNA, oligo(dT) stretches and the sequencing of specific genes, we conclude that many, but not all Dictyostelium genes contain a run of about 25 dA residues near

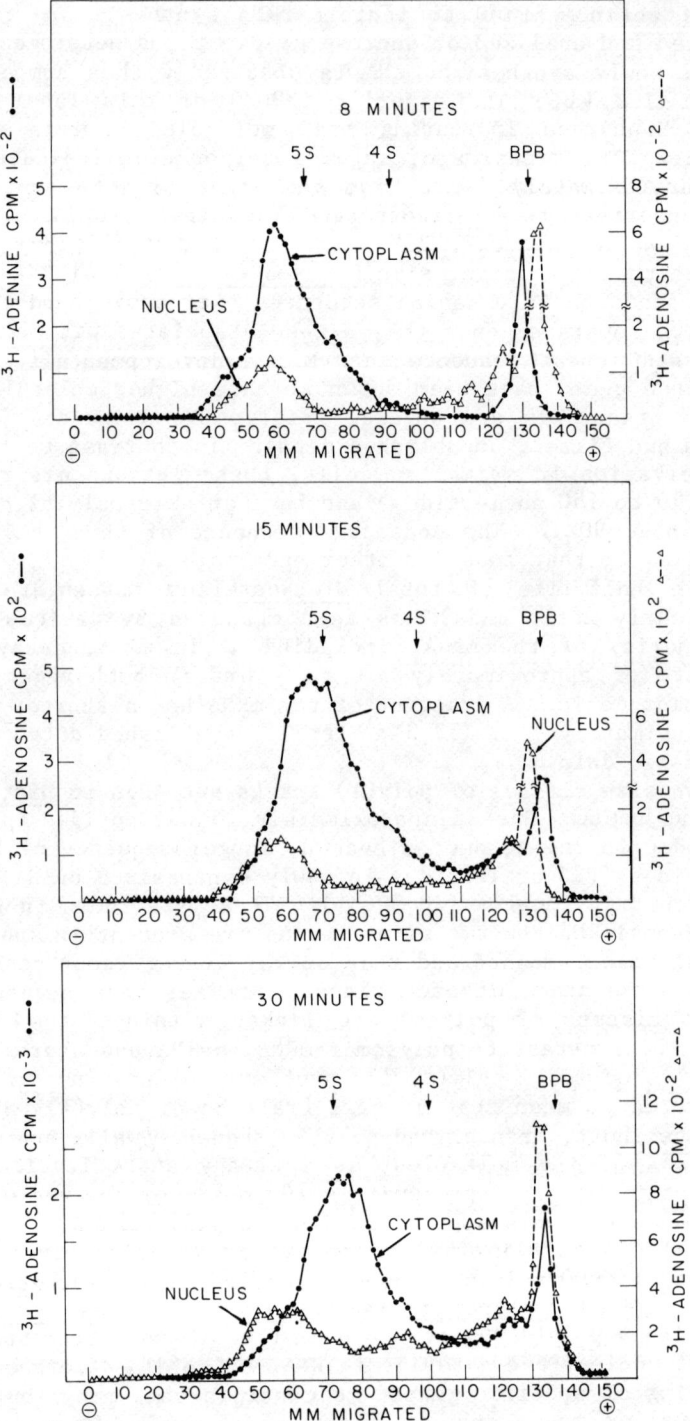

the 3' end, which is transcribed and found in both hnRNA and mRNA. The termination of transcription occurs several nucleotides past the poly(A)$_{25}$ and the poly(A)$_{125}$ is added post-transcriptionally (McKeown and Firtel, unpublished data).

A large amount of work has been done to determine what fraction of the genome is transcribed throughout the life cycle (6, 36-38). RNA excess hybridization of cellular RNA to single-copy DNA tracer and cDNA indicates that approxiately 50 to 60% of the double-stranded genome (25 to 30% single-strand) is transcribed during Dictyostelium development. From these data, we also conclude that there are four abundance classes of poly(A)$^+$ RNA. The first 3 components, comprising approximately 10, 30 and 60% of the total, contain approximately 50% of the complexity of the nuclear transcripts. The class representing about 10% of the mass consists of a few highly abundant mRNAs such as actin, with each present as between 0.5 and 2% of the total mRNA mass. The middle abundance class, representing about 30% of the mass, consists of approximately 300 different mRNAs each found at approximately 300 copies per cell. The third class consists of approximately 4×10^3 1.2 kb sequences found at 1 to 10 copies per cell. The fourth class, representing only 2% of the mass of poly(A)$^+$ RNA, consists of about 4×10^3 individual sequences, found in nuclear RNA, partially represented in cytoplasmic RNA, but not present on polysomes. It is unclear whether this fourth class represents specific transcription of genes whose products are needed at extremely low levels or represents an extremely low-level transcription of genes which are normally expressed at a higher level at some other time in development. It is possible that there is nonspecific transcription of various regions of the Dictyostelium genome (6,36).

With RNA isolated at various times in development, we have shown that there are approximately 2.5 to 3.5×10^3 new mRNAs transcribed during development that are not found on polysomes in vegetative cells (6; Firtel, unpublished data).

More recent results by Blumberg and Lodish (37,38) indicate that a higher fraction of the genome (80 to 85%) is transcribed throughout development; approximately 5,000 genes are transcribed during differentiation that are not expressed in exponentially growing vegetative cells. While there are differences between the way the cells are grown in Lodish's laboratory and ours, it

---

←Figure 6. Size of nuclear and cytoplasmic poly(A) from Dictyostelium heterodisperse RNA. Vegetative cells labeled for the indicated times with [^{3}H]adenosine were fractionated into nuclear and cytoplasmic poly(A)$^+$ RNA (6,10,28,29). Poly(A) was purified from the fractions and analyzed on 10% polyacrylamide gels. ^{32}P-Dictyostelium 4S and 5S RNA and bromphenol blue were the markers. The nuclear and the cytoplasmic gels were superimposed and plotted on the same figure.

is improbable that these can account for all the differences in the transcription pattern observed. However, results from both laboratories are qualitatively similar. We have only examined a very small fraction of the Dictyostelium genome by DNA sequencing, but it is difficult to imagine from the length of the regions between mapped genes and the extremely low G+C content of these regions that 80 to 85% of the single-copy DNA is expressed (unpublished data).

ORGANIZATION AND REGULATION OF EXPRESSION OF REPEAT GENE FAMILIES

The two gene families encoding actin and discoidin I represent two different types of gene families with regard to regulation during Dictyostelium development. Actin is encoded by a family of 17 genes and has a relatively complex pattern of synthesis during development, as some family members are regulated differentially. Discoidin I is encoded by a 3 to 4 member gene family which is expressed coordinately.

The study of repeated genes is focused on several questions: 1) How are members of a family organized? 2) What is the relationship between linked genes? 3) What is the sequence homology between gene members in the protein coding regions and in the 5' and 3' noncoding regions? 4) Is there a correlation between homology and coordinate gene regulation? 5) When during development are different members expressed and to what relative level?

ACTIN GENES

Actin appears to be essential in all eukaryotic cells. It is important in both muscle and nonmuscle cellular motility. Cytoplasmic microfilaments of actin are essential for cell shape changes involved in gastrulation and organogenesis in early embryos (for review, see ref. 39), cytokinetics (40,41) and amoeboid movement (see refs. 42,43). Actin filaments are also involved in the acrosome reaction of sperm and in cone growth in neuroelongation (44). Nonmuscle actin is associated with the plasma membrane in both growing and nongrowing cells (45-47). Its presence in the nonhistone component of chromosomal protein suggests that it is also involved in either chromosome movement during mitosis or meiosis or in chromosome condensation (48,49).

In Dictyostelium, actin represents approximately 5% of newly incorporated ^{35}S-methionine in vegetative cells (50,51; Firtel, unpublished data). Actin fibers are found in the cytoplasm in close association with the plasma membrane and are probably involved in amoeboid movement of vegetative cells (52,53). We have also seen the presence of actin in association with Dictyostelium nonhistone chromosomal protein (Firtel, unpublished data). However, in these studies, we cannot be absolutely sure that the actin present does not result from a small amount of cytoplasmic contamination. Actin from vegetative cells has been sequenced (54) and its polymerization and interaction with mammalian meromyosin has been examined (55).

Isoelectric Forms of Actin

When in vivo or in vitro synthesized actins are analyzed by 2-D gel electrophoresis, one major spot is observed plus more acidic and more basic spots (see Figure 7) (56,57). All of these spots are synthesized in response to purified actin mRNA and all bind to DNase I, which has a strong affinity for actins. The same isoelectric forms are synthesized throughout Dictyostelium development. However, there appears to be some quantitative change in the relative amount of each spot. During germination, when actin synthesis goes from an undetectable level ($<1/10^5$) to about 1% of newly synthesized protein, all of the isoelectric forms appear coordinately (56,57).

Vanderkerchove and Weber (54) have sequenced purified Dictyostelium actin and have found only one major actin form. No minor forms are detected at a level of >5% of the total actin utilized in their procedure. An analysis of newly synthesized actins has also been made by Rubenstein who has concluded that the more acidic minor form represents an unacetylated actin (58).

The discrepancies in the number of isoelectric forms of actin have not been resolved. One of the minor species could be a product of the M6 gene (see below), which has an amino acid change of a histidine to a tyrosine at amino acid 40; this probably results in a slight shift in the isoelectric point.

Developmental Regulation of Actin

A developmental time course of ^{35}S-methionine incorporation into newly synthesized protein is shown in Figure 8. Actin is one of the major proteins synthesized in vegetative cells, but it is synthesized at a several fold lower level during culmination (50,51,56,57,59-61; Firtel, unpublished data). A summary of the relative level of actin synthesis during development is shown in Figure 9. Actin expression has been examined during germination as well as during the multicellular differentiation stages. Dictyostelium spores can be synchronously germinated. Approximately 4 hrs after activation of the spores, amoebae emerge from the spores and enter into the vegetative growth phase in the presence of food. During the first hour after spore activation, no actin synthesis is observed (at a level of <0.01%) although a new set of germination-specific proteins can be observed. After 1 hr,

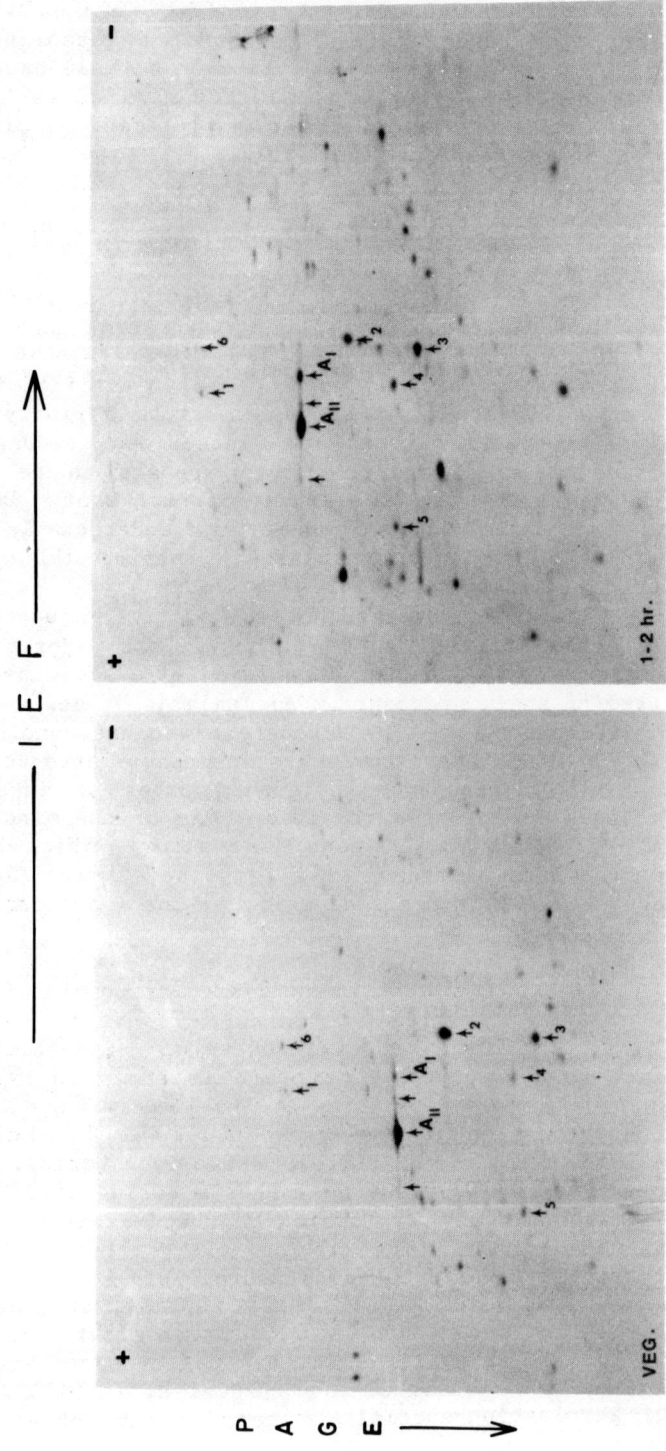

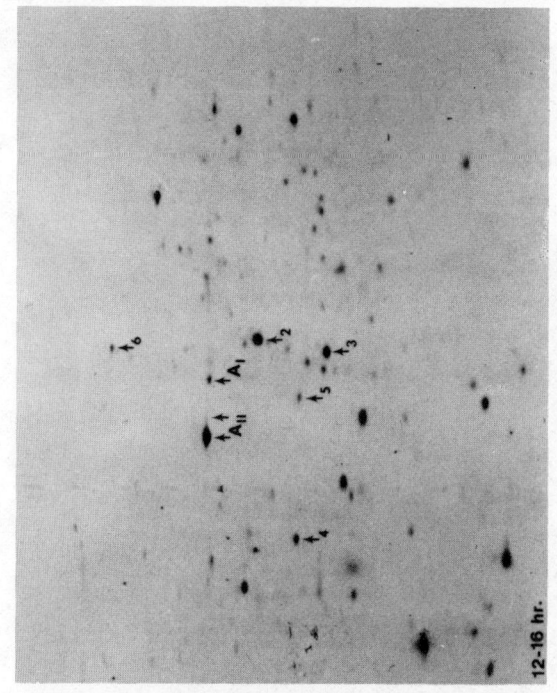
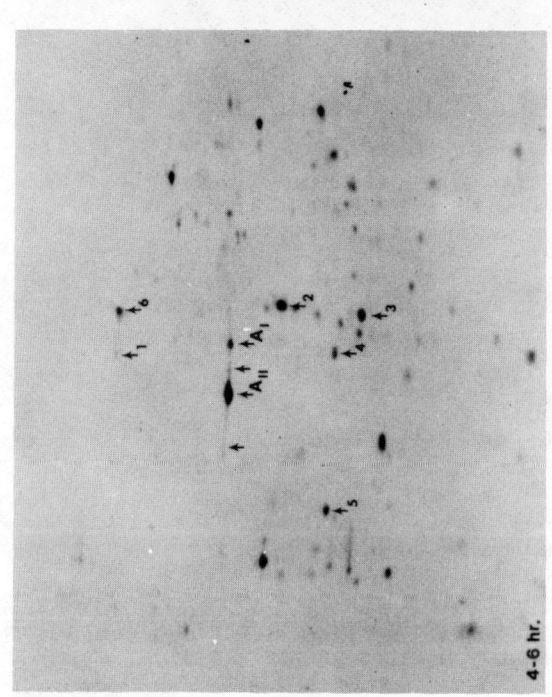

Figure 7. Two-dimensional gel electrophoresis of Dictyostelium actin. Cells were labeled with 35S-met during the Dictyostelium life cycle for the time periods shown and analyzed on 2-D gels (see refs. 56,57,59,63). AI and AII represent two of the more prominent actin isoforms. Adjacent arrows point to minor isoforms which are always present.

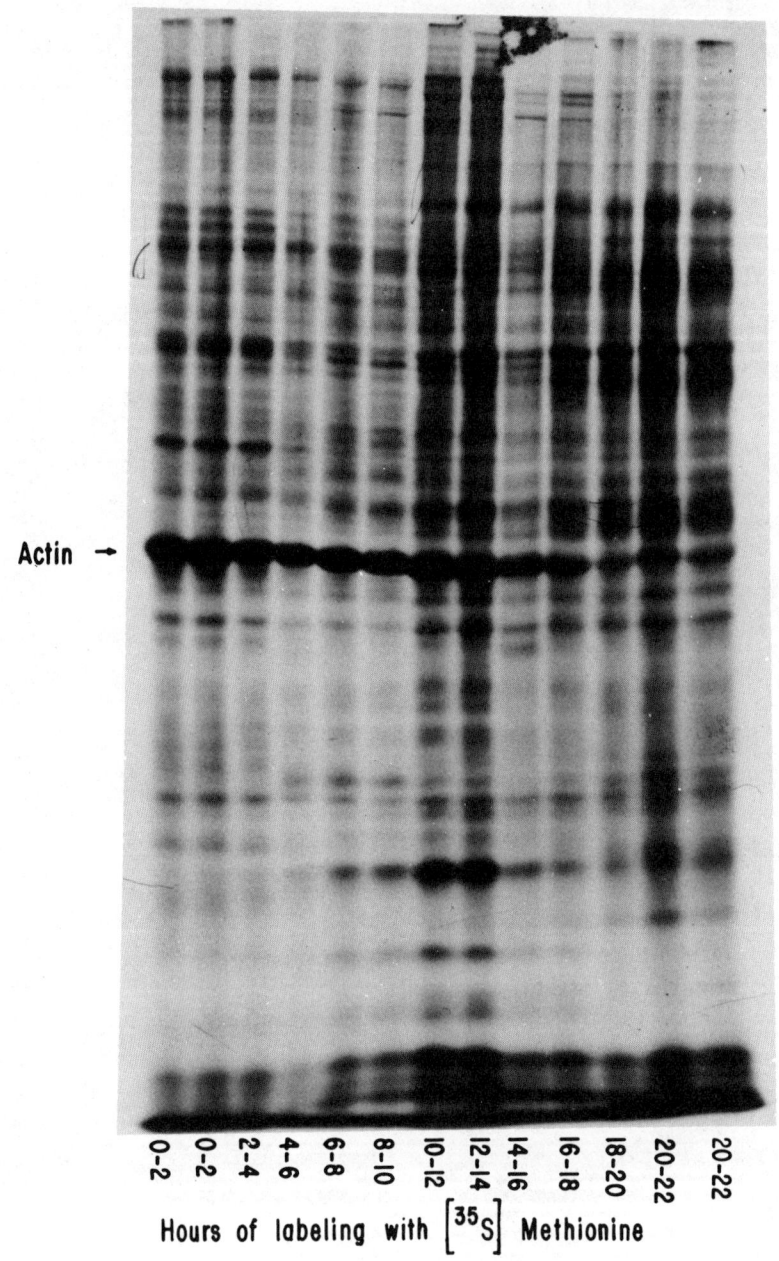

Figure 8. In vivo labeled proteins during Dictyostelium development. Dictyostelium cells were plated for development and labeled with ^{35}S-met for the time periods shown (10,51). Samples were electrophoresed on a 10% acrylamide gel and fluorographed. The major band corresponding to actin is labeled.

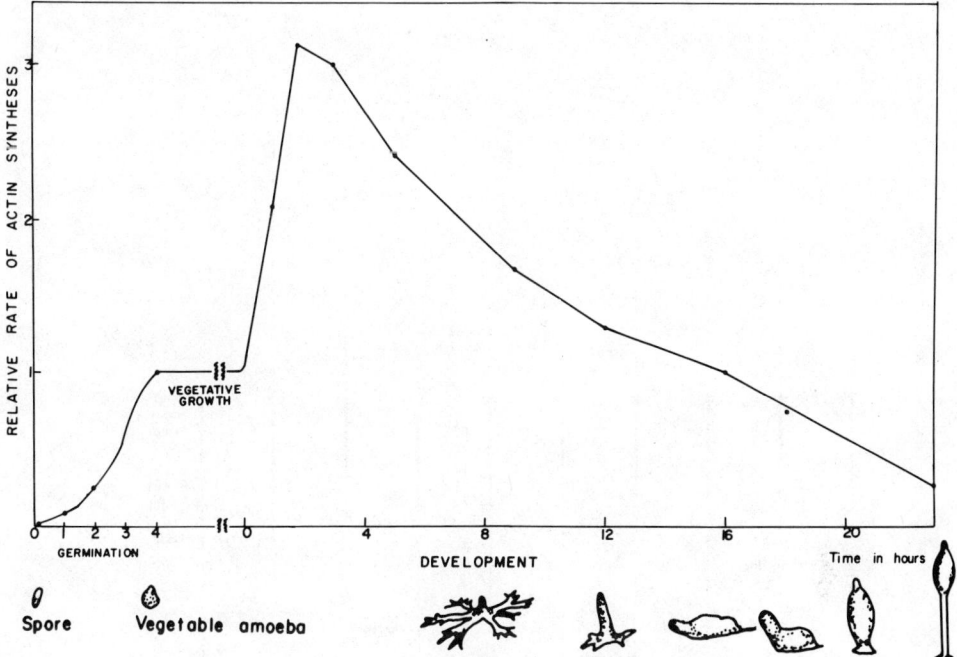

Figure 9. Developmental time course of actin synthesis. Shown is the relative rate of actin synthesis during Dictyostelium development. (Summarized from refs. 50,51,56,57,59-62; Firtel, unpublished data).

actin synthesis can be detected and the relative level increases to the vegetative level by the time of emergence. When newly-synthesized proteins are examined on 2-D gels, all isoelectric forms of actin observed appear coordinately (56).

Cells are plated for development and actin synthesis increases several fold over the next 2 to 4 hrs representing 15 to 20% of the ^{35}S-Met incorporation (50,51,57,59,60). After 6 hrs of development, the relative rate of actin synthesis decreases through the remainder of development and represents only several percent of the protein synthesis by late development. These results suggest that actin mRNA synthesis increases to a relatively high rate during the initial stages of development and then decreases shortly thereafter. It is possible that this increase in actin synthesis is necessary for the increased level of cell migration which takes place during this time in the developmental cycle. The relative rate of actin mRNA synthesis has been determined by hybridization of pulse-labeled poly(A)$^+$ RNA to cloned actin cDNA and measurement of the relative level of actin mRNA by in vitro protein synthesis with newly-synthesized mRNA purified by differential binding to poly(U)-Sepharose or oligo(dT) cellulose (56,57,59-61). The amount of actin mRNA has been quantitated by translation of isolated RNA and by RNA excess hybridization

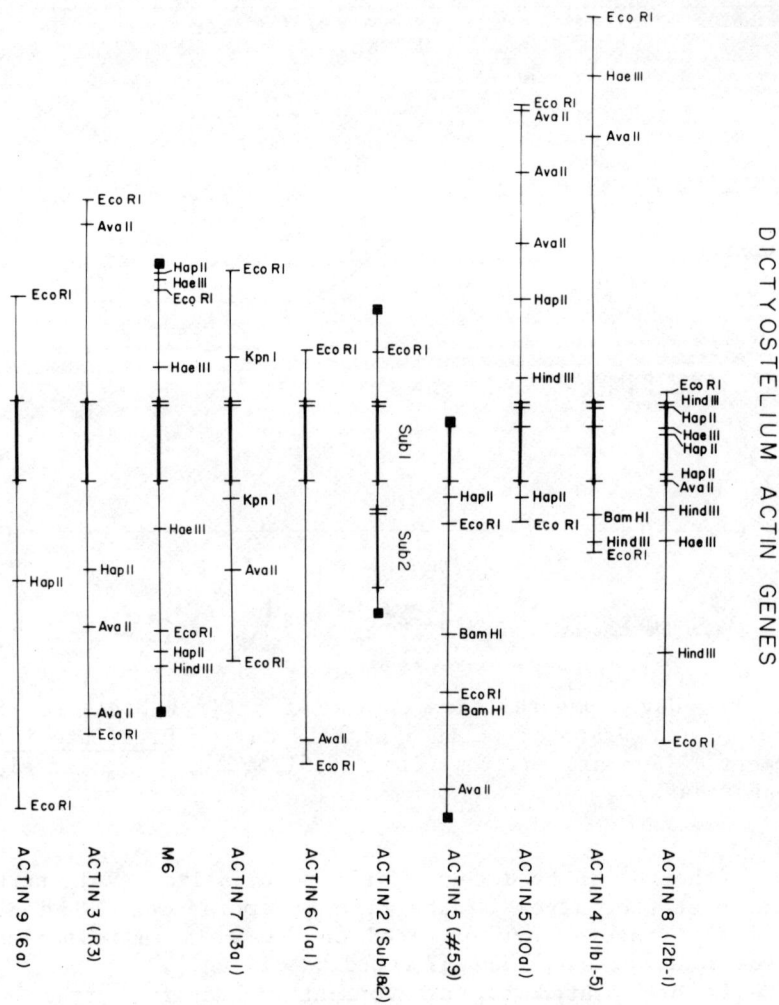

Figure 10. Restriction maps of cloned actin genes. Chimeric plasmids containing actin genes and surrounding chromosomal regions were restriction mapped by the standard methods of single and multiple digestion followed by electrophoresis. The actin genes are aligned and shown as thicker bars. The <u>Hind</u>III site in all genes lies at the 5' end of the protein coding region (see refs. 16,63-65,73 for details).

(56,57,59-63). These results show that the increases and decreases in actin synthesis during <u>Dictyostelium</u> development are accompanied by increases and decreases in the relative amount of actin mRNA. These changes in actin mRNA are due to the expected increases and decreases in the relative rate of actin mRNA synthesis.

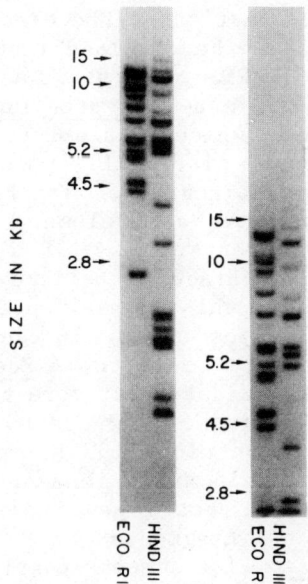

Figure 11. Southern blot hybridization of actin gene probe to Dictyostelium nuclear DNA. Dictyostelium DNA was digested with EcoRI or HindIII restriction endonucleases. Samples were split in half and fragments separated by electrophoresis. One-half of each sample was run until the bromophenol blue marker had moved 80% of the length of the gel; the other was run twice as long. Actin sequences were visualized by hybridizing ^{32}P-labeled actin cDNA to a Southern blot (98) made from the gel (73).

Structure of Actin Genes

In 1976, Kindle isolated a recombinant plasmid carrying a genomic Dictyostelium DNA insert complementary to actin mRNA. Analysis of this clone showed that actin mRNA was highly abundant and that the actin mRNA complementary region was reiterated between 15 and 20 times in the Dictyostelium genome (63-65). Subsequently, we have isolated recombinant plasmids containing 10 separate Dictyostelium actin genes plus a large number of recombinant plasmids carrying actin cDNA (16,66).

The restriction maps of these actin genes are shown in Figure 10. All isolated actin genes have a HindIII site very near the 5' end of the coding sequence and an AvaII (Sau96I) site very close to the UAA termination signal. In addition, the genes contain a series of HapII (HpaII) and HaeIII sites in the coding region. Actin 8 has the largest number of these restriction sites. The presence and absence of these sites may indicate the evolutionary relationship among the different genes.

A good deal of information on the organization and homology of various actin genes can be obtained from experiments such as that shown in Figure 11. Figure 11 shows a Southern blot hybridization experiment in which a Dictyostelium genomic DNA is digested with either EcoRI, which does not cut in any of the known Dictyostelium actin genes, or HindIII, which cuts once near the 5' end of all isolated actin genes. The blots were then probed with pcDd actin B_1, an actin cDNA clone. Hybridization to the HindIII restriction fragments shows 17 individual restriction fragments (see ref. 66). Since we believe that each actin gene has a single HindIII site, this number represents the actual number of actin genes in Dictyostelium. In contrast, the hybridization probe is complementary to 13 to 14 EcoRI restriction fragments. This result may indicate that more than one actin gene is found on the same EcoRI fragment. This is known to be the case for the two genes found on clone pDd actin 2. Other DNA blot data suggest that the single actin gene on plasmid pDd actin M6 is within 7 kb of another actin gene, although they would be located on different EcoRI fragments (65). The size heterogeneity of both EcoRI and HindIII fragments complementary to the actin probe strongly suggests that the genes are not arranged in tandem array with a common, homogeneous length intergenic spacer as found in the sea urchin, Drosophila histone genes and the ribosomal genes of most eukaryotes (see refs. 67,68). If all the actin genes were linked on one chromosome, they would require a minimum of 130 kb of DNA.

Examination of the intensity of hybridization of the probe to various restriction fragments shows that certain fragments hybridize much more intensely than others. For some of the EcoRI restriction fragments, this is probably due to the presence of more than one gene. In the case of the HindIII restriction fragments, the restriction fragments that hybridize more intensely than others may contain genes that show a greater sequence homology to the gene used as a probe. Those showing a weaker hybridization represent genes that are more diverged in nucleotide sequence. This conclusion can also be seen in Figure 12 in which probes from three Dictyostelium genes were hybridized to Southern DNA blots. Each gene hybridizes most strongly to the restriction fragment carrying that gene and shows a range of intensity of hybridization to other genes. With different probes, the same restriction fragments hybridize but the intensity of hybridization is different. This difference is most obvious with probe 2 (actin 2-Sub 2), which shows a strong hybridization to only a couple of restriction fragments and relatively weak hybridization to the others. The amount of nucleotide sequence divergence between these three genes was measured from thermal denaturation profiles (see Figure 13). Gene M6 and actins 2-Sub 2 have the highest amount of sequence divergence (approximately 10%) while M6 and actin 2-Sub 1, or actin 2-Sub 1 and Sub 2 have approximately 5% sequence mismatch between them. Anywhere from 2

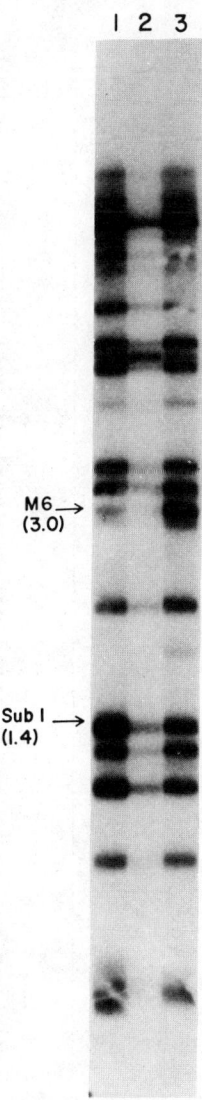

HapⅡ

Figure 12. Homologies between actin genes. A DNA blot filter was prepared from a HapII digest of total Dictyostelium DNA; the filter was sliced into one-quarter inch strips and each strip was hybridized to a different ^{32}P-labeled probe. After thorough washing, the filter strips were realigned and autoradiographed at -70°C. The ^{32}P-labeled probes hybridized to each strip were the actin gene containing restriction fragments from 1) pDd actin 2-Sub 1; 2)PdD actin 2-Sub 2; 3) M6 (see refs. 63-65).

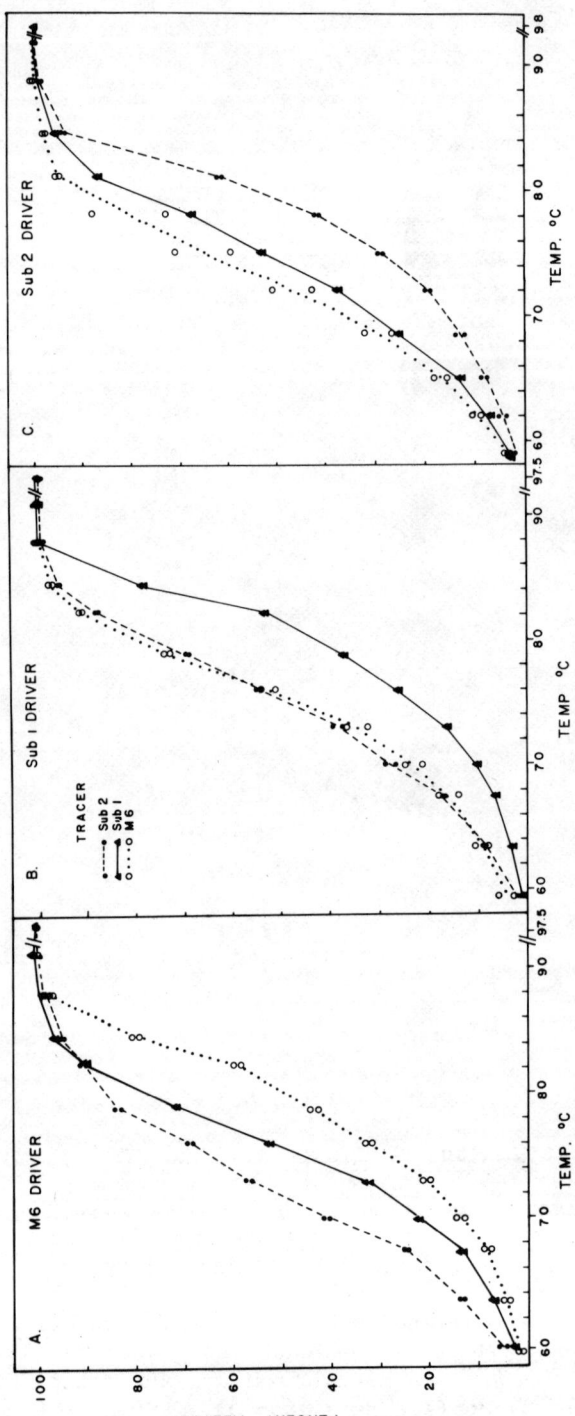

Figure 13. Melting curves for heteroduplexes formed between various actin genes. Plasmids M6, pDd actin 2-Sub 1 and pDd actin 2-Sub 2 were randomly sheared by mild depurination and hybridized in vast excess (greater than 200-fold) to ^{32}P-nick-translated actin gene containing restriction fragments. Hybridization was at 55°C in 0.12 M phosphate buffer (BP) for at least 20 times the calculated Cot1/2 for the unlabeled driver DNA. The samples were then bound to 1 ml hydroxylapatite columns in 0.12 M PB at 55°C and washed with 2x1 ml of preheated 0.12 M PB. All columns for a given unlabeled driver DNA were contained in the same variable temperature water bath. Unbound DNA was eluted with 2x1 ml of 0.12 M PB after 5 min of equilibration at the temperature of the indicated points. The number of counts not eluted at 55°C but eluted at temperatures up to 97.5 to 98°C inclusive was taken as 100% of the amount bound. This number was at least 30,000 Cerenkov cpm in all experiments. (A) Unlabeled M6 driver; (B) unlabeled pDd actin 2-Sub 1 as driver; (C) unlabeled pDd actin 2-Sub 2 as driver. (o---o) ^{32}P 1.7 kb HaeIII-Hap fragment of M6 as tracer. (▲---▲) ^{32}P 1.4 kb HindIII fragment of Sub 1 as tracer. (●---●) ^{32}P 1.3 kb HindIII fragment of Sub 2 as tracer (see ref. 63).

to 10% base mismatch within those sections of the protein coding regions which have been sequenced is observed (see below).

The extent of homology of the genes and intergene regions was initially analyzed by EM heteroduplex mapping. The mapping between genes M6, actin 2-Sub 1 and actin 2-Sub 2 shows a region of homology of approximately 1.1 kb -- the distance sufficient to encode actin protein (64,65). Heteroduplexes between two cDNA clones, pdDd actin B_1 and pdDd actin A_1, and genomic clones showed the absence of any large intervening sequences. Subsequent fine structure restriction mapping of various cDNA clones and their corresponding genomic clones combined with DNA sequencing and S1 nuclease mapping indicate that none of the isolated actin genes contains intervening sequences. In contrast, a single intervening sequence has been observed in the single yeast actin gene (69,70) as well as in the multiple Drosophila actin genes (71,72; N. Davidson, personal communication).

Heteroduplex mapping of cDNA clones and restriction mapping of cDNA and genomic clones indicate that there are no intervening sequences within the protein coding regions. S1 nuclease mapping and DNA sequencing of cDNA and genomic clones (see below) also indicate that no intervening sequences are found in the 5' and 3' untranslated regions of the genes cloned. As shown in Figure 10, all genomic and cDNA clones isolated to date contain both the HindIII and Sau96I restriction endonuclease sites located very near the 5' end of the protein coding regions, respectively. As a method of examining actin genes which have not yet been isolated, Dictyostelium DNA was digested with various restriction endonucleases including a combination of HindIII and Sau96I. If all the actin genes contain one and only one site for HindIII and Sau96I restriction endonucleases and if no intervening sequences are present, then digestion of genomic DNA with both of these enzymes will produce a single fragment of approximately 1.0 kb that will hybridize intensely with a probe from the protein coding region. The results of such an experiment are shown in Figure 14. There is an extremely intense band of hybridization at a DNA fragment size of 1.0 kb (marked with an arrow). There are a small number of higher molecular weight bands which show a low level of hybridization. Restriction endonuclease digests with HindIII, Sau96I and EcoRI are also shown. The higher molecular weight bands observed on the HindIII-Sau96I digests are generally less intense than most of the bands observed on the Sau96I or the HindIII digests alone. There are several possible explanations for these low intensity hybridizations. It is possible that some of the Dictyostelium genes that have not been isolated are missing a HindIII or a Sau96I restriction site, thus producing a higher molecular weight fragment. If this were the case, these would only be genes with a relatively low level of homology to B_1 since the higher molecular bands observed show a less intense hybridization than most of the HindIII restriction fragments carrying only a single Dictyostelium actin gene. Another possibility

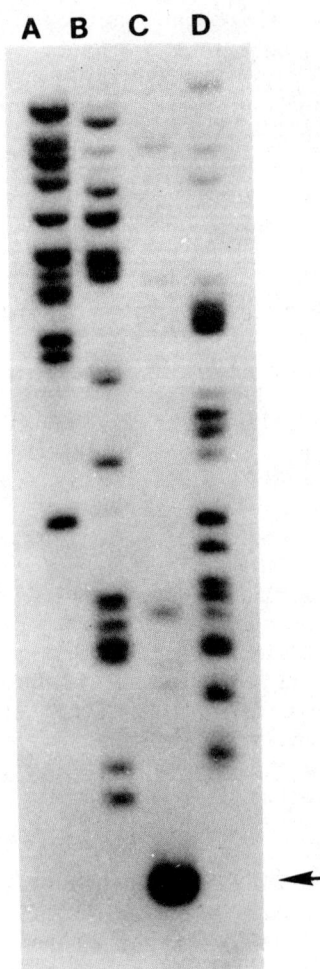

Figure 14. Internal organization of actin genes. Dictyostelium DNA was treated with restriction nucleases EcoRI, HindIII, HindIII + Sau96I and Sau96I and the fragments were separated by electrophoresis on an 0.8% agarose gel. ^{32}P-labeled pcDd actin B$_1$ DNA was hybridized to a Southern blot made from the gel. A. EcoRI; B. HindIII; C. Sau96I + HindIII; D. Sau96I.

is that there may be some other sequences in the Dictyostelium genome that have a poor sequence homology to the probe used, some nonspecific homology to the G-C tail or a low level of homology between the AUG and the HindIII sites to show hybridization in a genomic DNA blot. The last possibility is that some of the Dic-

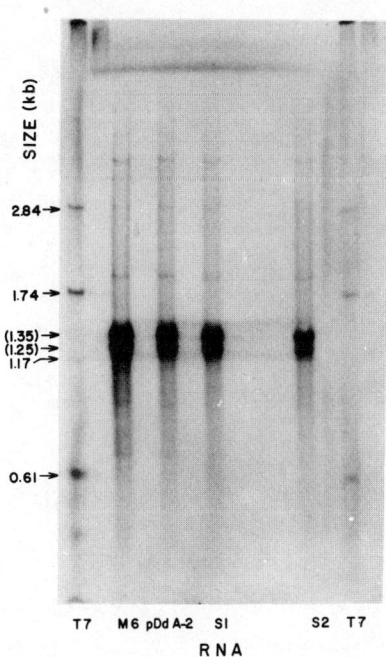

Figure 15. Mobility of actin mRNA in a methyl mercuric hydroxide-containing agarose gel. Poly(A)-containing in vivo labeled RNA from vegetative cells was hybridized to nitrocellulose-bound DNA, eluted and electrophoresed on a 1.8% agarose gel containing 5 mM methyl mercuric hydroxide. ^{32}P-labeled T7 early RNA was used as molecular weight markers. M6, pDda-S1 and S2 refer to mRNA complementary to M6, pDd actin 2, pdD actin 2-Sub 1 and pDd actin 2-Sub 1 DNA (see ref. 65).

tyostelium actin genes may have intervening sequences. This possibility is not likely for the very high molecular weight fragments. The reason for this claim is that the intervening sequences would have to be extremely large and all introns in Dictyostelium examined thus far in the M3 and M4 genes are relatively short (90 to 150 nucleotides in length) and the nuclear precursors are only between 150 and 400 nucleotides larger than the mRNA. If a Sau96I or HindIII restriction site were in one of the intervening sequences, it would produce fragments of lower molecular weight than the 1.0 kb fragment assuming that both the 5' and 3' ends of these genes contained the observed HindIII and Sau96I sites. Isolation of the remainder of the Dictyostelium actin genes should clarify these data.

The length and location of heteroduplexes between actin genes suggest that the only region of homology between the indi-

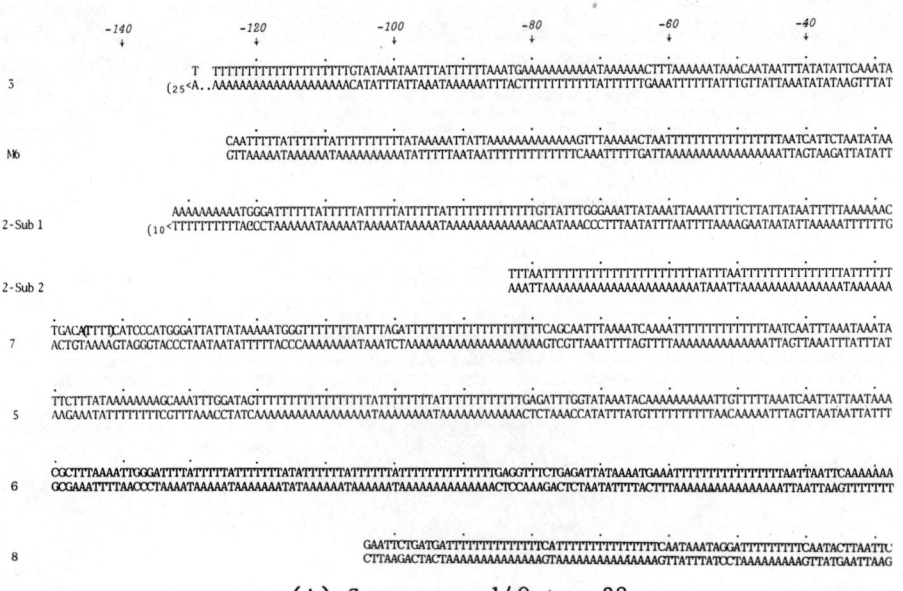

(A) Sequence −140 to −28

(B) Sequence −27 to +66

```
                                    90                        120                                150
                                     ↓                         ↓                                  ↓
         Ala Gly Asp Asp Ala Pro Arg Ala Val Phe Pro Ser Ile Val Gly Arg Pro Arg His Thr Gly Val Met Val Gly Met Gly Gln Lys
         GCU GGU GAU GAU GCC CCA CGU GCU GUU UUC CCA UCU AUU GUU GGU CGU CCA AGA CAC ACU GGU GUU AUG GUC GGU AUG GGU CAA AAA GA
B₁       CGA CCA CTA CTA CGG GGT GCA CGA CAA AAG GGT AGA TAA CAA CCA GCA GGT TCT GTG TGA CCA CAA TAC CAG CCA TAC CCA GTT TTT CT

         Ala Gly Asp Asp Ala Pro Arg Ala Leu Phe Pro Ser Ile Val Gly Arg Pro Arg Tyr Thr Gly Val Met Val Gly Met Gly Gln Lys
         GCU GGU GAU GAC GCU CCA CGU GCU CUU UUC CCA UCU AUU GUU GGU CGU CCU AGA UAU ACU GGU GUU AUG GUC GGU AUG GGU CAA AAA GA
3        CGA CCA CTA CTA CTG CGA GGT GCA CGA GAA AAG GGT AGA TAA CAA CCA GCA GGA TCT ATA TGA CCA CAA TAC CAA CCA TAC CCA GTT TTT

         Ala Gly Asp Asp Ala Pro Arg Ala Leu Phe Pro Ser Ile Val Gly Arg Pro Arg Tyr Thr Gly Val Met Val Gly Met Gly Gln Lys
         GCU GGU GAU GAU GCC CCA CGU GCU CUU UUC CCA UCU AUU GUU GGU CGU CCU AGA UAU ACU GGU GUU AUG GUU AUG GGU CAA AAA GA
M6       CGA CCA CTA CTA CGG GGT GCA CGA GAA AAG GGT AGA TAA CAA CCA GCA GGA TCT ATA TGA CCA CAA TAC CAA CCA TAC CCA GTT TTT CT

         Ala Gly Asp Asp Ala Pro Arg Ala Val Phe Pro Ser Ile Val Gly Arg Pro Arg His Thr Gly Val Met Val Gly Met Gly Gln Lys
         GCU GGU GAU GAU GCU CCA CGU GCU GUU UUC CCA UCA AUU GUU GGU CGU CCA AGA CAU ACU GGU GUU AUG GUC GGU AUG GGU CAA AAA G
2-Sub 1  CGA CLA CTA CTA CGA GGT GCA CGA CAA AAG GGT AGT TAA CAA CCA GCA GGT TCT GTA TGA CCA CAA TAC CAG CCA TAC CCA GTT TTT C

         Ala Gly Asp Asp Ala Pro Arg Ala Val Phe Pro Ser Asn Val Gly Arg Gln Arg Tyr
         GCG GGU GAU GAU GCU CCA CGU GCU GUU UUC CCA UCU AUU GUU GGU CGU CAA AGA UAU
2-Sub 2  CGC CCA CTA CTA CGA GGT GCA CGA CAA AAG GGT AGA TTA CAA CCA GCA GUU UCA ATA

         Ala Gly Asp Asp Ala Pro Arg Ala Val Phe Pro Ser Ile Val Gly Arg Pro Arg His Thr Gly Val Met Val Gly Met Gly Gln Lys Asp
         GCU GGU GAU GAC GCU CCA CGU GCU GUU UUC CCA UCU AUU GUU GGU CGU CCA AGA CAU ACU GGC GUU AUG GUU GGU AUG GGU CAA AAA GAU
7        CGA CCA CTA CTG CGA GGT GCA CGA CAA AAG GGT AGA TAA CAA CCA GCA GGT TCT GTA TGA CCA CAA TAC CAA CCA TAC CCA GTT TTT CTA

         Ala Gly Asp Asp Ala Pro Arg Ala Val Phe Pro Ser Ile Val Gly Arg Pro Arg His Thr Gly Val Met Val Gly Met Gly Gln Lys Asp
         GCU GGU GAU GAU GCU CCA CGU GCU GUU UUC CCA UCA AUU GUU GGU CGU CCA AGA CAC ACU GGU GUU AUG GUC GGU AUG GGU CAA AAA GAU
5        CGA CCA CTA CTA CGA GGT GCA CGA CAA AAG GGT AGT TAA CAA CCA GCA GGU UCU GUG UGA CCA CAA UAC CCA UAC CCA GUU TTT CTA

         Ala Gly Asp Asp Ala Pro Arg Ala Val Phe Pro Ser Ile Val Gly Arg Pro Arg His Thr Gly Val Met Val Gly Met Gly Gln Lys Asp
         GCU GGU GAU GAU GCC CCA CGU GCU GUU UUC CCA UCA AUU GUU GGU CGU CCA AGA CAC ACU GGU GUU AUG GUU GGU AUG GGU CAA AAA GAU
6        CGA CCA CTA CTA CGG GGT GCA CGA CAA AAG GGT AGT TAA CAA CCA GCA GGT TCT GTG TGA CCA CAA TAC CAA CCA TAC CCA GTT TTT CTA

         Ala Gly Asp Asp Ala Pro Arg Ala Val Phe Pro Ser Ile Val Gly Arg Pro Arg His Thr Gly Val Met Val Gly Met Gly Gln Lys Asp
         GCU GGU GAU GAU GCC CCA CGU GCU GUU UUC CCA UCA AUU GUU GGU CGU CCA AGA CAC ACU GGU GUU AUG GUC GGU AUG GGU CAA AAA GAU
8        CGA CCA CTA CTA CGG GGT GCA CGA AAG GGT AGT TAA CAA CCA GCA GGT TCT GTG TGA CCA CAA TAC CAG CCA TAC CCA GTT TTT CTG AGT

         Ala Gly Asp Asp Ala Pro Arg Ala Val Phe Pro Ser Ile Val Gly Arg Pro Arg His Thr Gly Val Met Val Gly Met Gly Gln Lys Asp
         GCU GGU GAU GAU GCC CCA CGU GCU GUU UUC CCA UCA AUU GUU GGU CGU CCA AGA CAC ACU GGU GUU AUG GUC GGU AUG GGU CAA
IEL-1    CGA CCA CTA CTA CGG GGT GCA CGA AAG GGT AGT TAA CAA CCA GCA GGT TCT GTG TGA CCA CAA TAC CAG CCA TAC CCA GTT
```

(C) Sequence +67 to +156

Figure 16. DNA sequence of 5' end of actin genes: coding region and sequence to nucleotide -150. The A of the AUG initiation codon of actin is labeled nucleotide 1. Nucleotides 5' to the AUG are denoted by negative numbers. The derived amino acid sequence is given above each DNA sequence. Underlined amino acids are those that are different from the Physarum actin sequence. Nucleotides in italics show changes in the structural sequence relative to pcDd actin B_1. The plus (top) strand of the structural sequence of the genomic clones and of the entire pcDd actin B_1 is written as ribonucleotides. The 13 dCMP residues at the 5' end of the pcDd actin B_1 represent the poly(dC) tail used for cloning. The parentheses indicate that the exact number of residues in these stretches is not known, but the values are probably correct to within one or two residues. Genes are abbreviated: B_1, pcDd actin B_1; M6, pDd actin M6; 2-Sub 1, pDd actin 2-Sub 1; 2-Sub 2, pDd actin 2-Sub 2; 3, 5 and 7, pdD actin 3, 5 and 7 (16).

vidual actin genes is the protein coding region. The 3' and 5' untranslated regions show little if any homology stable under heteroduplex conditions. When actin mRNA is isolated by hybridization selection to actin gene clones and sized on denaturing gels or analyzed by Northern RNA blots, two molecular weight species of 1.25 and 1.35 kb are observed (see Figure 15). These same species, in approximately the same ratio, are found in newly synthesized RNA during development. Subtracting approximately 50 nucleotides for the steady-state length of the poly(A) in Dictyostelium, a gene length of 1.2 and 1.3 kb is obtained for the two species (65). The 1.1 kb observed by heteroduplex analysis was reproducibly shorter than at least the 1.3 kb mRNA size (64).

Regions at the N terminus and the C terminus have been sequenced for eight genomic clones and several cDNA clones (16,66, 73,74). Figure 16 shows that most of the genes examined encode the same protein in the N terminal end. Sequence analysis of approximately 70 codons in the C terminal region also indicates that the genes with no amino acid substitution in the N terminal region show no substitution in the C terminal region. In all, we have sequenced between 40 and 60% of the protein coding region of several genes. These genes appear to encode the major Dictyostelium actin sequenced by Vanderchevkove and Weber (64) despite the absence of complete sequencing data. However, there is some third position nucleotide substitution between the various genes.

Three genes, actin M6, actin 3 and actin 2-Sub 2 show amino acid substitution in a variable number of positions. Actin M6 shows three amino acid substitutions in positions 2, 31 and 41 while actin 3 shows the same substitutions plus two additional ones. Actin 2-Sub 2 shows the largest number of substitutions with 10 occurring within the first 41 codons (16,73). There is a large amount of evidence indicating that actin 2 is a pseudo-gene (see below).

The region 5' to the AUG initiation codon is extremely A+T rich and there are no extensive homologies between the actin genes except for a run of A residues immediately preceding the AUG translation initiation codon. While the number of As preceding the AUG codon is variable, they are present in front of all actin genes.

Transcriptional Analysis of the Actin Genes

Since there is extensive homology between the genes, it is not possible to use differential hybridization with an entire gene probe to examine the transcription of individual actin genes. In order to circumvent this problem, we have developed a procedure that makes use of the homology between the AUG and the HindIII site and the nonhomology between the translational start codon and the 5' end of the messenger RNA. We have been able to determine the relative expression of particular actin genes during Dictyostelium differentiation. The general protocol is shown in Figure 17. In this procedure, a single-stranded DNA probe complementary to the 5' untranslated region is hybridized to poly(A)$^+$ RNA and is then treated with S1 nuclease to remove unhybridized single-strand tails. These fragments are then sized on polyacrylamide gels next to appropriate restriction fragment mar-

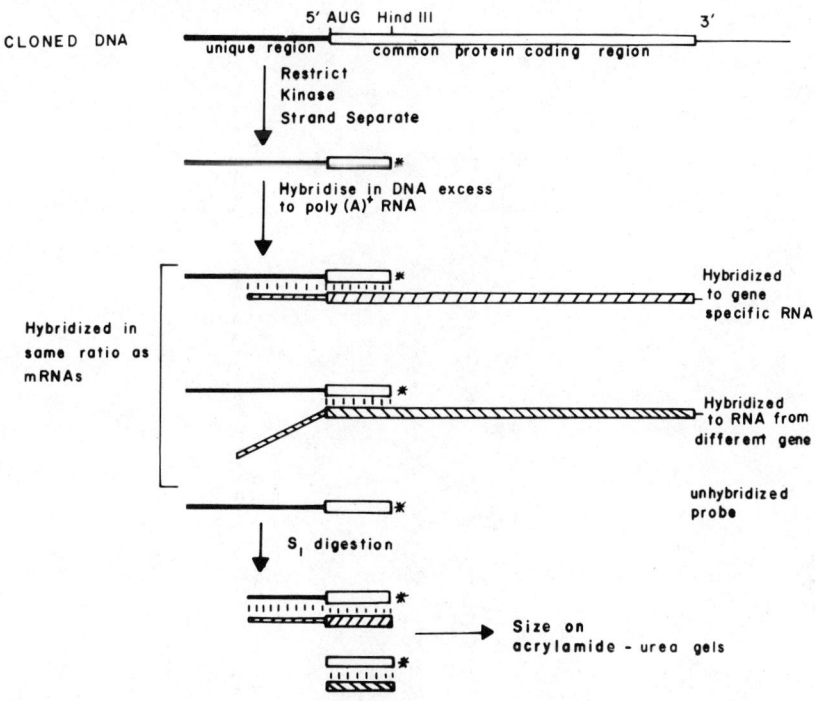

Figure 17. Localization of mRNA ends by S1 nuclease treatment. Actin plasmids were cut with HindIII and 5' phosphates replaced with [^{32}P] using γ-[^{32}P]ATP and polynucleotide kinase. Single-stranded, labeled DNA was isolated from strand separation acrylamide gels and hybridized to poly(A)$^+$ RNA. After S1 nuclease treatment, fragments protected by RNA were separated on 12% acrylamide-urea gels (see ref. 73).

kers or against a DNA sequencing ladder of the particular probe used. The results of an experiment with actin 8 are shown in Figure 18. Two molecular weight ranges of fragments are observed. The molecular weights differ slightly depending upon the S1 nuclease concentration used. Even with the highest S1 concentration, there is a series of bands in the region of the AUG and in the region presumably the 5' end of the messenger RNA. The presence of these larger fragments is evidence that actin 8 is transcribed. The heterogeneity around the AUG may be due to several factors including partial melting within this very A+T rich region and the variable number of A residues found adjacent to the AUG in the various genes sequenced. There is also heterogeneity in the length of fragments we expect to map at the 5' end of the actin 8 messenger RNA. This also could be due to partial melting of the end of the fragment due to the high A+T content or to possible heterogeneity of the start sequence. At present, the

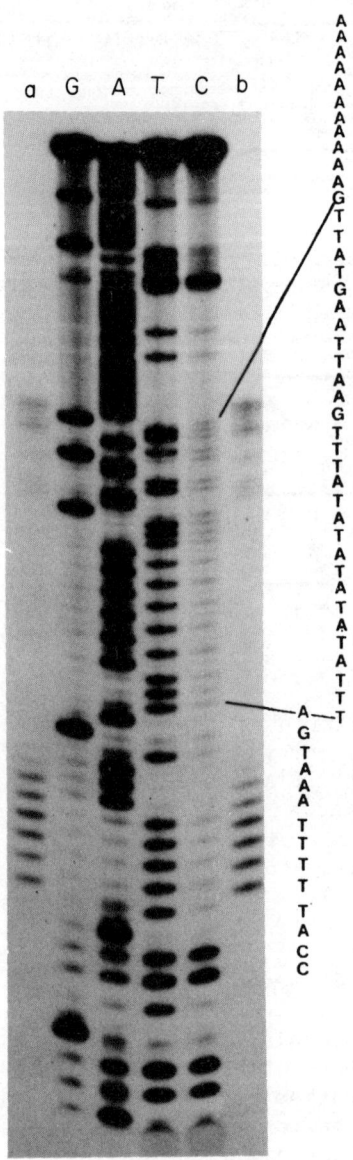

Figure 18. 5' ends of actin mRNA. Actin 8 DNA was digested with HindIII and EcoRI and treated as described in Figure 17. The DNA protected from S1 digestion was then sized by comparison to a Maxam/Gilbert sequencing ladder of the same fragment. Fragments in the sequencing ladder are 1.5 nucleotides farther down the gel than equivalent fragments treated with S1 due to the fact that the sequencing reaction eliminates the base and retains a 3' phosphate while S1 retains the base but with a 3' OH group (99).

DICTYOSTELIUM MULTIGENE FAMILIES

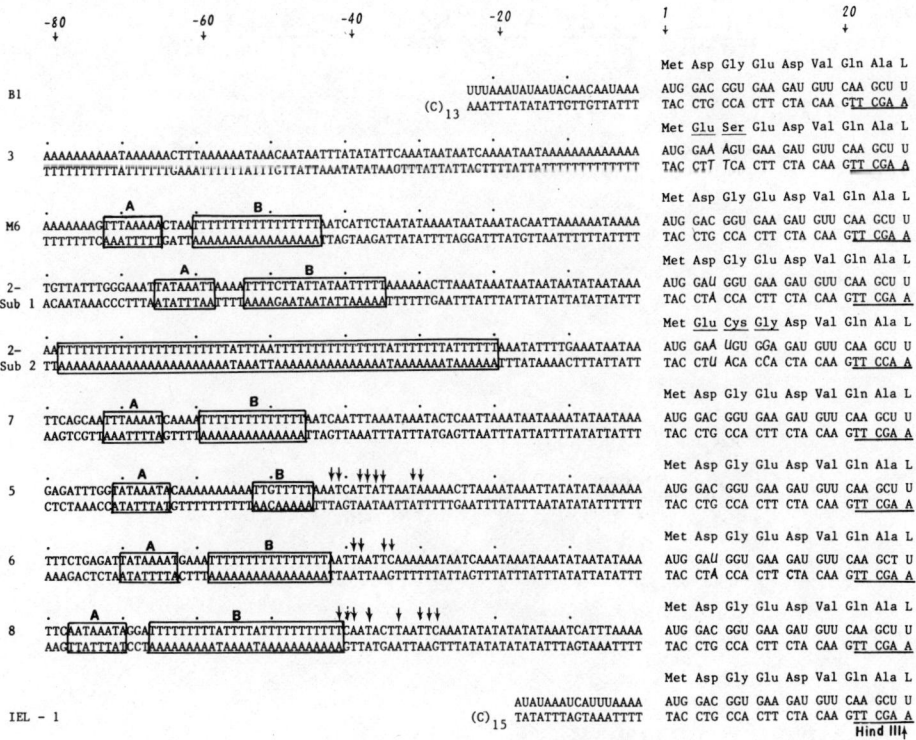

Figure 19. 5' ends of actin mRNAs. The potential 5' ends of the mRNAs derived from actin 5, actin 6 and actin 8 were determined in experiments similar to those shown in Figure 16. The vertical arrows indicate potential starts. The boxed region A corresponds to a potential Goldberg-Hogness box (96). The boxed region B corresponds to the homopolymer T stretch (see ref. 73).

data cannot rule out either possibility. It has not been possible to isolate actin 8 specific messenger RNA for 5' end sequencing adjacent to the CAP site. The conclusion that actin 8 is transcribed is corroborated by the fact that a cDNA complementary to the actin 8 messenger RNA has been identified by sequence of the 5' and 3' untranslated regions as well as a portion of the protein coding region.

Similar analyses have been done for actin M6, actin 5, actin 6 and actin 2-Sub 2. Actin M6, 5 and 6 all show hybridization in the region of approximately 30 to 40 nucleotides 5' to the AUG which suggests that these genes are transcribed. No expression is observed at the level of approximately 0.5% of the total actin messenger RNA with actin 2-Sub 2. These results suggest that actin 2-Sub 2 is not transcribed. Figure 19 shows the positions of 5' ends of actin 5, 6 and 8 as mapped by S1 nuclease hybridi-

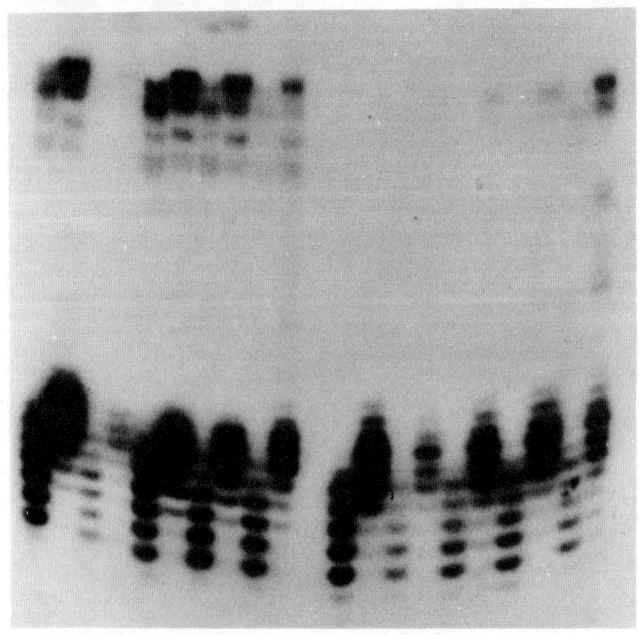

Figure 20. Actin gene expression through development. Actin 6 and actin 8 were treated as described in Figure 17 except RNAs from different stages in development were hybridized to excess labeled DNA prior to S1 treatment. Two different S1 concentrations were used for each hybridization. The higher S1 concentration was three times the lower. NC-4 are wild-type cells grown on bacteria. Ax-3 is an axenic strain grown in HL5 liquid medium.

zation. The 5' end of M6 is in a homologous position (data not shown) (see below). Potential starts are clustered in a region approximately 30 to 40 nucleotides 5' to the AUG. The analysis of this sequence is presented in a later section.

In order to quantitate the relative expression of the different actin genes during <u>Dictyostelium</u> development, an identical hybridization experiment is performed except that the probe is hybridized in DNA excess such that all actin mRNA sequences hybridize. By quantitating the number of counts in the fragments migrating in the region of the proposed 5' end versus the total number of counts hybridized, the expression (or level of mRNA) of a given actin gene relative to other actin genes can be determined. Figure 20 and Table 1 show an experiment and a compilation of data from a series of experiments on the expression of

Table 1

Expression of Actin Genes During Development

This table summarizes the relative expression of 4 of the actin genes during several stages of Dictyostelium development. The expression is given as percent total cellular actin mRNA. RNA was isolated at the developmental times indicated. Ax-3 are axenic cells grown in liquid culture. NC-4 are wild-type cells grown on K. aerogenes.

	Ax-3		NC-4	
	0 hr	3 hr	8 hr	20 hr
Actin 8	17±1	28±1	29±1	22±4
Actin 6	23±1	9±1	6±2	2±0.5
Actin 5	~7	~7	~7	~7
M6	1-3	1-3	1-3	1-3

the 4 genes. The different actin genes examined are not expressed at the same level. Actin 6 and actin 8 appear to be expressed at different levels at different times during Dictyostelium development while some genes are expressed at low levels at all times examined. M6, which encodes an actin with three amino acid substitutions relative to actin 5, 6 and 8, is transcribed at a relatively low level. The level is sufficiently low to prevent the quantitation of change in expression at different stages in development. The half-life of actin mRNA is essentially the same throughout development (61,62). If this result is true for specific actin mRNAs as well as for the bulk of actin mRNA, experiments may show that the relative level of transcription of these genes changes during Dictyostelium development and that expression of different actin genes is controlled mainly at the level of transcription.

At the present time, only these 4 actin genes plus actin 2-Sub 2 have been analyzed. DNA sequencing of large sections of actin 5, actin 6 and actin 8 suggests that they all encode the major isoelectric form of Dictyostelium actin sequenced by Vanderkerchove and Weber. Why should there be three genes with different patterns of expression for the same protein? It is possible that certain genes may be expressed at varying levels at different times in development because of different translational properties. It is not known why actin 8 has a high level of expression throughout development.

Actin 2-Sub 2: A Pseudo-gene

There is evidence to suggest that actin 2-Sub 2 is not transcribed but is a pseudo-gene. Transcriptional studies cannot

```
                5'                                                                                       Ter                                                                                   3'

pcDd Actin ITL-1       UAAAC AAAUAAUUAAAAC U AGUGAUCAAA GUCCUCUCACAAACAAUUAUUGUAAAAUAUAUAAUAAA
pDd Actin 5            UAAAC AAAAAAaAAAAC cgAGUGAUCAAA GcGCUGUCACAAA A UUAUGaAAAAUAUuUAAUAgu
pDd Actin 6            UAAACuAAACAAUUAAAAC c AGUGAUCAAAuGU CUUCUCACAcuuAAcaAUaUAAuAUuUAUAguau
pDd Actin 2-Sub 1      UAAAu uAAUuAaAAAAuuU AGUGAUCAAA GUCCUCUCACA CAAaaAUuauuauauauguacaau
pcDd Actin III-12      UAAAucA                UGAUGAAAuGUCCUUCaCAuAAAaAuaauaauaauaauaacaaua

pcDd Actin ITL-1(cont.) AUACAUUAUUU AA UCAUUUUUAUUUUUCUUUUAGUUCGUCGAUCUCUUUAUCCGACUAUU    UAAAAUUAAUUCU poly(A)
pDd Actin 5(cont.)      AUA AUaAUUuAAaUCuUUUUUAUUUUU UUUAGUCGUG UCUUUAUCCGACUuUaaaaaUAAAa AAUUGU (A)11 in genome
pDd Actin 6(cont.)      auaauaaaucucaauaaauauaaauucuaauuauuuugaauggugugucuuauccagccauc (A)33 in genome
pDd Actin 2-Sub 1(cont.) aauaacaauaaaaccccauaaaaaauauaaacuuuuucungauagucgugaucuuauucgaccuuu (A14) in genome
pcDd Actin III-12(cont.) aaaauauuuaaagugauauaauaaaauuaauaacuuuuuuuuaagguuguugaucuuuauccgaccuu poly(A)
```

A. ITL-1 is taken as the standard and the other 4 3' ends are compared to it. A best fit sequence homology was done by eye. The capital letters in the other 4 sequences refer to nucleotides homologous to ITL-1.

```
                5'                                        Ter                                                    3'

pcDd Actin III-12      UAAAUcA        UGAUGAAAGUCGUUCaCAuAaAAAuAAUaaUAaUA AUAU aACAA
pDd Actin 2-Sub 1      UAAAUuAaauuaaauuaaaauuuagUGAUGAAAGUCGUUCcCAcAcAAa AUuAU UAuAUAUguACAA
pDdDd Actin III-12(cont.) UAAUAuAuuUAAAuguauAAUAAAAUuUAAuuAC UUUUUuUUuAuGGUuGUCGAUCUUUAUCCGACCUU poly(A)
pDd Actin 2-Sub 1(cont.) UAAUAAcAa UAAAaacccAAUAAAAAUaUAA AcuUUUUcUUUgAUaGUcGUUcAUCUUUAUCCGACCUUU (A14) in genome
```

B. A similar homology between III-12 and Actin 2-Sub 1.

detect any RNA complementary to actin 2-Sub 2 at a level of less than 0.5% of total Dictyostelium actin mRNA (73). In addition, within the first 41 codons there are 11 amino acid substitutions (16). Some of these lie in the region which is conserved in all the known actin proteins. Numerous amino acid substitutions are also present in the C terminal part of the protein coding region (74). Actin 2-Sub 2 also lacks an appropriate Goldberg-Hogness or TATAA box, a region common to all expressed Dictyostelium mRNA encoding genes (see below). There is an extremely long poly(T) stretch that extends well into the region that contains the TATAA box and the 5' untranslated sequences of genes known to be transcribed. It is not known whether any of the other actin mRNA complementary sequences are pseudo-genes. Actin 3 also shows a high degree of amino acid substitution and has a 5' end that is slightly different from other actin genes (16). Transcriptional analysis of actin 3 has not been performed.

Sequence Homology in the Actin 3' Untranslated Region

There is a variable amount of sequence homology among the 3' untranslated regions of the actin genes that have been sequenced. As shown in Figure 21A, the sequence homology ranges from quite extensive to only a short distance 3' to the UAA termination codon. The homology between genes pcDd actin ITL-1 and actin 5 extends almost to the poly(A) while genes pcDd actin III-12 and Actin 2-Sub 1 show relatively little sequence homology with gene ITL-1. However, when genes actin 2-Sub 1 and pcdD actin III-12 are compared (see Figure 21B), a strong sequence homology is observed throughout the 3' untranslated region except for a deletion of 16 nucleotides in pcDd actin III-12. To examine whether there may be subfamilies of actin genes with 3' end homologies, the 3' region from the AvaII site through the poly(A) tail of plasmid pcDd ITL-1 was isolated, labeled and used as a probe against a Southern DNA blot of EcoRI digested genomic DNA. After hybridization to this probe, the counts were eluted and the same Southern strip was then hybridized to an entire pcDd actin B_1 probe, which is complementary to the protein coding region. Figure 22 shows that the ITL-1 3' untranslated region probe hybridizes 5 EcoRI fragments which also hybridize to the B_1 probe while showing essentially no hybridization to the other actin gene-containing bands. The hybridization with the ITL-1 probe is not just to bands that hybridize intensely with the B_1 coding

←Figure 21. the 3' ends of actin mRNA and actin genes. The sequences of the 3' noncoding regions of several actin genes and cDNAs are shown. All of the cDNAs are derived from the 1.35 kb actin mRNA. A second isolate of the III-12 cDNA has a poly(A) of about 25, a single T residue and a longer poly(A) region.

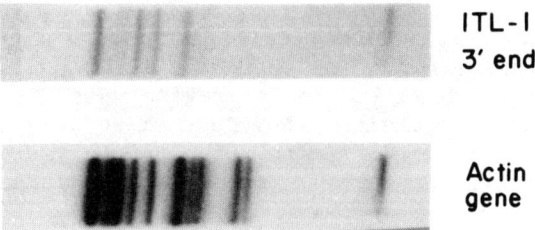

Figure 22. Hybridization of the pcDd actin ITL-1 3' end to EcoRI digested DNA. pcDd ITL-1 was digested with Sau96I restriction endonuclease and labeled with [^{32}P] by kinasing. The 3' end fragment was isolated and hybridized to a Southern filter of EcoRI digested Dictyostelium DNA. For comparison, labeled actin DNA was later hybridized to the same filter.

region probe, suggesting that it is specific to homologous 3' ends. From these results, we propose that ITL-1 represents a member of a subfamily of five actin genes which shown an evolutionary sequence relationship in the 3' untranslated region.

DISCOIDIN I GENES

Properties of Discoidin

When different species of Dictyostelium are mixed together they form co-aggregates in response to cAMP. However, the cells of the various species will eventually sort into species-specific aggregates and then differentiate into fruiting bodies containing cells of only one species (see ref. 1). This suggested that the molecules involved in aggregation are similar in all Dictyostelium species examined while those involved in cell-cell cohesion are species-specific. In Dictyostelium discoideum several proteins thought to be involved in cell-cell cohesion have been identified. Among these are several lectins and glycoproteins, including Contact Site A and gp150 (75-78).

In 1972, Barondes and co-workers identified two D. discoideum lectins, discoidin I and discoidin II, whose biological characteristics suggested they play a role in the cell-cell cohesion process (79). (These proteins were also designated CBP-26 (for carbohydrate binding protein) and CBP-24, but the names discoidin I and II are now used by most research groups (80).) Subsequently, Barondes and co-workers also identified analogous molecules in other species of Dictyostelium and in Polyspondilium (81,82). These proteins are isolated as multimers of molecular weight

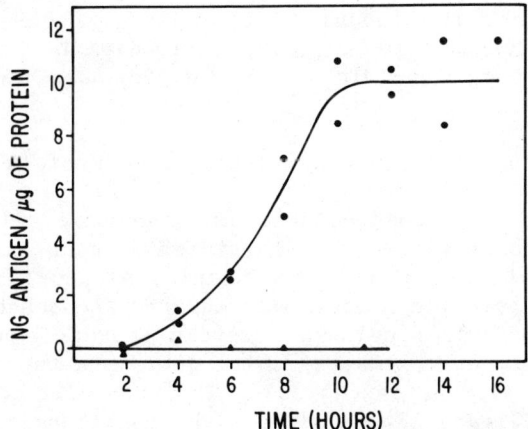

Figure 23. Quantitation of total carbohydrate-binding protein (CBP) with radioimmunoassay. NC-4 cells were harvested from bacterial growth plates for development. Cells were collected at 2 hr intervals and the soluble protein fractions were used for radioimmunoassays. The noncohesive mutant WL3, which was shown to lack CBP both on the surface and in the cytoplasm, was used as the control. The results of two experiments with NC-4 were plotted together and each experimental point represents the average concentration of CBP in the soluble protein fraction calculated from the values obtained with different dilutions of the protein sample. Cells at 0 and 2 hr had no inhibitory effect on the antibody even when 250 µg of soluble protein/sample were used (see ref. 80 for details).

>100,000. Discoidin I and II are isolated as tetramers whose subunits do not appear to have any post-translational modifications such as covalently bound sugar residues (83). Discoidin I and II are lectins with different hierarchies of affinities for various carbohydrates. Discoidin I binds to galactose and modified galactose residues, with the highest affinity for N-acetyl-galactosamine while discoidin II has a high affinity for lactose. Both discoidins can be purified by affinity chromatography on Sepharose (80,83,84). The two activities can be differentiated and assayed biologically by the agglutination of specific types of red blood cells (RBC). Discoidin I and II will agglutinate fixed rabbit RBC while only discoidin I will agglutinate fixed sheep RBC. Thus with a dilution assay of semi-purified protein

or cell lysates, the amount of biologically active discoidin I can be quantitated. In addition, antibodies have been prepared against purified discoidin which can be used in quantitative assays.

Function and Localization of Discoidin I

Discoidin I is undetectable in vegetative cells feeding on bacteria. When development is initiated by starvation, discoidin I mRNA synthesis can be detected within one hour and discoidin I protein by 3 hrs. The protein accumulates and reaches a level of about 1% of the total cellular protein (10^6 molecules per cell) by about 10 hrs, approximately coincident with the initial formation of tight aggregates (85,86). The accumulation of discoidin I is shown in Figure 23. Studies with inhibitors of RNA and protein synthesis indicate that prior RNA and concomitant protein synthesis are necessary for the appearance of the protein (unpublished data). Changes in the level of functional Discoidin I mRNA, assayed by in vitro translation and immunoprecipitation, coincide with changes in the relative rate of in vivo Discoidin I synthesis (87).

Studies with ^{125}I and fluorescent antibodies for cell surface labeling of Dictyostelium cells during development have shown that some discoidin I is found on the cell surface (85,86,88). Interestingly, less than one-tenth of the cellular discoidin I is found on the cell surface at any time in development while the remainder is in the soluble part of the cell. The protein appears to be loosely associated with the cell membrane. The surface discoidin I can be stripped off with N-acetylgalactosamine, which is subsequently replaced by discoidin I molecules from the soluble pool inside the cell (89). The physiological role of the internal discoidin I during development may be as a storage source of the protein.

Biochemical and immunological studies with lectins isolated from various species of cellular slime molds strongly suggest that these molecules are involved in species-specific cell-cell cohesion (90). The fact that the increase in the amount of cell surface discoidin I is directly correlated with the ability of shaking Dictyostelium cells to form aggregates supports this view.

Additional evidence that discoidin I is involved in cell-cell cohesion comes from the analysis of the D. discoideum mutant strain HJR-1, an agg⁻ (aggregation minus) mutant of D. discoideum (91). During development, the mutant cells chemotax but never form a tight cell mass and thus never culminate. The mutant accumulates wild-type levels of immunologically cross-reactive discoidin I, but this discoidin I does not bind to sugars or agglutinate sheep RBC and thus seems to be biologically inactive. The discoidin II retains wild-type properties. Dictyostelium parasexual genetics shows that the mutant maps to linkage group II. Revertants of the mutants were isolated by selecting cells capable of forming tight aggregates and culminating (92). Three classes of aggregation-competent revertants were obtained. Class I revertants form normal aggregates and differentiate indistinguishably from wild-type cells. Class II revertants form less defined aggregates whose development is delayed. Class III revertants form poor aggregates whose differentiation is appreciably delayed and form very few fruiting bodies (92). When the level of discoidin I was assayed radioimmunologically, the original mutant and all three classes of revertants had the same amount of protein as wild-type cells. However, when the biological specific activity (ability to agglutinate sheep RBC per mass of discoidin I) of discoidin I was assayed, the revertant in Class I had the same specific activity as wild-type cells while the other two had specific activities which were proportional to their qualitative ability to form aggregates and to further differentiate. These results strongly suggest that the original mutation affects the biological activity of discoidin I but has no effect on the accumulation of the protein.

Three equimolar isoelectric forms are detected by two-dimensional gel analysis of purified discoidin I or immunologically precipitated in vivo and in vitro synthesized discoidin I (Poole and Firtel, unpublished data). A 2-D gel analysis of discoidin I isolated from the mutant HJR-1 shows spots that run coincident with wild-type discoidin I, suggesting that the mutant may have a change in an uncharged amino acid. The mutant does not affect the relative ratios of the three isoelectric forms. The spot intensities appear to have the same relative ratios of the three isoelectric forms. This result agrees with radioimmunoassays indicating there is no change in the amount of discoidin I in the mutant. At present, the revertants have not been analyzed.

The genetic and biological data suggest that a mutation in one of the discoidin I genes can completely destroy the biological activity of discoidin I in vivo and in vitro. If there were three equally expressed genes making equivalent subunits, a mutant in one gene would decrease the activity to 20% of wild-type discoidin I levels. Since all the activity is destroyed in the mutant, we suggest that the gene that is mutated is either a required subunit of the lectin tetramer or affects an unidentified modification of the discoidin I proteins. Presently, we support the hypothesis that discoidin I is probably an obligatory heterotetramer. In vitro translation of hybrid-selected discoidin I mRNA showed that the immunologically precipitated translation product has the same molecular weight as in vivo labeled discoidin I protein (93,94). This suggested that there is probably not an excised leader sequence on this molecule.

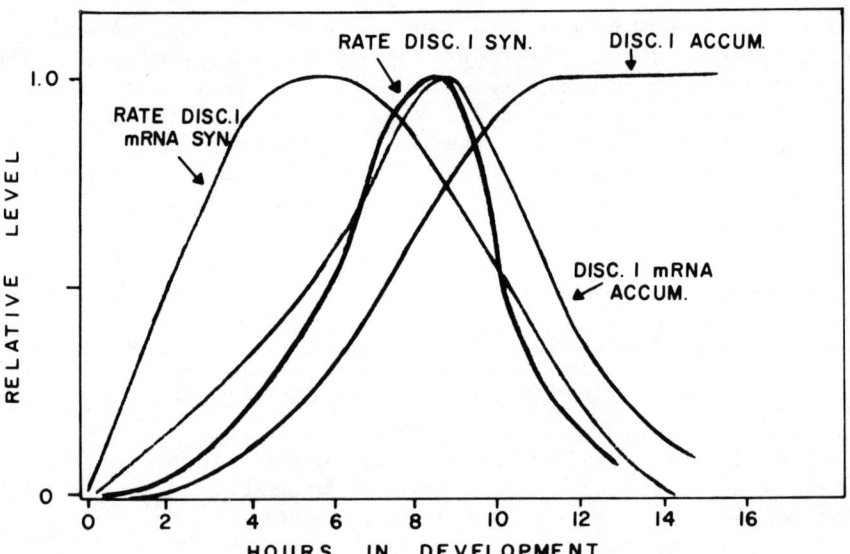

Figure 24. Regulation of discoidin I gene expression. Shown are the relative levels of discoidin I mRNa and relative rate of synthesis of discoidin I mRNA and protein synthesis through <u>Dictyostelium</u> development (data summarized from refs. 87,93,94; Poole, Rowekamp and Firtel, unpublished data).

Developmental Regulation of the Discoidin I Genes

With cloned cDNA probes, DNA excess hybridizations to RNA pulse-labeled <u>in vivo</u> during different stages in <u>Dictyostelium</u> development have been done to determine the transcriptionally active period of the discoidin I genes (94). Also, RNA excess hybridization experiments were done to determine the relative amounts of discoidin I mRNA at various times during development. Newly-synthesized discoidin I mRNA can first be detected very shortly after the cells are starved. The level of transcription reaches a peak at approximately 6 hrs and thereafter falls. A summary of the expression of the discoidin I is shown in Figure 24. Williams and co-workers, using a cDNA clone complementary to discoidin I, have also determined the transcriptionally active period of discoidin I genes with <u>in vitro</u> transcription experiments in isolated nuclei (95). Their results qualitatively agree with our <u>in vivo</u> transcription studies, though the developmental timing is somewhat advanced. This may be a result of their using an axenically grown strain (Ax-2) for their studies in contrast to the bacterially grown NC-4 strain. RNA excess hybridization experiments can detect about one copy of discoidin I mRNA per cell during vegetative growth on bacteria. The level increases between 500- and 1000-fold by aggregation and then falls 50-fold

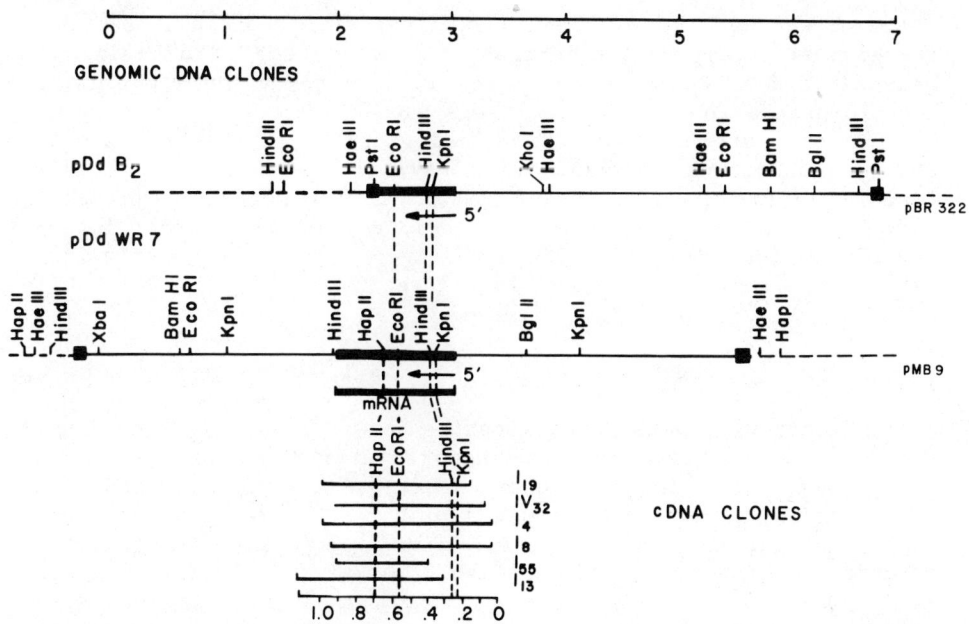

Figure 25. Restriction maps of the genomic and cDNA discoidin I recombinant plasmids. The figure shows the restriction endonuclease cleavage maps of recombinant plasmids carrying genomic and cDNA sequences complementary to discoidin I mRNA. pDdDisc-B_2 contains a random shear fragment inserted into the PstI site of pBR322 by poly(dG)-poly(dC) tailing (93). pDdDisc-WR7 contains a 5.6 kb HaeIII fragment inserted into the EcoRI site of pMB9 by poly(dA)-poly(dT) tailing (17). The cDNAs were inserted into the PstI site of pBR322 by poly(dG)-poly(dC) tailing. Vehicle sequences are represented by dashed lines, sites of insertion by solid boxes and Dictyostelium DNA inserts by thin solid lines. the thick solid lines indicate regions of the genomic clones complementary to mRNA. The vehicle sequences for the cDNA clones are not shown (see ref. 94 for details).

by 16 hrs in development (94). These results are in general agreement with earlier results in our laboratory assaying discoidin I mRNA by in vitro translation, although some of the in vitro translation data using RNA isolated from later time points during Dictyostelium development are not in total agreement (87). Northern RNA blot hybridization suggests that some of the in vitro translation data with RNA isolated after 10 hrs of development resulted in an overestimation in the amount of discoidin I mRNA (unpublished data). Asynchrony of cells used in these earlier experiments may be a possible explanation. It is not clear if the discoidin I genes are transcribed in bona fide vegetative cells or whether the extremely low level of discoidin I mRNA observed represents the transcriptional activation of discoidin I genes in a few Dictyostelium cells in a microenvironment equivalent to starvation.

Because of the interesting biological properties of discoidin I, we screened a library of cDNA clones made from mRNA isolated at 6 hrs in development to identify a series of plasmids carrying cDNA complementary to discoidin I mRNA. A restriction map of the isolated cDNA clones and two of the subsequently isolated genomic clones is shown in Figure 25.

Organization of Discoidin I Genes

DNA excess renaturation kinetics showed that the discoidin I cDNA is complementary to sequences reiterated about 4 times in the genome. When the discoidin I cDNA probe was hybridized to a Dictyostelium genomic Southern DNA blot filter, it was clear that discoidin I is encoded by a small multigene family. An analysis of genomic restriction endonuclease cleavage fragments hybridized to 5' (i.e., N-terminus of the protein) or 3' (C-terminus) specific discoidin I probes indicates the presence of approximately four discoidin I cDNA complementary regions in the genome (see Figure 26). Moreover, the blotting data indicated that at least two of the genes are tightly linked, since a 1.4 kb fragment was generated by several restriction enzymes and this fragment hybridized to both the 5' and 3' specific probe. Similar results suggested that two genes may also be linked on a 5.6 kb EcoRI fragment (see below). Unfortunately, the exact number of discoidin I genes cannot be determined from these results since more restriction fragments hybridized to the 5'-end probe than hybridized to the 3'-end probe. The exact nature of this discrepancy is still unclear and will require the isolation of all complementary restriction fragments and subsequent sequence analysis to clarify this. (See ref. 94 for details.)

Several genomic clones containing portions of three discoidin I genes have already been isolated in recombinant plasmids and phage (14,94). One of these contains a 1.4 kb EcoRI fragment carrying the 5' half of one discoidin I gene and the 3' half of another thus confirming that two of the discoidin I genes are tightly linked and in the same orientation. A different recombinant clone contains a 5.6 kb EcoRI fragment carrying the 3' half of a discoidin I gene and also a region which hybridized to a 5' specific probe. Initial sequence analysis from the EcoRI site on the side hybridizing to a 5' discoidin I cDNA probe does not show any direct sequence homology with other cloned discoidin I 5' sequences. At the present time, the nature of the cross-hybridization cannot be explained and must await further sequencing. We also have isolated a recombinant plasmid with the 5' half of a single discoidin I gene and one that contains a differ-

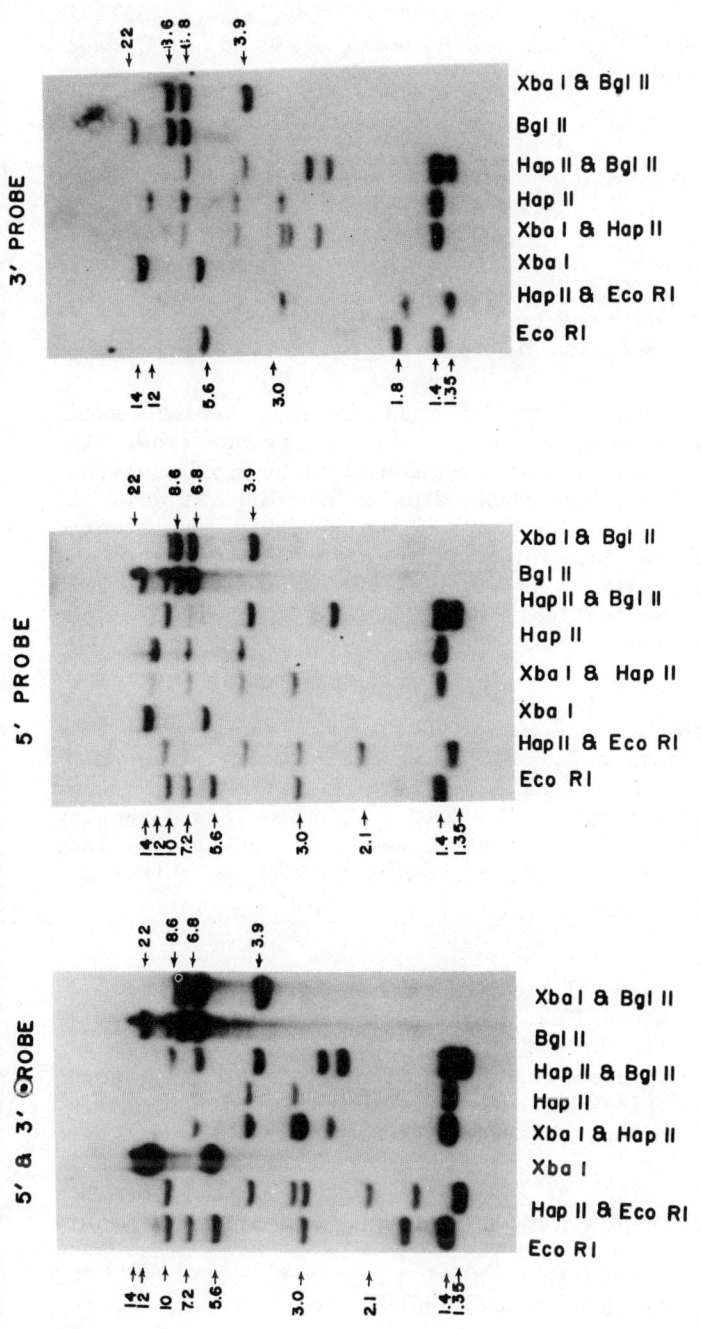

Figure 26. Genomic restriction endonuclease fragments complementary to a discoidin I mRNA probe and to 5' and 3' discoidin I gene probes. (A) Dictyostelium DNA was digested with various combinations of restriction endonucleases and the fragments were size-fractionated by electrophoresis on three identical 0.8% agarose gels and transferred to nitrocellulose. The filters were hybridized individually with nick-translated 32P-labeled 5'-specific, 3'-specific probe or 5' plus 3' probes. The 5'- and 3'-specific probes were the 0.5 kb and 0.35 kb EcoRI-PstI fragments, respectively, of pcDdDisc-IV$_{32}$. The restriction endonuclease digestions are shown on the sides of the Figure. Sizes shown on the sides of the DNA blots are in kilobases. Note: in the blot hybridized with the 5' plus 3' probe, the larger fragments in the HapII and XbaI plus HapII lanes did not transfer well; they can be seen in the blots hybridized to 5' and 3' probes (see ref. 94 for details).

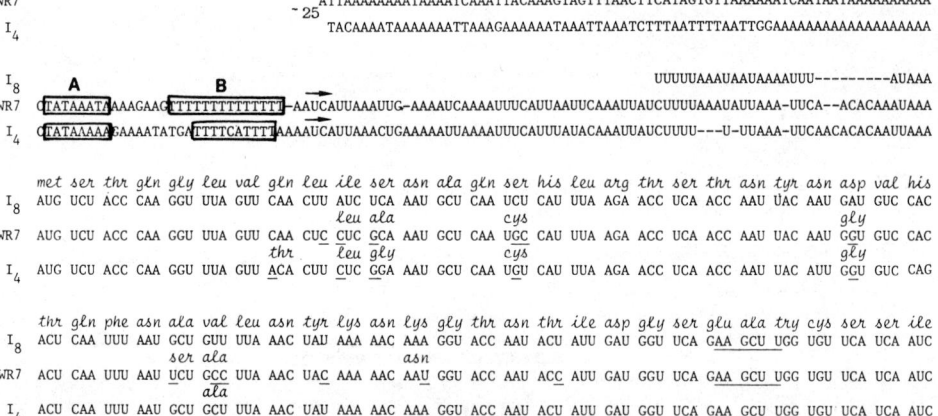

Figure 27. DNA sequence of the 5' ends of the discoidin genes WR7, I_4 and I_8. The mRNA start region is shown by an arrow. The nucleotide sequence of I_8 has been completed to approximately the same region as I_4 and WR7 and shows similar homology as does WR7 and I_4 up to the B region (data not shown). The A and B boxed regions correspond to potential Goldberg-Hogness box and T stretch. The HindIII site is underlined. The derived amino acid sequence is shown. Nucleotide changes versus I_8 are underlined (see ref. 14 for details).

ent discoidin I gene plus flanking sequences (see Figure 25). This complete gene does not appear to be linked to any other discoidin I sequences within 10 kb on either side. Finally, recombinant DNA molecules carrying a 7.2 kb EcoRI fragment and showing hybridization to 5' discoidin I cDNA clones have not been isolated and it is not clear if a discoidin I gene lies on this fragment.

Structure of the Discoidin I Genes

It was clear from the restriction endonuclease mapping and cross-hybridization studies that the cloned discoidin I genes have considerable homology within their protein coding regions. Three 5' gene regions and three 3' gene regions have now been completely sequenced and possess some interesting features (93). The protein coding regions of all three genes are extremely homologous. Although several amino acid substitutions lie within the N-terminal portion of the protein also, all three genes show negligible nucleotide substitution in the third position. The majority of nucleotide changes between the individual genes result in amino acid substitutions. The sequence data on the 5' end of three discoidin I genes are shown in Figure 27.

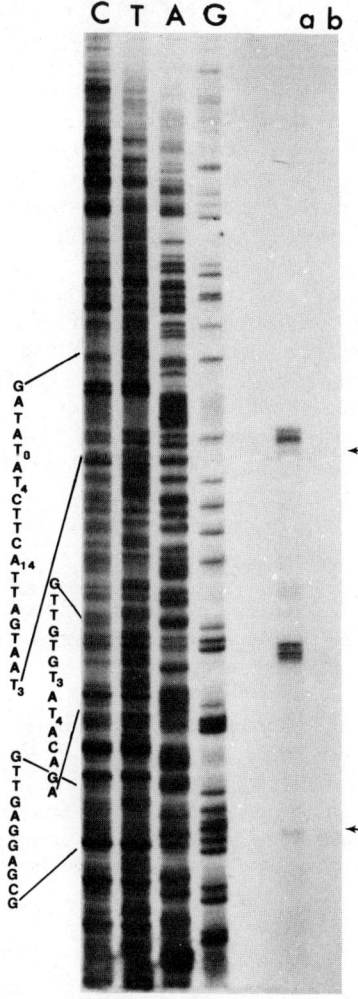

Figure 28. 5' end of discoidin genes. With the procedure described in Figures 17 and 18, the 5' end of discoidin gene WR7 (see Figure 25 for restriction map) was analyzed. The HindIII restriction site near the 5' end of the gene was labeled. The uppermost arrow represents the 5' end of the mRNA. The bands seen in lane a between the arrows map the deletions between WR7 and I_4 and I_8 (see Figure 27) in the 5' untranslated region. The lower arrow maps the nucleotide substitutions around codon 10 (see Figure 27). Lanes a and b are the experimentals; two different S1 nuclease concentrations were used. C, T, A and G are the sequencing ladder of the same fragment used in the hybridization (see ref. 14 for details).

With techniques identical to those used for the actin genes, we have shown that these three discoidin genes are transcribed in vivo and we have mapped the 5' end of the mRNAs (see Figure 28). By comparing sequence data on the 5' end of these three genes to that of cDNAs, we have shown that we have isolated the genomic clones encoding cDNAs I_4 and I_8. The 5' ends of the mRNA for all three genes map to the same relative site. The 5' untranslated region of all three genes shows extensive homology except for a few small deletions in some genes (14). The homology among discoidin I 5' ends contrasts the lack of homology among actin 5' ends. Since the discoidin I genes are coordinately expressed, probably at similar levels, it is possible that this 5' region may contain sequences important for controlling discoidin I expression at either the transcriptional or translational level.

Though the 5' ends and protein coding regions are highly conserved, the 3' ends of the discoidin I genes show little homology. Again, this is at variance with the actin family, where 3' subfamilies of sequence homology are seen. The lack of homology between various discoidin I 3' untranslated regions may indicate that these regions are not important for transcriptional or translational developmental control of this gene family. The site of the 3'-end of discoidin I-WR7 mRNA has been mapped and is approximately 380 nucleotides beyond the EcoRI site (14). The 3' ends of the other genes are as yet undetermined.

The sequencing data raise puzzling questions concerning the evolution of the discoidin I genes. While there is a relative paucity of silent third position changes in the codons and the highly conserved 5' ends, the 3' ends show no homology beyond the UAA termination codon. It is possible that the 3' ends have evolved much more rapidly than other regions of the genes. It is also possible that the 3' ends were not involved in the duplication of the discoidin I genes. Since the Dictyostelium genome is extremely A+T rich, it is possible that the genes duplicated into regions which could act as a 3' untranslated region.

OTHER DEVELOPMENTALLY REGULATED MULTIGENE FAMILIES

When cDNA clones, selected by their ability to hybridize preferentially to RNA isolated from developing cells over RNA isolated from vegetative cells, are used to probe the Dictyostelium genome by Southern blot hybridization, it is clear that several are encoded by multigene families (93). One of these families, M3, is presently being examined in detail in our laboratory (Brandis and Firtel, unpublished data). A combination of Southern blot hybridization and DNA sequencing analysis indicates that a cloned 7.2 kb EcoRI fragment contains two copies of a gene that encodes a developmentally regulated 2.0 kb mRNA. In a rabbit reticulocyte MDL, the mRNA is translated into a polypeptide of 50,000 daltons. Both genes show strong nucleotide sequence homology in their protein coding regions and are interrupted by several introns. These introns are relatively short (90 to 150 nucleotides) and contain approximately 60% T residues and 30% A residues. A similar size and G+C content has been observed for the introns in M4, a gene associated with a 5' interspersed repeat element (15,31,32). The splice junction sequences are very similar to the consensus sequence identified for higher eukaryotes. Transcription of the M3 genes is initiated upon starvation and reaches a maximum level during aggregation, approximately 1 to 2 hrs later than the discoidin I genes. When the cells are starved and shaken in liquid culture instead of being plated on buffer-saturated filters, the M3 mRNA is not accumulated. If, however, the shaking starved culture is pulsed with cyclic AMP, the M3 genes show a developmental pattern very similar to normal developing in vivo cells, although mRNA levels are reduced. The M3 protein has not been identified but these developmental kinetics imply that it may play a role in aggregation.

SEQUENCE ANALYSIS OF THE 5' AND 3' REGIONS

5' Ends of Dictyostelium Genes

At the present time, the 5' ends of four Dictyostelium actin genes, three discoidin I genes and gene M4, which is associated with a short interspersed repeat sequence, have been sequenced and the 5' ends mapped. Figure 29 summarizes these data. In comparing the regions immediately upstream to the 5' ends of the genes, we observe two regions that are homologous between all the genes. There is a TATAA or Goldberg-Hogness box (96) approximately 40 nucleotides 5' to the transcription start site. This region is approximately 10 nucleotides more 5' to the start sequence than in higher eukaryotes (97). There is also a homopolymer stretch of T residues that lies between the Goldberg-Hogness box and the mRNA start sequence (Kimmel and Firtel, unpublished data). The mRNA start lies extremely close to the 3' end of this T run. All transcribed Dictyostelium genes examined have this oligo(dT) sequence, including actin genes whose transcriptional properties are not known. Although all 5' untranslated regions are extremely A+T rich, we feel that the location of the potential Goldberg-Hogness box and T stretch are significant and may represent a promoter region for Dictyostelium genes.

It is quite possible that the T run aids the RNA polymerase since a homopolymer stretch of poly(dA)-poly(dT) has a lower melting point that poly(dAT). If this were true, there might be a direct correlation between the length or number of nucleotide substitutions in the T run with the level of transcription of a particular gene. However, for the three discoidin I genes this seems not to be the case. Two dimensional gel electrophoresis

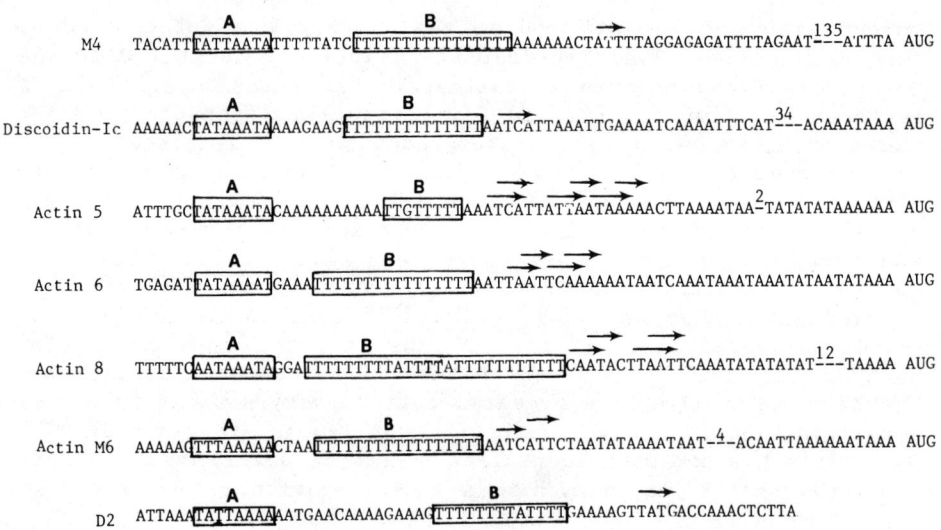

Figure 29. Sequence comparison of the 5' ends of Dictyostelium genes. The boxed A and B regions correspond to the TaTAA and T stretches. The arrows represent the region of the 5' end. The numbers above the dashed line represent the number of nucleotides deleted from the 5' untranslated region in order to line up the TATAA box and AUG initiation codon. For the protein coding genes, note the A residues preceding the AUG (see also Figures 16, 19 and 27). References: M4 (Kimmel and Firtel, unpublished data); discoidin Ic (WR7) (14); actin 5, 6, 8 and M6 (73). D2 is a potential gene for a low molecular weight nucleolar RNA complementary to rat U3 RNA (100).

has shown that there are approximately equal molar quantities of the three isoelectric forms of discoidin I. In addition, mRNAs transcribed from the three discoidin I genes have been identified and are present in approximately equal amounts. However, both the length and sequence purity of the T runs of these genes vary. Until surrogate genetics can be done in which specific deletions and nucleotide substitutions are made in both the Goldberg-Hogness box and T stretch, the importance of these sequences will be undetermined.

In the case of some of the actin genes, we know the A+T rich regions extend several hundred nucleotides beyond the 3' and 5' ends of the genes. From the absence of many restriction sites in regions 1 to 2 kb 3' and 5' from both the actin and discoidin genes we suggest that A+T rich regions may extend even further. The absence of restriction sites along long stretches of Dictyostelium DNA seems to be a general phenomenon. The relatively long poly(dA)-poly(dT) regions 5' to the region of initiation of transcription of the actin and discoidin genes are of interest but, at the present time, the function of these sequences is unknown. It is interesting to note that nucleosomes tend to form poorly on such homopolymer stretches. It is possible that these regions are involved in nucleosome phasing and thus involved with the formation of the chromatin structure around these genes.

Analysis of 5' and 3' Untranslated Regions

As described previously, the entire Dictyostelium genome has an extremely low (22%) G+C content. Nucleotide sequence analysis of the protein coding regions of the genes described in this review as well as other sequenced Dictyostelium genes shows a G+C content of approximately 30 to 40% although in silent third position nucleotide A or T is utilized more frequently than G or C. This G+C content is not remarkable when compared with the G+C content of other organisms. In contrast, both the 5' and 3' ends of all the genes examined plus the introns in the M4 and M3 genes are in general greater than 85% A+T (14). The significance of this is unknown.

CONCLUSIONS

The developmentally regulated multigene families in Dictyostelium are of considerable interest. All three of the discoidin I 5' gene regions isolated are transcribed and are most probably coordinately regulated. Two discoidin I genes are tightly linked while another is >10 kb away from other discoidin I genes. The discoidin mRNAs are all very homologous at the 5' untranslated region and possess very few third base changes while the 3' untranslated regions are totally different. The actin genes show very little amino acid substitution between the genes but about 5% divergence in the third codon position. In the case of the actin genes, the 5' untranslated regions are totally divergent while the 3' untranslated regions fall into sequence subfamilies. It is not known if these subfamilies are functionally distinct or only reflect evolutionary relationships between the genes.

We have also isolated and characterized the duplicated gene coding for the M3 protein, which is also developmentally regulated. At the present time, the function of this gene in Dictyostelium is not known.

It is hoped that with the development of a transformation system in Dictyostelium, we will be able to elucidate the importance of individual gene family members and the function of the sequences flanking the individual genes.

Acknowledgments: M.M. and S.P. are USPHS predoctoral trainees (GM07240). A.R.K. is the recipient of an American Cancer Society (California Division) Senior Fellowship (AC5 CA DIV-D365). J.B. is supported by an American Cancer Society postdoctoral fellowship (PF1639). R.A.F. is the recipient of an American Cancer Society Faculty Research Award. This work has been supported in part by grants from NIH (GM24279) and NSF (PCM78-21773) to R.A.F.

REFERENCES

1   Raper, K.B. (1935) J. Agr. Res. 50, 135-147.
2   Bonner, J.T. (1967) The Cellular Slime Molds, Ed. 2, Princeton University Press, Princeton, NJ.
3   Loomis, W.F. (1975) Dictyostelium discoideum: A Developmental System, Academic Press, New York, NY.
4   Sussman, M. (1956) Annu. Rev. Microbiol. 10, 21-36.
5   Jacobson, A. and Lodish, H.F. (1975) Annu. Rev. Genet. 9, 145-186.
6   Firtel, R.A. and Jacobson, A. (1977) in Biochemistry of Cell Differentiation (Paul, J., ed.), Vol. 15, pp. 377-429, University Park Press, Baltimore, MD.
7   Sussman, M. (1966) Current Topics Develop. Biol. 1, 61-83.
8   Sussman, M. and Brackenbury, R. (1976) Annu. Rev. Plant Physiol. 27, 229-257.
9   Roth, R., Ashworth, J.M. and Sussman, M. (1968) Proc. Nat. Acad. Sci. U.S.A. 59, 1235-1242.
10 Firtel, R.A., Baxter, L. and Lodish, H.F. (1973) J. Mol. Biol. 79, 315-327.
11 Sussman, R.R. and Rayner, E.P. (1971) Arch. Biochem. Biophys. 144, 127-137.
12 Firtel, R.A. and Bonner, J. (1972) J. Mol. Biol. 66, 339-361.
13 Jacobson, A., Firtel, R.A. and Lodish, H.F. (1974) J. Mol. Biol. 82, 213-230.
14 Poole, S., Rowekamp, W., Lamar, E.E. and Firtel, R.A. (1980) (submitted for publication).
15 Kimmel, A.R. and Firtel, R.A. (1980) Nucl. Acids. Res. (in press).
16 Firtel, R.A., Timm, R., Kimmel, A. and McKeown, M. (1979) Proc. Nat. Acad. Sci. U.S.A. 76, 6206-6210.
17 Cockburn, A.F., Newkirk, M.J. and Firtel, R.A. (1976) Cell 9, 605-613.
18 Cockburn, A., Taylor, W. and Firtel, R.A. (1978) Chromosoma 70, 19-29.
19 Frankel, G., Cockburn, A., Kindle, K. and Firtel, R.A. (1977) J. Mol. Biol. 109, 539-558.
20 Maizels, N. (1976) Cell 9, 431-438.

21. Firtel, R.A., Cockburn, A., Frankel, G. and Hershfield, V. (1976) J. Mol. Biol. 102, 831-852.
22. Karen, K.B. and Gall, J.G. (1976) J. Mol. Biol. 104, 421-453.
23. Engberg, J., Andersson, P., Leick, V. and Collins, J. (1976) J. Mol. Biol. 104, 455-470.
24. Vogt, V.M. and Braun, R. (1976) J. Mol. Biol. 106, 567-587.
25. Kimmel, A.R. and Gorovsky, B.A. (1978) Chromosoma 67, 1-20.
26. Blackburn, E.H. and Gall, J.G. (1978) J. Mol. Biol. 120, 33-53.
27. Firtel, R.A. and Kindle, K. (1975) Cell 5, 401-411.
28. Firtel, R.A., Kindle, K. and Huxley, M.P. (1976) Fed. Proc. 35, 13-22.
29. Firtel, R.A. and Lodish, H.F. (1973) J. Mol. Biol. 79, 295-314.
30. Lodish, H.F., Firtel, R.A. and Jacobson, A. (1974) Cold Spring Harbor Symp. Quant. Biol. 38, 899-914.
31. Kimmel, A.R. and Firtel, R.A. (1979) Cell 16, 787-796.
32. Kimmel, A.R., Lai, C. and Firtel, R.A. (1979) ICN-UCLA Symp. Mol. Cell. Biol. 14, 195-203.
33. Palatnick, C.M., Storti, R.V., Capone, A.K. and Jacobson, A. (1980) J. Mol. Biol. 140 (in press).
34. Margolskee, J. and Lodish, H.F. (1980) Develop. Biol. 74, 37-49.
35. Jacobson, A., Firtel, R.A. and Lodish, H.F. (1974) Proc. Nat. Acad. Sci. U.S.A. 71, 1607-1611.
36. Firtel, R.A. (1972) J. Mol. Biol. 66, 363-377.
37. Blumberg, D.D. and Lodish, H.F. (1980) Develop. Biol. 78, 268-284.
38. Blumberg, D.D. and Lodish, H.F. (1980) Develop. Biol. 78, 285-300.
39. Wessells, N.K., Spooner, B.S., Ash, J.S., Bradley, M.O., Luduena, M.A., Taylor, E.L., Wrenn, J.P. and Yasmada, K.M. (1971) Science 171, 135-142.
40. Schroeder, T.E. (1973) Proc. Nat. Acad. Sci. U.S.A. 70, 1688-1692.
41. Sanger, J.W. (1975) Proc. Nat. Acad. Sci. U.S.A. 72, 1913-1916.
42. Pollard, T.D. and Weihung, R.R. (1974) in Critical Reviews in Biochemistry (Fasman, G., ed.), Vol. 2, pp. 1-65, CRC, Cleveland, OH.
43. Durham, A.C.H. (1974) Cell 2, 123-135.
44. Tilney, L.G. and Setmers, P. (1975) J. Cell. Biol. 66, 508-522.
45. Bray, D. (1973) Nature 244, 93-95.
46. Gruenstein, E., Rich, A. and Weihung, R.R. (1975) J. Cell. Biol. 64, 223-234.
47. Lazarides, E. and Lindberg, U. (1974) Proc. Nat. Acad. Sci. U.S.A. 71, 4742-4746.

48. Douvas, A.S., Harrington, C.A. and Bonner, J. (1975) Proc. Nat. Acad. Sci. U.S.A. 72, 3902-3906.
49. Le Stourgeon, W.M., Faher, W.M., Yang, Y.Z., Bertram, J.S. and Rusck, H.P. (1975) Biochim. Biophys. Acta 379, 529-539.
50. Tuckman, J., Alton, T.A. and Lodish, H.F. (1974) Develop. Biol. 40, 116-128.
51. Alton, T.A. and Lodish, H.F. (1977) Develop. Biol. 60, 180-206.
52. Clarke, M. and Spudich, J.A. (1974) J. Mol. Biol. 86, 209-222.
53. Spudich, J.A. (1974) J. Biol. Chem. 249, 6013-6020.
54. Vandekerchlove, J. and Weber, K. (1980) Nature 284, 475-477.
55. Spudich, J.A. and Cooke, R. (1975) J. Biol. Chem. 250, 7485-7493.
56. MacLeod, C., Firtel, R.A. and Papkoff, J. (1980) Develop. Biol. 76, 263-274.
57. MacLeod, C. (1979) Thesis, University of California, San Diego, CA.
58. Rubenstein, P. and Deuchler, J. (1979) J. Biol. Chem. 254, 11142-11147.
59. Kindle, K. (1978) Thesis, University of California, San Diego, CA.
60. Kindle, K., Taylor, W., McKeown, M. and Firtel, R.A. (1977) in Developments in Cell Biology (Cappuccinelli, P. and Ashworth, J.M., eds.), Vol. I, pp. 273-290, Elsevier/North Holland Biomedical Press, Amsterdam.
61. Palatnile, C.M., Storti, R.V. and Jacobson, A. (1979) J. Mol. Biol. 128, 371-395.
62. Margolskee, J.P. and Lodish, H.F. (1980) Develop. Biol. 74, 50-64.
63. Kindle, K.L. and Firtel, R.A. (1978) Cell 15, 763-778.
64. Bender, W., Davidson, N., Kindle, K.L., Taylor, W., Silverman, M. and Firtel, R.A. (1978) Cell 15, 779-788.
65. McKeown, M., Taylor, W., Kindle, K., Firtel, R., Bender, W. and Davidson, N. (1978) Cell 15, 789-800.
66. McKeown, M. and Firtel, R.A. (1980) (submitted for publication).
67. Dawid, J.B. and Long, E.O. (1980) Annu. Rev. Biochem. 49, 727-764.
68. Kedes, L.H. (1979) Annu. Rev. Biochem. 48, 837-870.
69. Gallevitz, D. and Sures, J. (1980) Proc. Nat. Acad. Sci. U.S.A. 77, 2546-2550.
70. Ng, R. and Abelson, J. (1980) Proc. Nat. Acad. Sci. U.S.A. 77, 3912-3916.
71. Tobin, S.L., Zulanf, E., Sanchez, F., Craig, E.A. and McCarthy, B.J. (1980) Cell 19, 121-131.
72. Fryberg, E.A., Kindle, K.L., Davidson, N. and Sodja, A. (1980) Cell 19, 365-378.
73. McKeown, M. and Firtel, R.A. (1980) (submitted for publication).

74. McKeown, M. and Firtel, R.A. (1980) (submitted for publication).
75. Gerisch, G. (1980) in Current Topics in Developmental Biology (Moscana, A. and Monroy, A., eds), Vol. III, pp. 157-197.
76. Muller, K., Gerisch, G., Fromme, I., Mayer, H. and Tsugita, A. (1977) Eur. J. Biochem. 99, 419-426.
77. Gelotsky, J.E., Birdwell, C.R., Weseman, J. and Lerner, R.A. (1980) Cell 21, 339-345.
78. Gelotsky, J.E., Weseman, J., Bakke, A. and Lerner, R.A. (1979) Cell 18, 391-398.
79. Rosen, S., Kafka, J., Simpson, D. and Barondes, S. (1973) Proc. Nat. Acad. Sci. U.S.A. 70, 2554-2557.
80. Sui, C., Lerner, R., Ma, G., Firtel, R. and Loomis, W. (1976) J. Mol. Biol. 100, 157-178.
81. Rosen, S., Reitherman, R. and Barondes, S. (1975) Exper. Cell Res. 95, 159-166.
82. Barondes, S.H. and Haywood, P.L. (1979) Biochim. Biophys. Acta 550, 297-328.
83. Frazier, W., Rosen, S., Reitherman, R. and Barondes, S. (1975) J. Biol. Chem. 250, 7714-7721.
84. Simpson, D.L., Rosen, S. and Barondes, S. (1974) Biochemistry 13, 3487-3493.
85. Sui, C.H., Loomis, W.F. and Lerner, R.A. (1978) in The Molecular Bases of Cell-Cell Interactions (Lerner, R.A. and Bergsma, D., eds.), pp. 439-458, Alan R. Liss, New York, NY.
86. Reitherman, R., Rosen, S., Frazier, W. and Barondes, S. (1975) Proc. Nat. Acad. Sci. U.S.A. 72, 3541-3545.
87. Ma, G. and Firtel, R. (1978) J. Biol. Chem. 253, 3924-3932.
88. Chang, C.M., Reitherman, R., Rosen, S. and Barondes, S. (1975) Exper. Cell Res. 95, 136-142.
89. Springer, W., Haywood, P. and Barondes, S. (1980) J. Cell. Biol. (in press).
90. Barondes, S.D. (1975) in The Molecular Bases of Cell-Cell Interaction (Lerner, R.A. and Bergsma, D., eds.), pp. 633-636, Alan R. Liss, New York, NY.
91. Ray, J., Shinnick, T. and Lerner, R. (1979) Nature 279, 215-221.
92. Shinnick, T. and Lerner, R. (1980) Proc. Nat. Acad. Sci. U.S.A. 77, 4788-4792.
93. Rowekamp, W. and Firtel, R. (1980) Develop. Biol. (in press).
94. Rowekamp, W., Poole, S. and Firtel, R. (1980) Cell 20, 495-505.
95. Williams, J., Lloyd, M. and Devine, J. (1979) Cell 17, 903-913.
96. Goldberg, M. (1979) Ph.D. Thesis, Stanford University, Stanford, CA.
97. Gannon, F., O'Hare, K., Perrin, F., LePennec, J.P., Benoist, C., Cochet, M., Breatnach, R., Royal, A., Garapin, A., Cami, B. and Chambon, P. (1979) Nature 278, 428-434.

98 Southern, E.M. (1975) J. Mol. Biol. 98, 504-517.
99 Sollner-Webb, B. and Reeder, R.H. (1979) Cell 18, 485-499.
100 Wise, J.A. and Weiner, A.M. (1980) Cell 22, 109-118.

COMPUTER ASSISTED METHODS FOR NUCLEIC ACID SEQUENCING

T.R. Gingeras and R.J. Roberts

Cold Spring Harbor Laboratory

Cold Spring Harbor, New York 11724

INTRODUCTION

Eight years ago procedures such as pyrimidine tracts (1) and wandering spots (2) were the methods of choice for determining the sequence of short stretches of nucleic acids. The use of computers at that time to assist in the collection and analysis of these sequences would have seemed unnecessary. However, the development of more productive and technically simpler methods for DNA sequencing has caused a dramatic increase in the total volume of nucleic acid sequence data. For example, within a single journal (Nucleic Acids Research) published bi-weekly, newly determined sequence data, including more than 27,000 nucleotides, were published during the first five months of 1980. The accumulation of nucleotide sequence data in the near future will undoubtedly continue at an increased rate. Such large quantities of sequence data have provided the stimulus to establish centralized sequence storage banks.* The utilization of computers at these centers to store and catalog sequence data is one obvious but trivial way in which computer technology has been useful to those interested in nucleotide sequences.

---

*Two centralized nucleotide data banks are being established, one in the United States (for information write to Dr. E. Jordan, Genetics Program, National Institutes of General Medical Sciences, Bethesda, MD 20205) and the second in Europe (Dr. K. Murray, European Molecular Biology Laboratory, Heidelberg, Germany).

It is the intent of this review to describe other ways in which computers have been useful in the determination and analysis of nucleotide sequences. It is hoped that this description will result in an appreciation of those areas in which computer technology has been or can be applied during the sequencing protocol.

DNA SEQUENCING

Presequencing State

Two major methods exist for the sequencing of DNA molecules: the chemical scission method of Maxam and Gilbert (3) and the chain termination method of Sanger et al. (4). By far the most widely used technique has been the chemical method of Maxam and Gilbert. However, because of the numerous technical steps and time required to carry out these steps, the construction of detailed restriction enzyme maps is almost a prerequisite for the use of this method. The chain termination method developed by Sanger and his colleagues is less dependent upon the construction of detailed maps because sequence can be produced rapidly. Thus, the time used to construct restriction enzyme maps can be used to produce the required DNA sequence. Restriction maps can be generated as the sequencing proceeds with redundant, overlapping sequences providing confirmation of the final sequence.

Although the construction of restriction maps by conventional techniques is discussed elsewhere (5), that survey does not include the more modern approach that computers can provide. Two sets of computer programs have been written to assist in the construction of restriction enzyme maps. Stefik (6) has described an algorithm that uses a model-driven approach to construct such maps. The program, called GA1, requires only the sizes of all fragments produced by single or multiple enzyme digestions; it solves the mapping problem by inferring structures with the use of an exhaustive model generator. Most of the models are eliminated by a pruning process that derives its rules from data supplied by the user (i.e., the total number and sizes of fragments generated by each single, double or triple enzyme digest). The origin of this approach is derived from the algorithms used in a set of programs (DENDRAL) developed to predict the molecular structures of organic molecules (7).

A major difficulty plagues this method as well as the more conventional methods of mapping. Any mapping effort relies heavily on the accurate determination of the fragment lengths produced by each combination of enzyme digests. If these lengths are known exactly, then mapping is reduced to a simple matter of addition. Unfortunately, existing gel systems, upon which enzyme digestions are fractionated, allow for only rough estimates of fragment lengths. Another computer program has been devised to overcome these inherent difficulties. Schroeder and Blattner (8) have described a computer program that uses a "least squares" method to improve the accuracy with which the size of each restriction fragment is known. From these improved estimates, restriction enzyme maps can be deduced and, by an iterative procedure, the map can be continually refined so that the map position for each restriction site minimizes the sum of the squares of the fractional (rather than the absolute) deviation of each measured fragment size from its predicted value.

It would be misleading to infer that a unique restriction map for any enzyme on any substrate can be the guaranteed product of either one or both of these programs at the present time. Rather, the programs should be viewed as providing a small number of possibilities for each map by the use of some well-defined mathematical techniques. A subsequent set of selected experiments can then be used to decide among the candidate possibilities.

Collection and Assembly of DNA Sequences

Collection. Both the Maxam/Gilbert and the Sanger sequencing methods result in the familiar ladder-like pattern of autoradiograph bands displaying the four channels corresponding to each of the four deoxyribonucleotides. Two simple but serious sources of error are associated with the reading of these autoradiographs. The first of these errors involves the careless reading of the gel (e.g., mistaking channels or skipping bands) and the second occurs when the results of this reading are subsequently transcribed.

We have developed a method to overcome errors at this stage by automatically transferring data from the autoradiographs directly into the memory of the computer. Figure 1 depicts the scheme used in this method. Our approach uses a digitizing tablet that operates by sending the location of any point on the surface of the pad to the computer when the point is touched by a signal pen. The location is represented by a digitized set of x, y coordinates. The autoradiograph is placed on the pad and the channels are identified by touching each of their four corners with the pen. This defines a quadrilateral for each channel, such that any point within this quadrilateral subsequently touched by the pen is automatically assigned the appropriate base. The autoradiograph is then read by touching each band with the pen and the location is recorded on the corresponding base. By repeating this process several times, the readings can be compared and the discrepancies highlighted by the computer by means of a matching subroutine. This allows for immediate proofreading of the sequence. Other areas of the digitizing tablet that are not covered by the autoradiographs are used to encode keys that

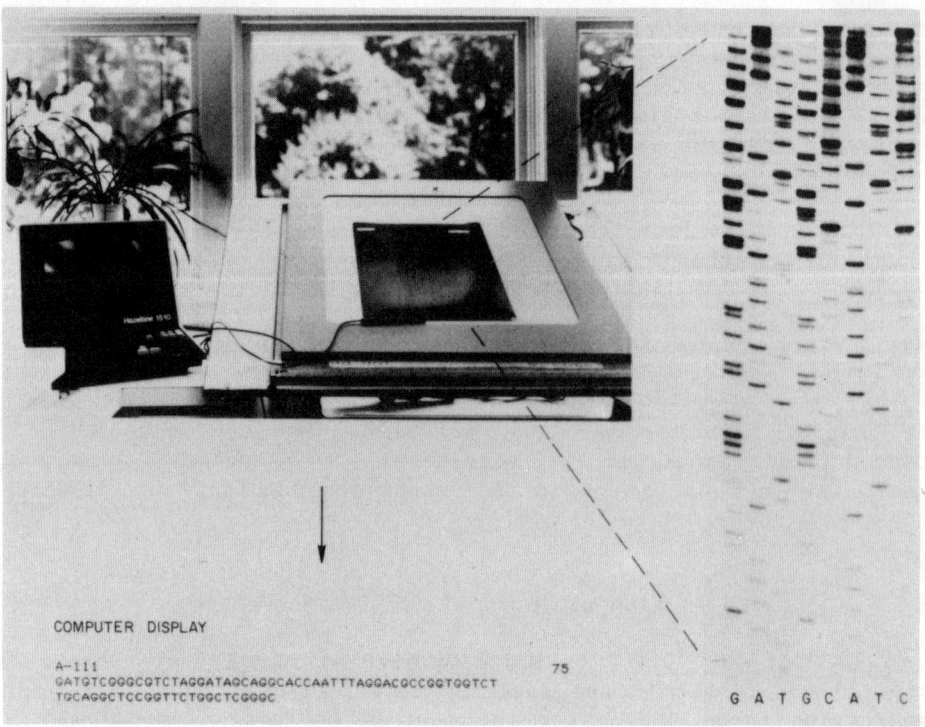

Figure 1. A semi-automated gel reading system. Such a system consists of an autoradiograph mounted on a translucent digitizing tablet that is connected to a computer and cathode ray tube terminal. The sequence is read from the autoradiograph (using the signal pen) directly into memory. The nucleotide sequence is stored and displayed on the terminal screen. The limit of resolution of the tablet is 0.1 mm. A menu of other functions, such as editing and homology searches, is encoded on the surface of the tablet in an area not covered by the autoradiograph. These functions can be activated by touching the designated areas with the signal pen.

signal the computer and invoke various useful functions. For instance, one key may call an editing subroutine to correct the sequence just read, or another key may invoke programs to compare the newly-entered sequence with other blocks of sequence already stored in memory.

It is clear that this approach permits the removal of those trivial errors that plague the present manual reading of autoradiographs. Another important feature of this scheme is that the actual reading still depends upon the experience and expertise of the experimenter. Those areas of the gel in which compression of bands occur (usually attributed to secondary structure) or those areas suffering from unequal migration down the gel ("smiling" effect, usually attributed to uneven heating of the gel surface) can be noted and interpreted by this simple method. These ambiguous areas become much more difficult to interpret if a completely automated gel reading system were to be employed.

It would seem that a fully automated gel reading system is the logical next step, considering the fact that greater and greater lengths of DNA molecules will eventually be sequenced, requiring larger numbers of sequencing gels. We have made preliminary explorations into this area and have noted the availability of highly sophisticated optical scanning devices that possess more than satisfactory resolution. The problem centers on the computer programs (software) needed to drive these devices. As yet, only initial efforts have been made (H. Lehrach, personal communication) in trying to encode in a computer program the decision-making abilities needed to interpret the artifacts (e.g., compression, unequal band intensities, unequal band migration) encountered with each sequencing gel. However, in view of the fact that a functional set of computer programs exists for the purpose of reading two-dimensional protein gels (9,10), it would seem that the difficulties in reading sequencing gels will eventually be solved.

<u>Assembly</u>. The product of each reading of an autoradiograph is a short segment of sequence that is unmapped relative to the intact genome. Consequently, the next stage is to order these segments by finding overlapping stretches of common nucleotides in order ultimately to join each segment into a long block of sequence that represents the primary structure of the original molecule.

Two collections of computer programs (11,12) have been written to accomplish this goal. Both sets of programs are highly interactive and help the user provide the required information. These programs include three important functions. First, all newly-acquired sequence data are transferred and stored in a computer file that acts as an informational archive. The second function allows the identification of those blocks of sequence with overlaps (i.e., homologous or complementary stretches of sequence). To accomplish this, the programs request that the user enter the stringency necessary to define the overlap (Figure 2). The programs compare the newly entered data with all blocks of previously entered sequences relating to a given sequencing project. The programs look for both direct homology and homology by complementarity. Both sets of programs print out the strings of nucleotides sharing overlapping sequence in a format which aligns the area of homology (Figure 2). The Staden programs (11) provide a useful feature during this stage. Characters other

Menu of ASSEMBLER Functions

```
WELCOME TO THE ASSEMBLER
WHAT REGION ARE YOU INTERESTED IN?
PLEASE KEY IN A LETTER A, B.......S
> N



THE ASSEMBLER HAS THE FOLLOWING CAPABILITIES:
1.   LISTING OF ENTIRE MASTER TABLE
2.   ENTRY OF NEW REGIONAL DATA
3.   LISTING OF PREVIOUS REGIONAL DATA
4.   ALIGNED DATA FOR A GIVEN REGION
5.   MELDING OF REGIONAL DATA
PLEASE KEY IN FUNCTION DESIRED (1...2..4)
> 4



DO YOU WANT TO ALIGN IN THE 3-5 DIRECTION?
> NO



HOW MANY IN A ROW FOR INITIAL OVERLAP?
> 11
WILL LOOK FOR 11
                        │ Results of options 4 and 5
                        ▼

OVERLAPPING SEGMENTS GM-1 (05/12/80)              AND    C-345 (05/10/80)
     3
    16    34   16
    33    50   19
    55    72   10
 *    *              *
 *    <              A  ANAACCAGGT AGGT>CAGAG AGGTAAACTG   *
 *   <TNGTTCTCG GGAAAGCGAT TGAACACGTG GGTA>CAGAG AGGTAAACTG *
 *                              *
 *   G<A>CGGATG AGCTGGGAGT AGA<CGG>CC TGGTCGTT<A GTAGAAGCTC *
 *   G< >CGGATG AGCTGGGAGT AGA<GGC>CC TGNTCGTT<G TAGA      *
 *
 *   TTGGAGTGCA CGGGCAACAG CTCGGCGCCC ACCACNGAAA GTTGCTGATC *
 *                                                          *
 *                                    *
 *   TGGCTCGTGG AGCGGAAGGT CAGGGGTCTT GCATCATNTC TGGCAAGACA *
 *                                                          *
 *
 *   TAGACTCTCG ACGAGTTAGT CA>
 *                           >
                   MELDED FRAGMENTS
 *   - ------- ---------    *                              *
 *   <TNGTTCTCG GGAAAGCGAA AGAACAAGGG AGGA>CAGAG AGGTAAACTG *
 *   -                              *         ------       *
 *   G<A>CGGATG AGCTGGGAGT AGA<CGC>CC TGGTCGTT<A GAAAAAGCTC *
 *   ---------- ----------  ---------- -----  ---- ---------- *
 *   TTGGAGTGCA CGGGCAACAG CTCGGCGCCC ACCACNGAAA GTTGCTGATC *
 *   ---------- ---------- ---------- -- ----------         *
 *   TGGCTCGTGG AGCGGAAGGT CAGGGGTCTT GCATCATNTC TGGCAAGACA *
 *   ---------- ----------  --
 *   TAGACTCTCG ACGAGTTAGT CA>
```

←Figure 2. A sample printout of the menu and results generated by the nucleotide sequence assembly program called ASSEMBLER (11). The menu indicates what options are available to the user. In this example, the program asks the user to identify a file name into which the data will be stored. Only a single letter response is required. In this case the ASSEMBLER program is asked to perform an alignment and meld function (options 4 and 5). Since the request is to align and meld, the user is questioned as to, a) what is to be searched and b) what stringency is required for the definition of the initial overlaps between any two sequences being compared.

The results of the alignment and meld functions are listed under the arrow. Overlapping sequences are identified by the table listed at the top of the sequences. Thus, within data GM-1 and G-345 there are three regions of overlap. At nucleotide 16 of GM-1 and nucleotide 34 of G-345 there exists an overlap of 16 nucleotides. This overlap is broken at this point but a new overlap takes over at position 33 of GM-1 and 50 of G-345. A detailed listing of the sequences present in GM-1 and G-345 is printed one over another in order to facilitate the easy recognition of such overlaps. Any areas of disagreement between the two sequences are placed in brackets. An asterisk (*) placed above the sequence indicates positions where N (an unidentified nucleotide) occurs in either or both sequence elements.

An example of melding the data present at GM-1 and G-345 into one sequence is also listed. If a dash appears over a base, the user is notified that there was no corresponding nucleotide at that position within the other set of data in order for a comparison to be made. Thus additional data are needed for these positions. A dot (˙) over the nucleotide indicates that two different readings exist at this position. This tells the user that this base requires reevaluation.

---

than the usual A, C, G, T can be used to represent those nucleotides that are difficult to determine. These ambiguities can be denoted by a special set of characters (e.g., R=A or G) and thus can be taken into account during the search for overlaps. The third important function supplied by these programs is that of melding any two strings of nucleotides containing overlapping sequences (Figure 2). Differences that appear at any position during a meld are denoted by a special set of diacritical marks placed over the nucleotides concerned. This serves to identify those positions which either require new data or reevaluation of the old data. The ultimate purpose of this feature is to join the original stretches of short sequence into longer and longer blocks of data until a total reconstruction is accomplished.

Verification of Sequence. An area in which computer programs can be particularly useful is in determining the accuracy of a newly-derived sequence. One way to do this is to confirm that a sequence contains all restriction enzyme sites mapped prior to the sequencing. A second way is to locate all of the restriction fragments that were utilized as primers during the sequencing operation. Finally, with the completed sequence, it should be possible to predict the number and sizes of the fragments observed upon digestion of the original molecule with a selected set of restriction enzymes. Because computer searches for the presence of a specific restriction site are both easily and rapidly done, this use of the computer represents a straightforward and efficient approach for verification of the accuracy of newly-derived nucleotide sequences.

ANALYSIS OF NUCLEIC ACID SEQUENCES

The primary goal of most sequencing projects is to relate the primary structure of a nucleic acid molecule with some functional property. Such analyses fall into two definite categories. The first category involves searches for sequences that possess proven functional properties. These programs consist of simple search and identification routines, similar to those used in finding restriction enzyme sites. For example, potential initiation and termination codons for translation can be quickly and accurately located with this simple type of analysis.

The second category of computer analysis examines nucleic acid sequences for more subtle signals. For example, DNA sequences capable of acting as promoters (for transcription by RNA polymerase) vary considerably, as do those sequences subject to RNA splicing. This second category of programs frequently operates in an inferential manner, storing large numbers of known active sequences and comparing these to known inactive but partially homologous sequences. Such comparisons are done in order to identify the subtle features that distinguish active from inactive sites. This second type of analysis represents the area of most promise. Future programs of this type will make use of elements from the artificial intelligence area of computer science in order to assist during this inferential process. However, it must be emphasized that such computer programs serve only as an experimental resource. The derivation of any new hypothesis by this means will still require experimental support.

Simple Types of Analysis

The desired product from this simple type of analysis is usually a list of the locations at which a specific pattern occurs. The output for this search may vary from a simple table to a detailed graphic display. Two collections of programs that possess these cataloguing features are available. The first set was developed by Korn and Queen (13,14) and is capable of printing sequences in a format that highlights the occurrences of specific short sequences, as directed from a standard list or by the user. The ability also exists to translate all or portions of an input sequence into its polypeptide equivalent. One useful feature of these search programs is that, although perfect matches with an input sequence are easily found, imperfect matches may also be located at the user's discretion. Furthermore, the mathematical probability of finding a given sequence by chance is assessed by the computer, thus giving an indication of the possible significance of nonrandom distributions of particular patterns. This last portion of the Korn and Queen program has been significantly modified (D. Brutlag, personal communication) so as to reflect the base compositon of the input sequence.

The original set of computer programs written to assist in a DNA sequencing project was developed by McCallum and Smith (15) during the sequencing of the ϕX174 genome. These programs were written in ANS COBOL, which is a business-oriented language designed for search operations. The programs were followed by an expanded collection written in FORTRAN (16,17). The later versions provided useful facilities for storage, editing and manipulation of long strings of sequence. An additional important feature of these programs is the fact that many of the subroutines were constructed as stand-alone modules. This feature is valuable because such modules can easily be transported and incorporated into more sophisticated programs.

The characteristic of constructing computer programs in this modular manner raises the issue of whether there is an optimal computer language for nucleic acid sequence analysis. It would be misleading to imply that because FORTRAN is in such wide usage within the scientific community it is the language ideally suited for sequence analysis. Rather, languages such as SNOBOL, which was designed to manipulate long character strings, or PASCAL, which was designed to be an easily understood and portable language, might be more suited for this type of analysis. Since such analytical programs should be viewed as another research tool available for molecular biology, it would seem that the transportability of any program should be a paramount feature. Consequently, it is probable that FORTRAN will continue to be widely used until such time as the utilization of these other higher-order languages becomes more widely accepted.

Inferential Types of Analysis

A second type of sequence analysis arises when the question posed by the user can be answered only by integrating the results from several simple types of searches. Often all possible occurrences of several specific types of patterns will need to be

examined. Against these results, some particular rules will be
applied. Those occurrences that obey the prescribed rules will
be retained, while the rest of the catalog will be discarded.
Further pruning can occur by the application of other known parameters or rules until only a few possibilities remain. These can
each be tested by further experimentation. A good illustration
is found in the computer programs written to generate secondary
structure models for RNA transcripts (18-23). These programs
invariably utilize the well-known base-pairing rules for polynucleotides and rely on previously-calculated, free-energy contribution for each base pair (24-26). Such models can be useful in
forming the basis for hypotheses relating secondary structure to
biological function if they can be shown to exist from additional
lines of evidence. A later section of this review will discuss
how it will be possible for computer programs to become autodidactic. Such programs would be capable of suggesting what other
lines of exploration could be pursued to test the validity of a
certain structure or hypothesis.

In addition to those computer programs that generate RNA
secondary structures, two other programs have been developed to
analyze nucleic acid sequences in a less straightforward manner.
We (27) and others (28) have attempted to utilize the known sequences of various viral genomes in order to predict the recognition sequences of new restriction endonucleases. This is accomplished by cleaving DNAs of known sequence with a new restriction
endonuclease and measuring the length of the fragments produced.
The computer is asked to produce a set of theoretical fragmentation patterns for that DNA sequence with all possible tetranucleotide, pentanucleotide and hexanucleotide sequence combinations. Many of these combinations of nucleotides already define
the recognition sequence of existing restriction enzymes (29).
After comparing the observed pattern with these theoretical patterns and eliminating those which differ significantly, a small
number of potential recognition sequences is left. In general,
the larger the number of fragments produced by the enzyme, the
greater the likelihood this program will arrive at a unique candidate sequence. The predicted sequence can be tested by either
cleaving another substrate of known sequence or perhaps by mapping one or more of the cleavage sites within the known sequence. The value of this computerized method is that new and
potentially useful recognition sites can be identified quickly
and, often, sample experiments can be devised to prove (or disprove) this predicted sequence.

Another computer program that analyzes nucleic acid sequences for additional subtle characteristics has been developed by
Staden (30). Because tRNA molecules are known to share a similar
secondary structure (cloverleaf), Staden has developed a program
to identify putative tRNA genes from within a long DNA sequence.
By applying this criteria of cloverleaf shape and a constant number of bases within particular portions of the molecule, two new

tRNA genes have been found within the human mitochondrial genome (31). It is worth noting that although these newly-discovered tRNA genes resemble other previously-described tRNA genes (32), they differ in their structure in several respects. These differences might have precluded their detection had the set of rules used by this program been overly stringent.

FUTURE APPLICATIONS

Table 1 provides a summary of programs currently available to assist in various aspects of nucleic acid sequencing. It has become apparent since the first compilation of this list (33) that recent emphasis is on the use of computers for sequence analysis. This is understandable since there is a wealth of nucleotide sequences now available but little in the way of understanding their significance.

A good example of this concerns the phenomenon of RNA splicing in eukaryotic cells. On the basis of existing data, the only common features that occur at all splicing sites are a GU dinucleotide at the 5' end of the intervening sequence and an AG dinucleotide at the 3' end of that sequence (34). It seems certain that other information, be it primary sequence or some structural feature dependent upon that primary sequence, is also necessary since not all GU, AG combinations are joined. We are approaching this problem by using the computer to generate potential mRNAs from a DNA sequence, making all possible pairwise combinations of GU and AG. The task then becomes one of finding a rule or rules that can be applied in order to distinguish correct splicing events from incorrect ones. The computer program that is under development approaches this task in two parts. The first part concerns trying to locate splice points at the nucleotide level when such sites can be approximated on the basis of electron microscopic measurements. Thus, these electron microscopic mapped positions are used to sort through the total population of possible spliced mRNA candidates produced by the computer. Each remaining mRNA is then translated and molecular weights of the predicted polypeptides used to further limit the candidates if, for instance, the actual size of the protein product made by the spliced mRNA is known. Further restrictions may be placed on the computer-generated candidates by requiring that certain sequences be present or absent from the final mRNA candidate because of the knowledge that specific tryptic peptides should or should not be present. The second part of this program will sort through the sequences around previously determined processing points in such a way as to find sequence elements that provide a unique environment for the two ends of a splice junction. This will allow the discrimination of actual splice points from the rest of the sequence.

Table 1

Computer Programs Used During Nucleic Acid Sequencing

Function	Program Name*	Program Language	Ref.
PRE-SEQUENCING PREPARATIONS			
Restriction enzyme mapping	"GA1"[a]	SAIL	6
	Least squares method for restriction mapping[b]	FORTRAN	8
Reverse translation	"REVTRANS"[c]	FORTRAN	unpub.
	Nucleic acid sequence analysis[d]		13
COLLECTION AND ASSEMBLY OF SEQUENCES			
Semi-automated autoradiograph reading	"READ"[e]	FORTRAN	unpub.
Automated autoradiograph reading	Scanning Program[f]		unpub.
Sequence assembly	"ASSEMBLER"[g]	FORTRAN	12
	Overlap-Meld[h]	FORTRAN	11
ANALYSIS OF NUCLEOTIDE SEQUENCES			
Printing, editing, storage and manipulation	Nucleic acid sequence analysis[d]	PL/1, SAIL**	13
Search Routines (restriction enzyme sites, direct repeats, true and dyad symmetries); Translation	DNA-Handling Program[h]	FORTRAN	15–17
Restriction enzyme recognition site predictions	"MONITOR"[g,l]	FORTRAN	27,28
tRNA Gene Prediction	"tRNA"[g]	FORTRAN	30
Measure frequency of occurrence of same elements at different distances from each other	DNA Confirmational Structure[m]	FORTRAN	unpub.
Optimal alignment and comparison of sequences	Sequence Homology and Alignment Search[i]	FORTRAN IV	unpub.
Computer-graphic matrix presentation of optimal sequence alignment	"REVEAL"[n]	H.P. BASIC	unpub.

Table 1 (contd.)

Computer Programs Used During Nucleic Acid Sequencing

Function	Program Name*	Program Language	Ref.
Secondary Structure Prediction	Secondary Structure Program[i]	FORTRAN	unpub.
	Secondary Structure Program[j]	APL, FORTRAN	20
Secondary Structure	Secondary Structure Program[o]	FORTRAN	23
Tertiary Structure Modeling	3D Molecular Modeling[k]	FORTRAN, BLISS	44, 45

*Refers to superscripts, which indicate sources of programs. Listing in quotations indicates titles and the remaining names are brief descriptions.

**This version is available from the Sumex System at Stanford University.

[a]Sumex System, Stanford University, P. Frieland, D. Brutlag, L. Kedes.

[b]University of Wisconsin, J.L. Schroeder, F. Blattner.

[c]Cold Spring Harbor Laboratory, R. Blumenthal, R.J. Roberts.

[d]National Institutes of Health, C. Queen, L. Korn.

[e]Cold Spring Harbor Laboratory, T.R. Gingeras, P. Rice, R.J. Roberts.

[f]European Mol. Biol. Lab., Heidelberg, Germany, S. Provencher, R. Vogel, V. Dovi, J. Lehrach.

[g]Cold Spring Larboratory, T.R. Gingeras, J. Milazzo, R.J. Roberts.

[h]MRC Laboratory of Molecular Biology, Cambridge, England, R. Staden.

[i]Syracuse University, G. Pavlakis, J. Vournakis.

[j]Univ. of California at Los Angeles, G.M. Studnick, G.M. Rahn, I.W. Cummings, W.A. Salser.

[k]National Institutes of Health, R.J. Feldmann.

[l]University of Wisconsin, C. Fuchs, E.C. Rosenvold, A. Honigman, W. Szybalski.

[m]Weizmann Institute, E.N. Trifonov.

[n]National Institutes of Health, J. Maizel.

[o]State University of New York at Stony Brook, R. Nussinov, A. Jacobson.

Such computer programs will have to be highly interactive since they are dependent on the user for information necessary for continuation. They must be capable of combining seemingly divergent pieces of information (e.g., electron microscopic mapping data, protein molecular weights, tryptic peptide profiles) in order to make a prediction about the most likely candidate(s). To accomplish this, the programs will be structured to become autodidactic.

Work in the field of artificial intelligence has been devoted to the search for computer-assisted methods that emulate the human mind during the solution of a problem. One interesting area in this field is the problem of interactive transfer of expertise. The goal of workers in this area is to establish a knowledge base (information bank) as well as a method (an interactive computer program) to solve relevant problems and update (self-correct) this information. The central theme of work in this area is the need for a program that not only uses its knowledge directly to solve a particular problem, but also examines its data base, abstracts it, reasons about it and directs its application. Such programs have been written for the diagnosis and treatment of infectious diseases (MYCIN) (35,36) as well as for the advice and selection of investments in the stock market (TEIRESIAS) (37). It seems clear that in the area of DNA sequence analysis, data bases could be constructed and a rule-based computer consultant (artificially intelligent program) developed to: 1) gather data banks (e.g., collect sequences of mRNAs, locate all GU, AGs, make all potential spliced RNA transcripts); 2) construct lists of rules (e.g., that splices occur only after GU or AG, that exon segments have 5' OH-group); 3) solve problems (e.g., predict splice points); 4) edit decision-making process (that a particular candidate site is unlikely since no splice site has been observed by means of electron microscopy); 5) suggest experiments (if splice occurs at nucleotide numbers X,Y, then the resulting protein product should be K daltons, with 2 methionine-containing tryptic peptides). Figure 3 illustrates in dialogue form how such a program could be applied to the problem of predicting RNA splice sites. Because the ultimate goal of workers interested in nucleic acid sequence analysis is to understand the information encoded within the nucleotide sequence, this frequently necessitates the application of information derived from several related but distinct areas of research. Consequently, the use of such autodidactic programs could be useful for constructing models and making predictions as a means of relating nucleic acid structure and functions.

Another and more challenging area where autodidactic programs could be applied is in the analysis of the role of secondary and tertiary structures in the function of nucleic acids. In reviewing long stretches of DNA sequence, it has become quite easy to conceive of DNA as a featureless one-dimensional string. The observation that $(dG-dC)_3$ crystallizes with a left-handed helical conformation (38) represents an extreme example of the fact that the DNA double helix is not perfectly smooth and regular. The experimental data concerning the sequences recognized by RNA polymerase (39) demonstrate clearly the importance of a three-dimensional approach and confirm concepts previously developed (40,41). In a similar vein, computer programs developed by Trifonov and Sussman (42; personal communication) address the question of DNA folding around the nucleosomes. They raise the possibility that patterns, existing within the primary sequence, have been designed to facilitate such folding.

As for the higher-order structures of RNA, computer modeling of single-stranded RNA molecules is, by now, a familiar exercise to those interested in nucleic acid sequences. We are slowly becoming accustomed to such two-dimensional representations of RNA. However, the need for better algorithms to predict secondary structures is also evident. Programs making use of autodidactic algorithms could improve the quality of the models by gathering data acquired from investigations such as X-ray crystallographic or single-stranded nuclease (S1 or T1 studies). These observations could be formulated into rules that could be used as guides during the model-building process. Pavlakis et al. (43) have attempted to correlate computer-generated secondary structure models for the 5' ends of both α- and β-globin RNAs with their sensitivity to S1 and T1 nucleases. Such nuclease studies are attempts to provide supportive evidence for the existence of computer-predicted structures and, in the process, indicate how such structures can influence function.

Computer programs that examine the three-dimensional properties of macromolecules have been devised by Feldmann (44,45). At present, these programs rely on X-ray crystallographic data to provide the basic parameters of a model and use computer-graphics to depict and transform the structures. Obviously, we are not yet able to define the rules that relate primary sequence to three-dimensional structure; more basic experimentation is necessary. However, if we are to understand the nature of the interactions controlling the behavior of macromolecules, it will be essential to expand both our experiments and our viewpoints into this third dimension. Computer graphics is likely to prove an important tool in these efforts.

CONCLUSIONS

Within a very short time, the alternative to using computer-assisted means for the collection, assembly and analysis of the ever-growing mass of nucleotide sequence data will be limited to that method illustrated in Figure 4. A variety of computer programs has been written to assist in the various tasks associated

I. FORMATION OF DATA BASES

1. *WHAT IS SEQUENCE SUBMITTED?*
   The DNA sequence of left-end 0-5% of Adenovirus-2 genome.
2. *HOW MANY CODING REGIONS ARE FOUND WITHIN THIS REGION?*
   One
3. *DO YOU KNOW WHERE (NUCLEOTIDE #) THIS AUG IS LOCATED?*
   Yes
   559
4. *HOW MANY SPLICE SPECIES OBSERVED?*
   3
5. *ARE THERE EM DATA FOR THE SPLICE SITES?*
   a) 3.1% and 3.7%
   b) 2.7% and 3.7%
   c) 1.7% and 3.7%
6. *ARE THERE ANY PROTEIN PRODUCTS OBSERVED TO COME FOR THIS REGION? (Y,N,)*
   Y
7. *HOW MANY PROTEIN PRODUCTS?*
   3
8. *WHAT ARE THEIR SIZES (DALTONS)?*
   a) 32,000
   b) 27,000
   c) 14,000

II. INFERENCE STEPS

Rules	Application
Rule 1	
1.1 Splice occurs only after GU and AG	Make all possible splice RNA species starting at nucleotide 559.
Rule 2	
2.1 1% Ad2 genome=360 bp	Map data to nucleotide number:
2.2 % genome x 360=nucleotide number	a. 1111 to 1225 (±50)
	b. 973 to 1225 (±50)
2.3 Error limits for EM are ±50 nucleotides	c. 610 to 1225 (±50)
Rule 3	These are candidates found within 50 nucleotides of positions indicated.
3.1 Locate all possible GU-AG pairs with 50 base pairs of positions given in rule 2.	
3.2 Construct all possible RNA splices using these GU-AG pairs.	
3.3 Translate these transcripts into amino acid sequences.	
3.4 Calculate molecular weight of protein coded for by each spliced RNA.	
3.5 Determine the number of limit tryptic, clostripain, V-8 S. aureus, cyanogen bromide cleavage products that may be produced by each encoded protein.	

Candidates	AUG	GT	AG	Size of Protein Coded	No. of Tryptic Peptides
1					
2					
3					
4					
5					
6					

III. INTERPRETATION OF RULES

9   Based on electron microscopic data <u>and</u> sizes of proteins observed from this region, I would predict the splices to be located at:
     a)   1111 and 1225 = 32K
     b)   973 and 1225 = 27K
     c)   610 and 1225 = 14K

10 I would suggest if these predictions are correct:
     a) that a tryptic peptide analysis of the 27K product will yield 2 methionine containing peptides;
     b) that the 32K product will be the only one to contain 5 cysteines.

Figure 3. A model dialogue resulting from the use of an autodidactic computer program designed to assist in predicting the sites of RNA splicing. This model makes use of data obtained from electron microscopic, protein chemistry and nucleic acid sequence experiments. It would rely on a basic data bank and a set of rules designed for the computer in order to generate a prediction. Also, it would be capable of generating new rules based on detecting features found at known splice points but absent at sequences known not be processing sites. Such programs are in use in both the medical (MYCIN) (34,35) and investment (TEIRESIAS) (36) fields)

with sequencing. The programs discussed in this review represent the first efforts in this area. These and future programs are new and promising tools for the molecular biologist interested in the structure and organization of nucleic acids. As these programs become more widely used, their level of sophistication will undoubtedly grow. However, it is important to conclude with the understanding that despite the levels of sophistication reached by such programs, they are only as accurate and imaginative as those who use them.

Figure 4.

This photo was supplied by Information Handling Services, Denver, Colorado (c) 1980.

REFERENCES

1. Shapiro, H.S. (1967) in Methods in Enzymology (Grossman, L. and Moldave, K., eds.), Vol. 12, pp. 205-414, Academic Press, New York, NY.
2. Brownlee, G.G. and Sanger, F. (1969) Eur. J. Biochem. 11, 395-399.
3. Maxam, A.M. and Gilbert, W. (1977) Proc. Nat. Acad. Sci. U.S.A. 74, 560-564.
4. Sanger, F., Nicklen, S. and Coulson, A.R. (1977) Proc. Nat. Acad. Sci. U.S.A. 74, 5463-5467.

5   Zabeau, M. and Roberts, R.J. (1979) in Molecular Genetics (Taylor, J.H., ed.), Part 3, pp. 1-63, Academic Press, New York, NY.
6   Stefik, M. (1978) Articifial Intell. 11, 85-114.
7   Buchanan, B.G. and Feigenbaum, E.A. (1978) Artificial Intell. 11, 5-24.
8   Schroeder, J.L. and Blattner, F. (1978) Gene 4, 167-174.
9   Garrels, J.I. (1979) J. Biol. Chem. 254, 7961-7977.
10 Taylor, J., Anderson, N.L., Coulter, B.P., Scandera, A.E. and Anderson, N.C. (1979) in Electrophoresis "79" (Radola, B.J., ed.), Springer-Verlag, Berlin.
11 Staden, R. (1979) Nucl. Acids Res. 6, 2601-2610.
12 Gingeras, T.R., Milazzo, J.P., Sciaky, D. and Roberts, R.J. (1979) Nucl. Acids Res. 7, 529-545.
13 Korn, L.J., Queen, C.L. and Wegman, M.N. (1977) Proc. Nat. Acad. Sci. U.S.A. 74, 4401-4405.
14 Queen, C.L. and Korn, L.J. (1980) in Methods in Enzymology (Grossman, L. and Moldave, K., eds.), Vol. 65, pp. 595-609, Academic Press, New York, NY.
15 McCallum, D. and Smith, M. (1977) J. Mol. Biol. 116, 29-30.
16 Staden, R. (1977) Nucl. Acids Res. 4, 4037-4051.
17 Staden, R. (1978) Nucl. Acids. Res. 5, 1013-1015.
18 Fitch, W.M. (1972) J. Mol. Biol. 1, 185-207.
19 Pipas, J.M. and McMahon, J.E. (1975) Proc. Nat. Acad. Sci. U.S.A. 72, 2017-2021.
20 Studnicka, G.M., Rahn, G.M., Cummings, I.W. and Salser, W.A. (1978) Nucl. Acids Res. 5, 3365-3387.
21 Nussinov, R., Pieczenik, G., Griggs, J. and Kleitman, D. (1978) SIAM J. Applied Math. 35, 68-76.
22 Phillipp, M., Ballinger, D. and Seliger, H. (1978) Naturwissenschaften 65, 388-389.
23 Nussinov, R. and Jacobson, A. (1980) Proc. Nat. Acad. Sci. U.S.A. (in press).
24 Gralla, J. and Crothers, D.M. (1973) J. Mol. Biol. 73, 497-511.
25 Tinoco, I., Uhlenbeck, O.C. and Levine, M. (1971) Nature 230, 362-367.
26 Tinoco, I., Borer, P.W., Dengler, B., Levine, M.D., Uhlenbeck, O.C., Crothers, D.M. and Gralla, J. (1973) Nature New Biol. 246, 40-43.
27 Gingeras, T.R., Milazzo, J.P. and Roberts, R.J. (1978) Nucl. Acids Res. 5, 4105-4127.
28 Fuchs, C., Rosenvold, E.C., Honigman, A. and Szybalski, W. (1978) Gene 4, 1-23.
29 Roberts, R.J. (1980) Nucl. Acids Res. 8, r63-r80.
30 Staden, R. (1980) Nucl. Acids Res. 8, 817-825.
31 Barrell, B.G., Bankier, A.T. and Droun, J. (1979) Nature 282, 189-194.
32 Sprinzel, M., Gruete, G., Spetzhaus, F. and Gauss, D.H. (1980) Nucl. Acids Res. 8, r1-r22.

33 Gingeras, T.R. and Roberts, R.J. (1980) Science 209, 1322-1328.
34 Breathnach, R., Mandel, J.D. and Chambon, P. (1977) Nature 270, 314-319.
35 Shortcliffe, E.H. (1976) in MYCIN: Computer-Based Consultations in Medical Therapeutics, American Elsevier, New York, NY.
36 Davis, R., Buchanan, B. and Shortcliffe, E.H. (1977) Artificial Intell. 8, 15-45.
37 Davis, R. (1979) Artificial Intell. 12, 121-157.
38 Want, A.H.J., Quigley, G.J., Kolpak, F.J., Crawford, J.L., van Boom, J.H., van der Marel, G. and Rich, A. (1979) Nature 282, 680-686.
39 Siebenlist, U., Simpson, R.B. and Gilbert, W. (1980) Cell 20, 269-281.
40 Wells, R.D., Blakesley, R.W., Hardies, S.C., Horn, G.T., Larson, G.E., Selsing, E., Burd, J.F., Chan, H.W., Dodgson, J.B., Jensen, K.F., Nes, I.F. and Wartell, R.M. (1977) CRC Crit. Rev. Biochem. 4, 305-340.
41 O'Neill, M.C. (1977) Nucl. Acids. Res. 4, 4439-4463.
42 Trifonov, E.N. and Sussman, J.L. (1980) Proc. Nat. Acad. Sci. U.S.A. 77, 3816-3820.
43 Pavlakis, G., Lockard, R.E., Varnvakopoulos, N., Rieser, L., RajBhandary, U.L. and Vorunakis, J.N. (1980) Cell 19, 91-102.
44 Feldmann, R.J., Bling, D.H., Furie, B.C. and Furie, B. (1978) Proc. Nat. Acad. Sci. U.S.A. 75, 5409-5412.
45 Bina, M., Feldmann, R.J. and Deelay, R.G. (1980) Proc. Nat. Acad. Sci. U.S.A. 77, 1278-1282.

INDEX

Acrolein, 224,239
Acrylamide-urea gel, 293
Actin developmental regulation, 277-283
　filament, 276
　genes in Dictyostelium, 61, 283-292,310,311
　　in Drosophila, 287
　　in Physarum, 291
　　in yeast, 61,68,78,287
　　isoelectric forms, 277
Actinomycin D, 274
ADH, see alcohol dehydrogenase
Aegilops caudata, 219,220
　comosa, 220
　columnaris, 220
Agarose gels, 39,71, see also gel electrophoresis
Agropyron, 218
Alanine transaminase, 268
Alcohol dehydrogenase
　assay, 224
　antibodies, 227
　Drosophila melanogaster, 239
　electrophoretic variants, naturally-occurring, 231-236
　maize, 223-264
　yeast, 77,239
Alkaline phosphatase, 268
Allelic exclusion, 159,171,172
Allyl alcohol selection, 224, 238-242
AluI, 34,45,47
Amber mutation, 7,10,11,19
Amoeba, 266,277
Amoeboid movement, 276

AMV, see avian myeloblastosis virus
Anaerobic polypeptides, 223,227-229
ANP, see anaerobic polypeptides
ANS COBOL, 327
Anther ear, 225
Anther wall protein synthesis, 229
Anthocyanin-conditioning genes, 256
Antibiotic resistance, transposons in, 50
Antibody diversity, 170
APL, 331
Arginine permease, 66,77
Arginosuccinate lyase, 77
ARS, 72
Artificial intelligence, 332
ASSEMBLER, 324,325,330
Autonomous replication sequence, 72
AvaI, 136,139,142
AvaII, 136,139,142,282,283,299
Auxin, 229
Avian myeloblastosis virus, 95
Avian sarcoma virus, 124

B cell, 157,159,173,174,176,178, 181,182
　lymphoma, 182
Bacterial genes in yeast, 58
BalI, 34
BALB/c mice, 157
BamHI, 38,39,41,149,162,198,270, 282,305
Barley, 229
BASIC, 330

Basic library copies, 145,146
Berkeley Fast, 227-230,260
Bevelac, 240
BglII, 27,100,136,149,305,307
BLC, see basic library copies
BLISS, 331
Bone marrow cell, see B cell
Bootstrapping, 60

Canavanine resistance in yeast, 66,74
Carbohydrate binding protein, 300
Carrot, 224,229
cDNA cloning, 92,93
Cell squashes, 208,209
Cellular immunity, 182
Cellulose for binding oligodeoxyribonucleotides, 3
Centromeres, 109
Charon 4A, 148
Chemical modification, 2
Chinese Spring, 213,217
Chloramphenicol resistance, 35, 36
Chorionic somatomammotropin, 190
Chromosome hybridization, 209
Citrobacter freundii, 40,43
Cloning, of human DNA, 190,191
Clustered and scrambled repeats, 111,112
C-mos, 97,98,101,103
Cointegration of plasmids, 37
Coleoptile protein synthesis, 229
ColE1, 114,119,120
Colony hybridization method, 94,117
Commerford ^{125}I procedure, 210
Complement fixation, 159,173
Complementation cloning, 63-67
Conalbumin DNA, 28
Copia, 114,117-123
Cyc1, 60
Cytochrome c, 60

δ, 72,73,79
David's Bread Loaf method, 239

Deletion, 33
  loop, 194,195
DENDRAL, 320
Density centrifugation, 268,269
Diabetes, 201
2,4-Dichlorophenoxyacetate, 229
Dictyostelium discoideum, 265-317
Dideoxyribonucleotide DNA sequencing, 25
Discoidin, 276
  genes, 300-311
Dispersed repeat DNA in Drosophila, 113-115,117-121
Diversity gene segments, 165-170
DNA polymerase I, see Pol I
DNA sequencing, 3,319-338
DNase I, 277
Dreyer-Bennett hypothesis, 159, 162
Drosophila melanogaster, 109-128
  circular DNA, 115,117
  heat shock, 227
  satellite, 217
  tRNA gene, 244
Drosophila simulans, 110,115,116

EcoRI, 38,39,41,60,66,71,97-99, 102,115,117,119,120,148,149, 190,191,270-272,282-284,287, 288,294,299,300,305-308
Edwardsiella tarda, 40,43
ELC, see expression linked copies
Elymus, 218
EMS, see ethyl methanesulfonate
Endosperm protein synthesis, 229
Enterobacter aerogenes, 40,43
Enterobacteriaceae, 42,48,49
Enterotoxin, 50
env of retrovirus, 91,93
Erwinieae, 41,42
Eschalier's Kco line, 117,118, 122
Escherichia coli C, 40,42
  K12, 40,42,44
  W, 40,42
  W3110, 40
Escherichieae, 40,42
Ethanol oxidase reaction, 224

INDEX

Ethidium bromide equilibrium
  centrifugation, 94, 117
  staining, 39,62,71,97
Ethyl methanesulfonate, 236,
  245-248,251
Exon, 161,163,172,174,177,180,
  181
3' Exonuclease of Pol I, 6,10,
  19
5' Exonuclease of Pol I, 5
Expression linked copies, 144-
  147,150

f1 phage, 123
fd vector, 28
Feline leukemia virus, 94
Feline sarcoma virus, 94,95
FeLV, see feline leukemia virus
FeSV, see feline sarcoma virus
Fetal hemoglobin syndrome, 192,
  193,195
Fluorescence-activated cell
  sorter, 162
Foldback DNA, 112,113
FORTRAN, 327,330,331
Fruiting body, 266,267
Fusidic acid resistance, 35,36

G4, 1
gag, 91,101
GA1, 320,330
Galactokinase, 59,77
Galactose-1-phosphate uridyl
  transferase, 59,77
Gastrulation, 276
Gel electrophoresis, 17,97,99,
  115,116,162,231,240,277-279,
  280,281
Gene competition hypothesis,
  233
Gene conversion, 81
Gene fusion, 82
Gene libraries, 95-98
  human, 191,200
  retrovirus, 103
  trypanosome, 148,149
Giant cell, 266
Giemsa stain, 210,215,216

Globin genes, human, 191,193,
  195-199
  RNA, computer analysis, 333
  yeast, 80
Glossina, 129
β-Glucosidase-2, 268
Glyceraldehyde-3-phosphate dehy-
  drogenase, 59,68,78
Glycogen phosphorylase, 268
Glycolytic cycle, 223
Glycoprotein of trypanosomes,
  129-155
Goldberg-Hogness box, 27,295,
  299,308,311,312
Group specific antigen, see gag

HaeII, 34,37,45,46,136
HaeIII, 8,16,17,34,37,282,283,
  286,305
Hairpin structure, 8
HapII, 282,283,285,286,305,307
Harvey sarcoma virus, 94
Haynaldia, 218
Heavy chain class switching,
  173-178
Hemoglobin Constant Spring, 197
Hemoglobin synthesis disorders,
  189,191
Hemophilia, 200
Henbane, 224
Heterochromatin, 109,110,212-215
  X chromosome, 110
Heteroduplex analysis, 89,100,
  194,287
HgiAI, 139,143
HhaI, 34,119
HincII, 136
HindIII, 18-20,39,41,100,116,
  119,120,136,139,140,144,149,
  198,201,270,282-284,286-289,
  292,294,305,308,309
HinfI, 34,37,45,46,120,136
Hirt extraction, 94
Histamine, 159
Histidine genes, yeast, 61,62,
  64,75,76,82
Histidinol dehydrogenase, 77
Histocompatibility markers, 200
Histone genes, Drosophila, 284
  yeast, 78

HLA, see histocompatibility
    markers
hnRNA, 273
Hordeum chilense, 218,219
HpaI, 100,120,198,201
HpaII, 34,45,46,136,283
HphI, 23,24
Hull-less Pop, 235
Human disease, 189-206
Human DNA, genetic defects, 189
    hemoglobin genes, 190
    libraries, 191
    maps, 190,200
Human growth hormone, 190,201
Hybrid-arrest protein synthesis, 89,93,133-135
Hybridization, in situ, 207-222
Hydroxylapatite chromatography, 52,210
Hysterix, 218

IgA, 160,176
IgD, 159,160,179,181,182
IgG, 158-160
IgM, 158-160,176,179,182
Illegitimate recombination, 33
Imidazole glycerol phosphate dehydratase, 61,77
Immunoglobulin
    constant region, 157-160
    heavy chains, 157,158
    human, 168
    hypervariable region, 157, 164,168,170
    light chains, 157,158
    mouse, 157-188
Insertion sequence, 33-55
In situ hybridization, 110,207-222
Insulin, human, 190,201
Intervening sequence, see intron
Intron, 67,70,81,101,150,163, 166,168,171,173,175,193,273, 289,311,313
IS1, 33-55
    nucleotide sequence, 34
    number of copies in bacteria, 37
IS2 in yeast, 62

Iso-1-cytochrome c gene, 60,61, 68-70,73,77
Iso-2-cytochrome c gene, 68
Isoposon, 50
Isopropylmalate dehydrogenase, 77
Iso-transposon, 50
Isovariable antigen types, 132, 133
IsoVAT, see isovariable antigen types
Inversions, 33
Inverted repeats, 51,71
    isolation, 51,52

Joining sequence, 164

Killer RNA, 124
King II, 215,216
Klebsiella aerogenes, 40,42,43, 265,297
Klebsielleae, 40,41,43
Klenow fragment, 5,12,13,16,19, 21,27
Knotted, 225,242
KpnI, 100,136,282,305

β-Lactamase gene, 135,136
Leloir parhway enzymes, 59
Leucine aminopeptidase, 268
Leucine genes in yeast, 62-64,71
Libraries, gene, see gene libraries
Ligase T4, 5,9
Long-period interspersion, 111
Lymphocyte, 159,172,179,182

Macrocyst, 266
Maize alcohol dehydrogenase, 223-264
    in situ hybridization, 212
α-Mannosidase, 268
Marker rescue, 4
Mast cells, 159
Mating type genes, 61,64,67-69, 76
    interconversion, 69
Maxam-Gilbert method, 41,294, 320,321
MboI, 139-141,145

Melting temperature,
  relationship with length, 4
Mercury resistance, 35,36
Mesocotyl ADH, 231
Mesocotyl protein synthesis, 229
Metaphase chromosomes, 207
  banding, 207,208
Mexican hat stage, 266
Middle repetitive DNA in Drosophila, 115-117
Miniplant, 225
Minus strand, 92
Mismatch detection, 20
Mismatch repair, 6,29
Mitochondrial DNA, 271
M-MuLV, see Moloney murine leukemia virus
MnlI, 18,192
Modifier genes, 261
Moloney murine leukemia virus, 93,94,100,124
Moloney sarcoma virus, 98-102
  integration, 99
MONITOR, 330
MSV, see Moloney sarcoma virus
Multigene families, 74,265-317
Mutagens, polynucleotides as, 2
Mutation efficiency
  oligo length, 10,12,16
  priming temperature, 12-14
MYCIN, 332,335
Myeloma, 162,171,172,174,175, 179

N-acetylgalactosamine, 302
N-acetylglucosaminidase, 268
Netropsin-CsCl gradients, 271, 272
Nomadic DNA in Drosophila, 117-125
  transcripts similar to retrovirus RNA, 125
  yeast, 123
Nonsense suppressors, 74
Northern blot, 291
N-(5'-phosphoribosyl) anthranilate isomerase, 78
Nucleolar organizer, 212,217
Nucleotide analog, 2

ochre suppressor, 74,81
Oligodeoxyribonucleotide for
  mutant selection, 20-24
  synthesis, 3
Oncogenic transformation, 91, 101
Onion, 229
Orotidine-5-phosphate decarboxylase, 69,78

P1 phage, 35,37
Pachytene chromosomes, 207
Papago Flour, 235
Paramutation, 243
Parasitemia, 130,131,147
PASCAL, 327
pBR322, 28,75,93,134
Pea, 229
Peromyscus, 147
Petunia, 224
Phaseolus, 208
3-Phosphoglycerate kinase, 62,78
Phosphoribosyl-AMP cyclohydrolase, 77
Phosphoribosyl-AMP pyrophosphorylase, 77
Phosphotriester method of synthesis, 3
Physarum, 269,291
Piñon pine, 229
Plaque blot screening, 94
Plasma cell tumor, see myeloma
2μ plasmid, 64,71,79
Plus strand, 92
Point mutations, construction, 2
pol of retrovirus, 91
Pol I, 5,12,13
Pollen, maize, 232,239
  stains for ADH, 249
Pollen mother cells, 208
Polyacrylamide gel electrophoresis, 227, see also gel electrophoresis
Poly(dA)-(dT) method, 120,305
Poly(dG)-(dC) method, 135,305
Polynucleotide kinase, 41
Polynucleotide phosphorylase, 3
Polynucleotides as mutagens, 2
Polysomal RNA, 273
Polyspondilium, 300

Polytene chromosomes, 114,118
  suspensor cells, 208
Potato, 224
Prephenate dehydrogenase, 78
Priming efficiencies, 8
Promoter, 311
Propidium di-iodide equilibrium centrifugation, 94
Proteeae, 41,42
Proteus mirabilis, 41,45
  morganii, 41
Provirus, 89-91
  cloning of integrated, 95-98
  cloning of unintegrated, 94, 95
Pseudo-gene, 292,297,299
Pseudoplasmodium, 266,267
PstI, 34,37-39,41,100,136,139, 270,305,307
PvuI, 136,139
PvuII, 100,149

R1, 35
R6, 35,48
R100, 35-37,41,48
Rabbit reticulocyte system, 134,163,223,227,229,311
r-Determinant region, 35,36
READ, 330
Reassociation kinetics, 109
Reciprocal effect, 234,255
  translocations, 256
Recombinant DNA, mutation introduction, 24-27
Red Pop, 235
Renaturation kinetics, 269
Repeated sequences, 207
  Dictyostelium, 276
  Drosophila, 109-128
  maize, 212
  rye, 212-215
  wheat, 212-215,217,218
  yeast, 72
Resistance plasmids, 35
Restriction endonuclease mapping, 190
Restriction endonucleases, new, determination of recognition sequences, 328
Reverse transcriptase, 91,92

Reverse transcription, 89,90,93
retroviridae, 89
Retrovirus as cloning vehicles, 101
  gene library, 103
REVTRANS, 330
Ribosomal genes, spacer, 284
  wheat, 212
  yeast, 70,71,78,79
Ribosome binding sequence, 19,20
Rice, 229
R-loop analysis, 89,100,101,163, 194
RNA polymerase I, 70
  II, 81
  III, 70,74,75,81
RNA processing, computer prediction of, 335
  mouse, 180
  Xenopus, 70
  yeast, 68,70
Root ADH, 231,234
Root protein synthesis, 229
Root-tip preparations, 208
Rous sarcoma virus, 93,94
RPC-5 chromatography, 97,99
RSV, see Rous sarcoma virus
RTF, 37
Rye, 212-216

S1 nuclease, 5,6,10,16,19,21,89, 112,292,293,296,299,309
  mapping, 287
SacI, 100
Saccharomyces carlsbergensis, 71
Saccharomyces cervisiae, see yeast
SAIL, 330
SalI, 75,139,142,270
Salivary gland nuclei, 115
Salmonella, 168
  typhimurium, 40,45
Sanger sequencing method, 320, 321
sarc, 98
Sau3A, 140,145
Sau96I, 283,287-289,300
Schneider's line, 2,117
Schroeder-Blattner program, 321
Screening for clones, 59-67

Scutellum ADH, 234
  protein synthesis, 229
Sea urchin DNA, 284
Secale cereale, 211,215,218-220
Seedling protein synthesis, 229
Sequence analysis, 326-329
Sequence storage banks, 319
Sequencing nucleic acids, 3, 319-338
Serodeme, 136,137
Serratia marcescens, 41-43
Shigella, 42,43
  boydii, 40,42,48,49
  dysenteriae, 40,43-49
  flexneri, 40,44,45,48,49
  sonnei, 40,43-45,49
Sickle cell anemia, 191,198,201
Site-specific mutagens, 1-32
Site-specific mutagenesis, 89
Slime molds, 265
SmaI, 100
Smiling effect, 323
SNOBOL, 327
Solanacae, 224
Solid-phase support synthesis, 3
Somatic mutation theory of antibody diversity, 170
Sorghum, 229
Sorus, 266
Southern blot, 37,41,51,60,66, 68,71,75,76,94,97,99,116,118, 138-148,171,174,183,190,283-285,299,306,307,310
Spacer DNA, 165,166,168,284
Spleen necrosis virus, 124
Squash, 229
src, 98
Staden programs, 323,324
Starch gel electrophoresis, 231,240, see also gel electrophoresis
Streptomycin resistance, 35,36
Sulfanilamide resistance, 35,36
SUP4, 28,72,73,75
Super Gold Pop, 228-230,235
Suppressor E. coli, 9
SV40 construction of deletion, 1
Switch recombination, 151

T cells, 182,183
T4 ligase, 5,9
Tandem duplications, 37
TEIRESIAS, 332,335
Teosinte, 229,230
Terminal transferase, 3,135
Termination, 7,11,15
Tetracycline resistance, 35,36
Tetrahymena, 269,270
Tetrasome, 259
Tetrazolium, 224
Thalassemia, 192,193,195-198, 200,201
Thermal denaturation, 269
Threonine deaminase-2, 268
Tn3, 50
Tn9, 35
Tn204, 35,36
Tn1681, 35
Tobacco, 224
Tomato, 229
Transition, 6,9,10,14,19,20
Transposable element, see transposon
Transpositions, 33
Transposon, 33,50,91
  similarity of bacterial and retrovirus provirus, 95,101
  Drosophila melanogaster, 121
Transvection, 243
Transversion, 8,14-19,27
Trehalose phosphate synthetase, 268
Triticum aestivum, 211
  dicoccum, 217,218
  spelta, 211
tRNA computer program for gene prediction, 330
  genes, assignment to mitochondrial DNA, 1
  genes, Xenopus, 68
  genes, yeast, 67,68,74,75,78, 79
Trypanosoma, 129-155
  brucei, 129,130,147,150,151
  congolense, 129
  equiperdum, 129,131,147
  gambiense, 129,130
  ploidy, 151
  procyclic, 133,151

Trypanosoma vivax, 129
Trypanosomiasis, 130,131
Tryptophan synthetase, 78
Tsetse fly, 129
Tubulin genes, 150
Ty1, 72,73,75,76,79,123
Tyrosine aminotransferase, 78
Tyrosine transaminase, 268

UDPgal polysaccharide transferase, 268
UDPG-pyrophosphorylase, 268
UDPH-epimerase, 268
Unequal crossover, 71,249
Uridine diphosphogalactose-4-epimerase, 59,77

Variant surface glycoprotein, see VSG
VASSA, see VSG
Virion, 90,91
V-mos, 97,98,100,101,103
VSA, see VSG
VSG, 129-155
  mRNA, 132-138

W3110, 40,44
Wheat, 212-215,217

X chromosome, 110
XbaI, 100,305,307
Xenopus borealis 5S gene, 244
Xenopus RNA processing, 70
XhoI, 100,139,142,305

Y chromosome, 110
Yeast cloning, 57-88
  genes expressed in E. coli, 58,61
  genome size, 58
  histones, 63
  ribosomal proteins, 63
  shotgun in E. coli, 64
  transformation, 58,64,65
  tRNA genes, 67,74

Zein gene, 212